IRRIGATION AND DRAINAGE ENGINEERING

IRRIGATION AND DRAINAGE ENGINEERING

Editor

Svenja Breisacher

scitus
academics

Irrigation and Drainage Engineering

Edited by **Svenja Breisacher**

This Book is Printed on Acid-Free Paper. ♾

ISBN: 9781681179131
Library of Congress Control Number: 2017948941

SCITUS Academics LLC,
616, Corporate Way, Suite 2, 4766,
Valley Cottage, NY 10989

www.scitusacademics.com

Notice

Printed in United States of America.

Preface

Irrigation and drainage has made a major contribution to facilitate increasing food production in order to meet the needs of an ever-growing world population. Irrigation, as the biggest water user in most regions of the world, is facing significant challenges in balancing social, economic and environmental needs for water. The water is basic source for all living in the world and it covers two third of the earth, it is the lifeblood of plants and its permanent or temporary shortage can cause serious damage to plants yield and quality. Irrigation has an important place in terms of water usage, its environmental impact and reuse of urban waste water. Irrigation; firstly begin with people becoming sedentary and agricultural practices, it has been made to improve the quality and quantity of crop production. In most countries agriculture represents by far the largest use of water and worldwide, it represents about 70% of total withdrawal and 90% of water consumption. Especially since the 1950s, people living in the cities started to be the owner of a garden, studies conducted in the urban landscape water is needed in these areas with different types of grass and ornamental crops, in order to ensure the sustainability of irrigation, which resulted in entering of irrigation into urban life. The sustainability of irrigation and drainage in the face of many variants and constraints like availability of water as a resource, ecological balance, socio-cultural impacts, and climate change effects lies in the strategies adopted and systems emplaced. Thus, sustainable irrigation and drainage management will be a frontier field in the future. This raises a number of questions; how can we increase the efficiency and productivity of water use under the circumstance of increasing uncertainties of flooding and drought? How can we reduce the nutrients output from agro-field by using irrigation and drainage technique, to reduce the pollution to groundwater and surface water?

This volume 'Irrigation and Drainage Engineering' aims to explore frontiers of knowledge in coining sustainable strategies and systems direly needed in managing the quality and quantity of water required for crop irrigation, surface and root zone drainage and flood management using available tools of research and development. The contribute chapters are written by well-known authors possessing various professional backgrounds and proficiency have dealt with these issues concerning the strategies and systems of irrigation and drainage. This book will be of invaluable for crop growers, agricultural engineers, water resource managers, academicians and graduate students alike.

Table of Contents

Chapter 1	A Spatial Analysis of Irrigation Technology	1
Chapter 2	Irrigation and Drainage in Agriculture: A Salinity and Environmental Perspective	29
Chapter 3	Estimation of Crop Coefficient and Water Requirement of Dutch Roses (Rosa hybrida) under Greenhouse and Open Field Conditions	53
Chapter 4	Wastewater Reuse for Irrigation — Practices, Safe Reuse and Perspectives	75
Chapter 5	Inhibition of Root Penetration in Subsurface Driplines by Impregnating the Drippers with Copper Oxide Particles	99
Chapter 6	Challenges and Constraints in Irrigation and Drainage Development: A World-Wide View	113
Chapter 7	A Study on Planned and Applied Irrigation Modules in Irrigation Networks: A Case Study at Büyük Menderes Basin, Turkey	119
Chapter 8	Diurnal Effects on The Efficiency of Drip Irrigation	133
Chapter 9	Potential of Waste Water Use for Jatropha Cultivation in Arid Environments	165
Chapter 10	Comparison of Different Methods to Estimate Mean Daily Evapotranspiration from Weekly Data at Patna, India	185

Chapter 11 Effect of Reference Conveyance Parameter Usage on Real Time Canal Performance: The Case of Fentale Irrigation Scheme in Ethiopia 201

Chapter 12 Optimizing Preplant Irrigation for Maize under Limited Water in the High Plains 215

Chapter 13 Assessment of Wastewater Quality of Drains for Irrigation 239

Chapter 14 Evaluating the Usefulness of Continuous Leaf Turgor Pressure Measurements for the Assessment of Persimmon Tree Water Status 257

Chapter 15 Effect of Deficit Irrigation on Water Productivity of Onion (Al lium cepal.)under Drip Irrigation 277

Chapter 16 The effects of applied water at various fractions of measured evapotranspiration on water relations and vegetative growth of Thompson Seedless grapevines 289

Chapter 17 Wheat Yield Response to Water Deficit under Central Pivot Irrigation System Using Remote Sensing Techniques 313

Chapter 18 Effect of Reference Conveyance Parameter Usage on Real Time Canal Performance: The Case of Fentale Irrigation Scheme in Ethiopia 327

Chapter 19 Determination of Evapotranspiration and Water Use Efficiency in Crop Production 341

Chapter 20 A Study on the Causes of Sediment Accumulation in the Drainage Systems 359

Chapter 21 Applications of SuDS Techniques in Harvesting Stormwater for Landscape Irrigation Purposes: Issues and Considerations 375

Index 399

CHAPTER 1

A Spatial Analysis of Irrigation Technology

Andrew Wright, Darren Hudson, Maria Mutuc

Department of Agricultural and Applied Economics, Texas Tech University, Lubbock, USA.

ABSTRACT

The nature of spatial spillovers in the adoption of irrigation technology is examined in this paper. Adopting a new technology is a decision that is based on economic and individual-specific factors. One of these individual factors might be communication with other users. It makes sense to expect that contact between users and non-users would follow a spatial pattern, and if knowledge spillovers are important to the adoption decision then resource managers need to be aware of their existence. Using counties in the Texas High Plains as the study area, the adoption of center pivot technology is examined using both Ordinary Least Squares and spatial regression models to determine if knowledge spillovers exist. Ultimately, no evidence was found that adoption practices in a county affects its neighbors; however, geographic location does matter to who adopts and when.

Keywords: Technology; Adoption; Water; Irrigation; Natural Resources

1. INTRODUCTION

In the production of a good, the value of a per unit input, and by extension how much of it is used at any given time, is determined in part by the technology related to the input's use. Ceteris paribus, technological progress decreases the costs faced by firms and increases how much of the input is used in the production process. When the input in question is a stock resource, technological advancement implies that the resource will be depleted at a faster rate.[1] It is, therefore, important for anyone whose responsibilities include resource management to understand the process by which technology is adopted by potential users and spreads through a population.

Technology adoption is important to resource management for any number of reasons. The introduction of a more efficient technology changes the incentives faced by producers; thus, any policies that aim at the resource's conservation may no longer achieve the desired behavior. Consequently, it is important for resource managers to understand which individuals are most likely to adopt the new technology and how the technology will spread over time in order to adjust rules regarding resource use accordingly. On the other hand, policy makers might view more efficient technology as a means to encourage conservation. In this case, understanding the diffusion process will make it easier to identify those individuals or regions that will have the most influence on how quickly the technology spreads.

A real world example of the importance of technology diffusion to resource management is the advancement of irrigation technology in arid regions such as the Texas High Plains. Agricultural production in this region of Texas is a part of the foundation of the region's economy, and part of the reason for this is the availability of groundwater from the Ogallala Aquifer for irrigation. Over time, producers have found more efficient means of extracting water from the aquifer. Early farms used furrow irrigation to water crops. Sprinklers began replacing furrow irrigation in the mid-20th century; early sprinklers were high pressure systems that required a large amount of energy to operate. In the late 1980s, low energy precision application (LEPA) technology was introduced, and is the primary irrigation technology used today.

As irrigation technology has become more efficient at extracting water from the aquifer, it has become economically viable to irrigate more acres (or irrigate the same acres more intensely). This extraction rate increase has led to concerns about the long-term viability of the aquifer as a resource for irrigation water and the development of plans and policies to curb extraction from the aquifer. In the midst of these concerns, a new irrigation technology has begun to gain prominence, sub-surface drip, which supplies water directly to the root zone of the plant as opposed to sprinklers that deliver water above the surface which is absorbed by the plant through the soil. As such, it is more efficient in its application of water and generally reduces the variable cost of irrigating. Producers, therefore, have an opportunity to increase the amount of water they can take from the aquifer where available, and as drip becomes more prevalent, decision makers concerned about conservation must consider how the spread of this new technology will change water use patterns.

The spread of a technology involves both adoption and diffusion. Adoption occurs when an individual chooses to accept and use the new technology. The decision to adopt can be thought of as an economic choice, where individuals seek to maximize utility or profits [1-4]. Individuals adopt a new innovation if it increases their net revenue from production. This idea has been expanded to acknowledge that individuals are heterogeneous in terms of their preferences as well as the physical and economic factors they must consider when choosing to adopt a new technology [1,2,5].

Diffusion can be thought of as the path that adoption takes through a population over time. Theoretically, the diffusion of technology from one person to another has been closely linked to interactions between individuals [6-11]. As users and nonusers interact, the users of a new technology communicate information about its benefits and its use. Nonusers take this information and use it to decide whether to adopt or not; thus diffusion can be thought of as a process of communication and imitation.

This transfer of knowledge through communication constitutes an information externality associated with the adoption of the new technology. When innovators and early adopters adopt new technology, they take on the costs associated with the risk of using the new technology. Other potential users benefit from the experience of early adopters through person to person transfer of the knowledge gained, and are able to adopt the new technology without the risk involved with early adoption. If the transfer of knowledge is informing adoption decisions, and consequently the diffusion of a technology, then it is necessary to understand the strength of this spillover relative to other factors.

This study examines the presence of information spillovers in the adoption of center pivot systems on the Texas High Plains. Center pivot technology is examined because, as opposed to drip systems, it has reached the end of its diffusion cycle, and there is more reliable data regarding its use. If the adoption of drip technology is determined by similar factors as center pivot systems, then the results of this study can be used as a benchmark for later studies involving drip adoption.

In the study, the number of center pivot systems in each county that is a part of the High Plains Underground Water Conservation District No. 1 (HPWD) is related to the physical qualities of the county such as the saturated thickness of the Ogallala Aquifer at different points in time and the county's geographic location. These physical characteristics are of particular interest, as it is likely that the area of the aquifer a county overlies will determine how many irrigation systems exist and how quickly this number grows. Furthermore, if there are counties that adopt first then it would be interesting to examine how much influence these counties have on the adoption level of their neighbors, or whether or not information spillovers exists in the adoption process. To this end, spatial econometric models are employed to account for any spatial dependence present in the number of center pivots in each county.

2. ELEMENTS OF THE ADOPTION/DIFFUSION PROCESS

The choice to adopt new technology can be thought of as an economic decision where individuals switch to a new innovation because it increases their net revenue from the production of a good. Thus, the decision to adopt will be driven in part by the costs associated with the new technology. References [8] and [10] separate the cost of adopting new innovations into "hardware" costs and "software" costs.

Hardware costs relate to the actual purchase of the technology and its maintenance, and technologies such as irrigation systems that exhibit large hardware costs often rely on a loan to fund the purchase. Thus, access to credit may help these innovations to spread [3,12,13]. Software costs relate to the cost of learning to use a technology, and the ability to learn is related to the concepts of human capital and education. Firms with high amounts of human capital are able to take advantage of their employees' capabilities and adopt earlier [3,4,13]. Individuals who are better educated can adopt earlier because they face a lower software cost, and due to their greater wealth are better able to afford the hardware costs [14]. Software costs can be reduced through personal experience and communication with existing users, so complicated technologies tend to spread from more skilled users to less skilled users as the stock of knowledge related to the technology increases [8].

Aside from the cost of a new technology, the decision to adopt may be influenced by a number of physical factors as well. While some physical factors, such as firm size, will affect technology adoption in general, these factors are especially important to agriculture due to the relation of agricultural production to physical qualities of the land itself.

Previous studies have found a relationship between the size of a firm and the ability to adopt new technology [15-17]. These authors argue that larger firms are better able to absorb the costs related to adopting a new innovation; thus, there is a lower limit to the size of a firm that can adopt an innovation, and the critical size increases along with the amount of fixed costs involved with the decision. Technology then diffuses from large firms to small firms.

In studies related to irrigation adoption, the effects of land quality, well depth, and water price on adoption have been studied in a variety of frameworks. Land quality, and related factors such as soil type, as well as well depth will determine whether traditional (furrow) irrigation systems or modern (sprinkler or drip) irrigation systems are used [1]. Specifically, the authors show that modern technology is more likely to be adopted on lower quality land, or in areas with deeper wells. Similarly, [18] examines irrigation technology adoption in an exhaustible resource framework and finds that individuals will not adopt modern technology on their highest quality land, but do adopt modern technology as water price increases. Another study of irrigation adoption by [5] confirms the relation between water price and the adoption of modern technology, in this case drip. The authors also find that the adoption of center pivot technology is less sensitive to land quality than that of drip due to the dependence of drip irrigation on field slope [5].

The decision to adopt a new technology is ultimately an individual decision based in part on both economic and physical factors. The difficulty, then, of modeling the adoption process is in collecting individual level data for analysis. The amount of information that must be collected is large, and the time to do so normally intensive. By studying adoption at the aggregate level, as this study proposes to do, the data collection process is much less effort intensive, but many of the micro level variables that affect the adoption decision must be dropped from the study; however, other important features, such as the existence of knowledge spillovers between regions can be explored. Identifying and measuring spillovers requires the use of econometric methods that model spatial dependence between observations. The next section explains how and why these spatial relationships exist and how they might affect a county level analysis.

3. SPATIAL RELATIONSHIPS AND IRRIGATION ADOPTION

Spatial location and technology adoption can be linked through the physical location of users and through the geographic relationship between current users and potential adopters. The relationship between physical location and the decision to adopt a new irrigation technology has been explained in detail in the previous section. To summarize, physical characteristics related to an individual's land will determine where adoption is economically viable. When neighboring counties share similar characteristics, for instance if the saturated thickness of an underlying aquifer is similar, it is reasonable to assume that they will share aggregate irrigation levels as well.

How the geographic relationship between current users and potential adopters affects technology adoption and diffusion is less clear, but not difficult to explain. Early concepts of the diffusion process, [7] for example, modeled the diffusion process as similar to a disease epidemic where the vector is contact between current and potential adopters [10]. As potential adopters come into contact with current users, they are "infected" with the new technology and become users themselves. While these early models did not account for proximity between current and potential users, it should be clear that the probability of contact becomes more likely when the two groups are close to one another. Thus, a technology should diffuse from current users to the individuals closest to them if spatial spillovers are present.

From the above explanation, spatial relationships are clearly important to consider when studying the adoption and diffusion of irrigation technology; furthermore, multiple empirical studies of technology diffusion provide evidence of spatial effects in the diffusion process [19- 21]. With this in mind, the growth of center pivot technologies in the HPWD area is analyzed using a set of spatial regression models. The next section describes the data and variables used in the estimation these models, and the structure of the models themselves.

4. DATA AND METHODS

4.1. Variables and Sources

This study assumes that adoption of irrigation technology will depend on a set of county specific physical factors faced by producers. Irrigation technology adoption is measured using center pivot counts for each county in the HPWD. Information on the number of center pivots systems was provided by the staff of the HPWD [22,23]. The water district collects this information on a semiannual basis only, thus this information was collected for the years 1986, 1990, 1993, 1995, 2005, and 2008, and seven cross sections were developed.

Two measures of saturated thickness are used as independent variables. The first of these is the saturated thickness of the aquifer in the current period. Saturated thickness is closely related to the cost of irrigation, which has been shown to affect the adoption choice [5,18]. It is expected that as the saturated thickness of the aquifer decreases, and the price of pumping water increases, that producers will switch to a more efficient irrigation system. The second variable is the change in saturated thickness over time, and is included to account for the fact that producers might take past experience into account along with current conditions when making the choice to adopt a new technology. In other words, in counties where the reduction in saturated thickness over time is larger, producers might choose to adopt more efficient irrigation technology to extend the life of the aquifer.

Saturated thickness information was collected primarily from the Texas Tech Center for Geospatial Technology (TTUCGT) [24]. The staff at the TTUCGT has developed county average measures of the characteristics of the Ogallala aquifer, such as water storage and saturated thickness. The TTUCGT spans the years 1990-2008. For years prior to 1990, saturated thickness was estimated using the average change in well depth for each county, reported annually in the HPWD newsletter. This calculation started with the 1990 saturated thickness in county i as the baseline, and saturated thickness in the previous year, ST_{t-1}, was related to the current year saturated thickness, ST_t, and change in well depth reported in year t, w_t, so that $ST_{t-1} = ST_t - w_t$.

Other physical characteristics considered in this study are the distance between a county and an experiment station, the amount of irrigated production in a county, and the aridity level in a county. The distance from an experiment station is considered important to the adoption choice because of the role experiment stations play in introducing new technology. These stations act as the central source for information referred to in [6-8], and [10] that introduces the innovation to a population through demonstrations of the new technology. Attending these demonstrations requires that producers travel to the stations, and it is assumed that this will be easier for producers who are closer to a station. There are two experiment stations in the HPWD, so the distance from county to station was calculated as the average distance from the county seat to an experiment station.

The level of irrigated production in a county signifies how willing producers might be to pursuing new irrigation technology. In counties where the amount of irrigated production is high, producers should be more interested in learning about and adopting more efficient irrigation systems. To measure the level of irrigated agriculture practiced in each county, acreage amounts for corn, cotton, sorghum, and wheat where retrieved from the National Agricultural Statistics Service (NASS) quick stats data base [25]. The number of irrigated acres was divided by total acreage to calculate the percent of agriculture that depended on irrigation.

The aridity level in a county indicates how important irrigation is to agricultural production. Producers in counties that are hotter and dryer will rely more on irrigation to supplement natural precipitation, and should be more likely to switch to more efficient irrigation technology when it becomes available. To measure aridity, the Lang Index [26] was calculated for each county in each year. This index is calculated as P/T where P is the annual precipitation in centimeters and T is the average annual temperature in C°. Temperature and precipitation data was retrieved from the Western Regional Climate Center website [27].

Once the data is collected, it is sorted into seven cross sections, one for each year that the number of center pivots in the HPWD was counted. Each cross section had thirteen observations. There are fifteen counties in the study area; however, only small parts of Armstrong and Potter Counties are actually a part of the HPWD, and these counties practice low levels of crop agriculture. For this reason it is assumed that these counties would have little influence on center pivot adoption in neighboring counties, and would not be influenced by interactions with other counties in the study area. There is some obvious concern about the small number of observations in a given year and the effect on degrees of freedom; however, there is no easy way around this limitation. The accounting procedure for center pivot numbers is only semi-annual in the HPWD which limits the number of years with data available and prevents the use of panel estimation methods; furthermore, expanding the study area to include other water district introduces new concerns such as the effect of differing organizational structures, accounting practices, and data collection procedures.

Additional information collected for this study includes the market average prices reported by NASS and the interest rate for the 3-month T-bill [28]. Crop prices and interest rates may influence the adoption decision; however, while these variables may show variation over time, they do not change from county to county. As the data related to adoption is organized into cross sections, any effect that these variables might have on adoption would simply reduce to a fixed effect and enter the intercept; therefore, crop prices and the interest rate are not included in the spatial analysis, but were regressed against center pivots to determine their level of correlation with the number of center pivot in a county.

With the data for each county organized into cross sections, exploratory analysis was conducted to confirm the existence of spatial clustering in center pivot counts. Then, an Ordinary Least Squares (OLS) model was estimated, and tested for spatial dependence in the error term and dependent variable using the Moran's I statistic and Lagrange Multiplier (LM) tests. If these tests confirmed spatial dependence, the model was re-estimated using spatial regression techniques. The specification of these models is explained below.

4.2. The OLS Model

In all of the models estimated, the dependent variable is the log of the number of center pivots (CP) in each county. The independent variables in the OLS model include the log of saturated thickness (ST), the 5-year change in saturated thickness (chg), the average distance to an experiment station (dist), the level of irrigated agriculture (irr), and the aridity index (arid):

$$\ln(CP) = \beta_0 + \beta_1 \ln(ST) + \beta_2 chg + \beta_3 dist + \beta_4 irr + \beta_5 arid + \in . \quad (1)$$

These variables are chosen because they affect the center pivot count in a county and have a reasonable amount of variation at the county level of aggregation; however, the estimated parameters from this equation are not the main interest of this study. What is of interest is whether any spillovers exist from the center pivot counts in neighboring studies.

4.3. The Spatial Lag Model

If spatial dependence exists in county center pivot counts, it means that the number of center pivots in each county is dependent on the number of center pivots in neighboring counties and can be represented by a spatial lag model (SAR): $y = \rho Wy + X\beta + \varepsilon$.

In this specification, a spatially lagged dependent variable, ρWy, is included as an independent variable to account for the effect of neighboring observations. In the SAR model, W is a spatial weighting matrix and ρ is a spatial autoregressive parameter that measures the strength of spatial interactions between neighbors. A statistically significant ρ implies that spillovers related to center pivot counts exist between counties. To account for spatial dependence in center pivot counts, Equation (1) is modified so that:

$$\ln(CP) = \beta_0 + \rho W \ln(CP) + \beta_1 \ln(ST) + \beta_2 chg + \beta_3 dist + \beta_4 irr + \beta_5 arid + \in . \quad (2)$$

4.4. The Spatial Error Model

A second form of spatial dependence is spatial autocorrelation, or spatial dependence in the error term. When spatial autocorrelation exists, observations are grouped in such a way that measurement error depends on location. Examples of this would be if counties with high center pivot counts tended to be grouped together (positive spatial autocorrelation), or if counties with high center pivot counts tended to neighbor counties with low center pivot counts (negative spatial autocorrelation). The spatial error model (SEM) accounts for spatial autocorrelation by modifying the error term in the OLS specification, $y = X\beta + \varepsilon$, so that $\varepsilon = \lambda W\xi + \mu$, where ξ is the spatially correlated part of the error term and $\mu \sim (0, \sigma^2 I_n)$ is spatially uncorrelated. This specification is identical to an OLS model, save for the error term, which now exhibits spatial dependence. In the SEM model, λ is a spatial autoregressive parameter that measures the strength of spatial correlation between errors, and μ is the portion of the error term that satisfies the assumptions of a normal regression model.

While spatial autocorrelation does not imply the existence of spillovers between counties, the existence of spatial autocorrelation in center pivot counts would have important implications for the diffusion of irrigation technology in this region. In this context, the presence spatial autocorrelation would imply that a county's physical location, and perhaps its physical characteristics, is what drive the adoption decision. If this is the case then future attempts to understand the adoption decisions of producers would need to rely on a micro level approach that considers physical factors.

4.5. The Spatial Weights Matrix

Both of the spatial model specifications rely on the use of the spatial weights matrix to model spatial interaction. In the matrix, each element, w_{ij}, represents the spatial relationship between points i and j, and $w_{ij} = 0 \ \forall \ i = j$. The types of weights matrices are many, and there is no formal method of choosing one method of modeling distance over another [29]. For this analysis a row standardized inverse distance matrix is used where distance,

d_{ij}, is the Euclidian distance between two counties, and each element is standardized by dividing it by the number of neighbors, in this case 13, so the value of every off diagonal element is $w_{ij} = d_{ij}^{-1}/13$ To calculate the Euclidian distance between two counties, the geographic coordinates for the county center were taken from the Texas State Historical Society Website [30]. Assuming the geographic coordinates for two counties are the corners of a triangle adjacent to the hypotenuse, the distance between counties can be calculated using the Pythagorean Theorem.

Using the type of weighting matrix specified above assumes that all counties in the study area will have some affect on each other, but the effect will decrease as distance increases. Using a different type of weighting matrix could potentially change the nature of any spatial relationships discovered in the regression analysis. For example, using a k nearest neighbor (knn) matrix would only consider a k number of specified counties as neighbors, and only these neighbors would be considered when measuring spatial relationships. Considering the small study area, and that all of the counties included in the data are part of the same regulatory institution through membership in the HPWD, it is assumed that even individuals in the counties farthest away from each other would have some chance for contact with each other. Using an inverse distance matrix with no limit placed on distance captures these interactions.

5. RESULTS

5.1. Correlation with Price and Interest Rate

In the analysis of aggregated spatial effects, economic variables, such as price and the interest rate, are left out because they do not vary for each observation in the cross section; thus, these variables would reduce to a fixed effect and enter the intercept during the estimation process. To examine how these variables impact the number of center pivots in a county, a regression was performed using an index of crop prices and the interest rate as independent variables. The price variable (P) is an index of the average market price reported by NASS for corn, cotton, peanuts, sorghum, and wheat. The year 1986 was used as the base and was set at 100. The interest rate (i) used was the 90 day T-bill rate. The model took the following form:

$$\ln(CP) = \beta_0 + \beta_1 \ln(P) + \beta_2 \ln(i) + \in . \quad (3)$$

The results in **Table 1** show that while these variables may be important to the individual, they have no statistically significant correlation with the aggregate level of center pivot adoption, thus confirming that omission of these variables from subsequent regressions will not result in omitted variable bias on the basis of interest rates and prices.

5.2. Exploratory Analysis

To confirm the existence of spatial clustering in center pivot numbers, a global Moran's I test statistic was calculated for the dependent variable in each cross section. This test, proposed by [31] examines whether or not the distribution across space of observations with similar values for a variable of interest is random or not. The Moran's I statistic is a value of global spatial autocorrelation which is calculated as

$$I = \frac{\sum_{i=1}^{n}\sum_{j=i}^{n} w_{ij} Z_i Z_j}{S_0 m_2} \tag{4}$$

where w_{ij} is an element of the weighting matrix, $Zi = Yi - \overline{Y}$, Y_i is the value of the variable of interest at location i, $\overline{Y}$ is the mean of the variable of interest, $S_0 = \sum_i \sum_j w_{ij}$, and $m_2 = \sum_i Zi^2 / n$. The statistic is compared to the expected value of I, $E(I) = -1/(N-1)$ using a z-score, and a positive I value implies positive autocorrelation. The I statistic for each cross section is reported in **Table 2**.

Table 1. Regression of price and interest rate variables on center pivot systems.

	Coefficient	Std. Err.	t	p-value
Price index	0.490	0.906	0.54	0.589
Interest rate	−0.623	0.461	−1.35	0.18
constant	4.018	4.972	0.81	0.421
R-squared	0.06			
Adj. R-square	0.04			

Table 2. Moran's I values for the dependent variable (lnCP).

Year	I	$E(I)$	$sd(I)$	Z	p-value
1986	0.062	−0.083	0.046	3.161	0.002
1990	0.027	−0.083	0.046	2.410	0.016
1993	0.008	−0.083	0.046	1.991	0.047
1995	−0.031	−0.083	0.046	1.133	0.257
1998	−0.051	−0.083	0.046	0.703	0.482
2005	−0.064	−0.083	0.046	0.418	0.676
2008	−0.076	−0.083	0.046	0.160	0.873

The Moran's I values indicate the existence of spatial autocorrelation in only the first three cross sections. To augment the Moran's I statistic, Moran's scatter plots were created that give a visual representation of these results by plotting observed values of the variable of interest against the spatially lagged variable of interest. The Moran scatter plot for all years are shown in Figures 1-7. Points in the upper right quadrant indicate counties with high center pivot counts that are close to other counties with high center pivot counts. Points in the lower left quadrant indicate counties with low center pivot counts that are close to other counties with low center pivot counts.

The points of the scatter plots for the first three years have a positive skew, but the skew is decreasing from year to year. By 1995, the points are beginning to cluster more around the origin. The scatter plots for the years 1998-2008 show that the observations continue to cluster more tightly around the origin as time passes, with one exception. Randall County is an extreme outlier in the 1995-2008 scatter plots because it has the lowest number of center pivots of all the counties and is next to a county with one of the highest center pivot counts. While this may be evidence in support of dropping Randall from the rest of the analysis, the county was kept due to concerns about the already low number of degrees of freedom.

From a diffusion perspective, the reduction in spatial autocorrelation over the time period studied suggests that specific characteristics of a county determine where early adoption occurred at the individual level. In the 1986 cross section, for example, the two upper-rightmost ob servations are Bailey and Parmer counties. These two counties are situated next to each other in the region and

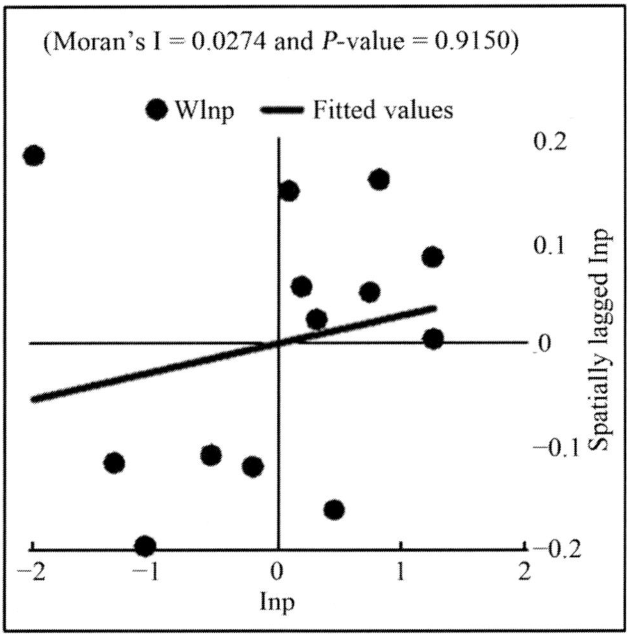

Figure 1. Moran's scatter plot of 1986 center pivots.

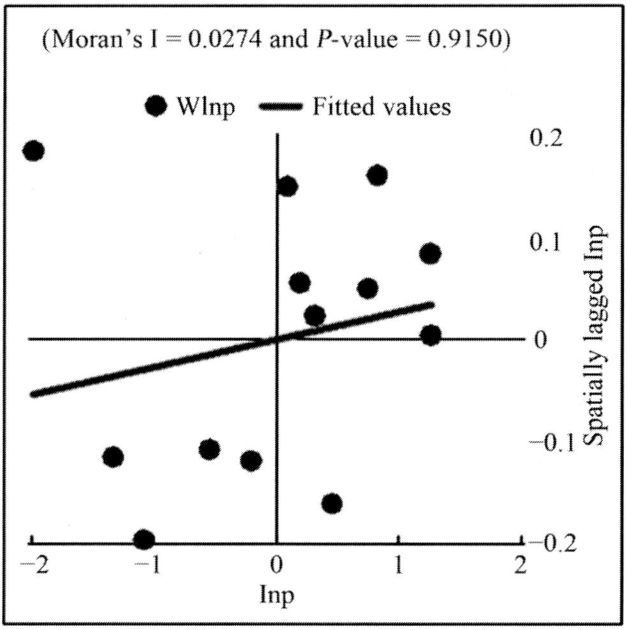

Figure 2. Moran's scatter plot of 1990 center pivots.

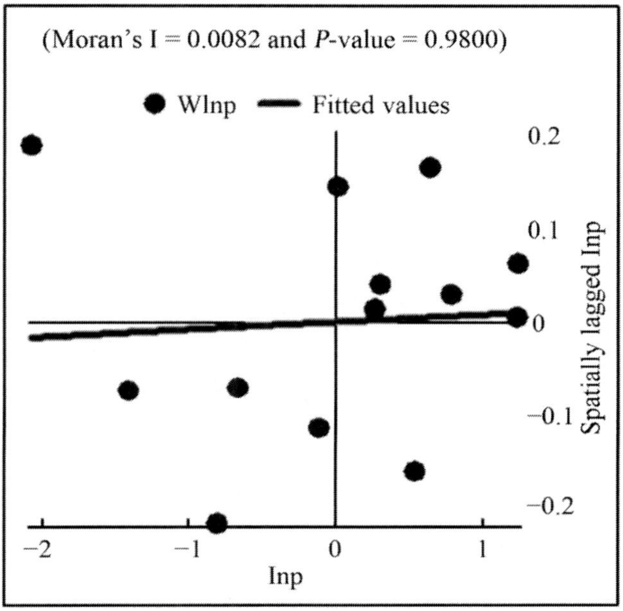

Figure 3. Moran's scatter plot of 1993 center pivots.

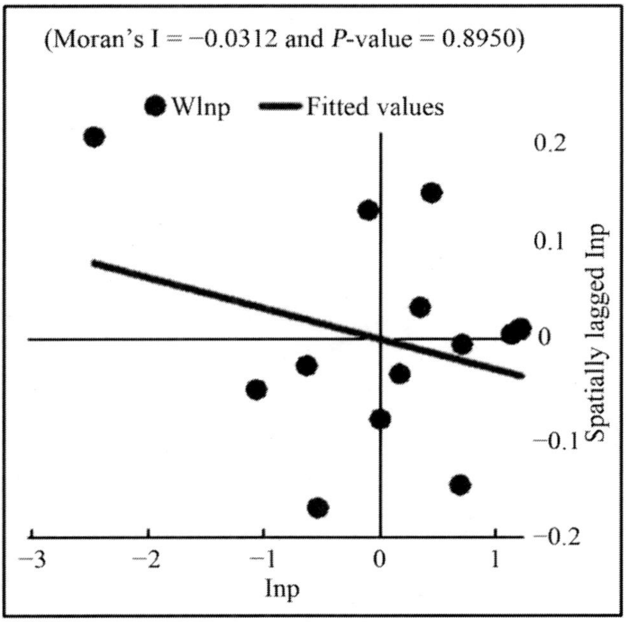

Figure 4. Moran's scatter plot of 1995 center pivots.

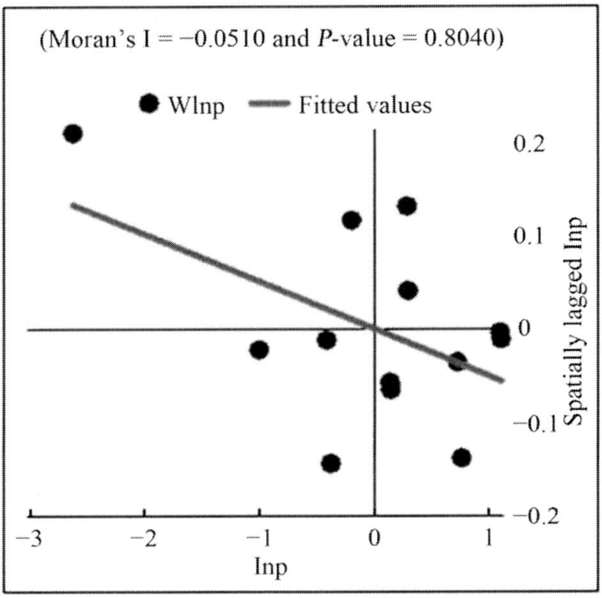

Figure 5. Moran's scatter plot of 1998 center pivots.

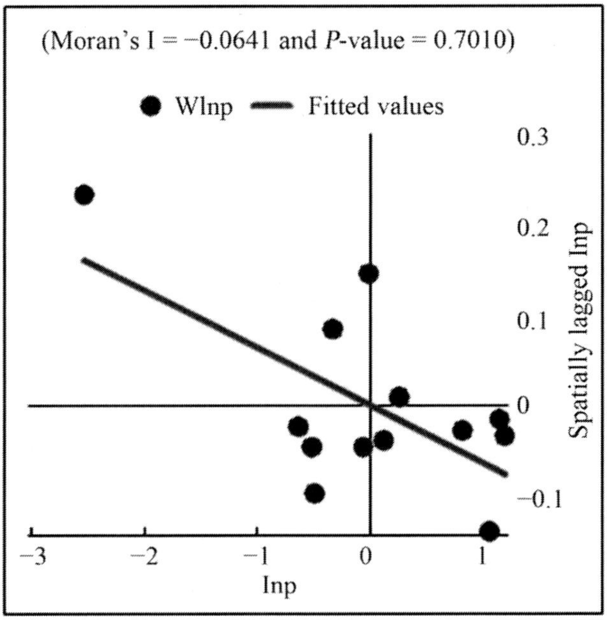

Figure 6. Moran's scatter plot of 2005 center pivots.

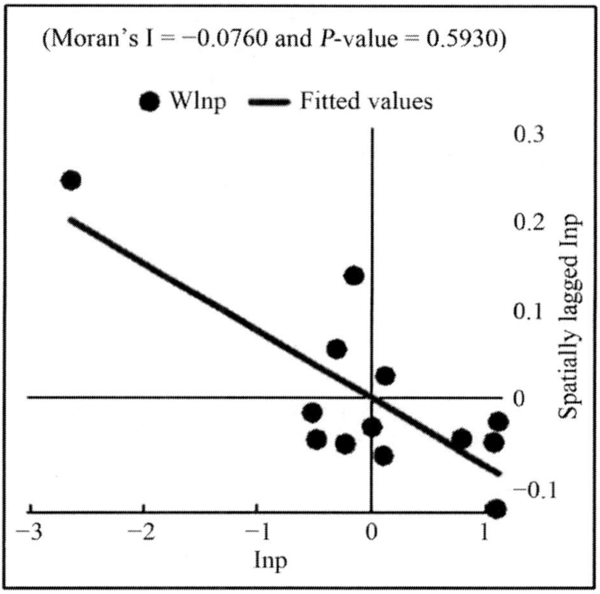

Figure 7. Moran's scatter plot of 2008 center pivots.

also share the trait of high saturated thickness in the aquifer. If this and other traits influence individual adoption decisions are spatially located then it would make sense for earlier years to show spatial autocorrelation, and for the autocorrelation to disappear as center pivot technology diffuses across the region.

5.3. Regression Analysis

The OLS and spatial models were estimated using STATA [32-34] for 1986, 1990, and 1993, and the results are reported in Tables 3-5. The first column in each table reports the results from the OLS regression. The second column pertains to the spatial error model (SEM) estimation and the last column reports results from the spatial lag model (SAR). At the bottom of the OLS column are the results from spatial diagnostic tests regarding the existence of spatial dependence. The first test is the Moran's I, which is significant at the 1% level in all three of these years and indicates that the residuals from the OLS regressions exhibit spatial autocorrelation.

The Lagrange Multiplier (LM) tests listed under the Moran's I statistic are used to determine how spatial dependence enters the model. For both spatial dependence in the error term and spatial dependence in the dependent variable, two test statistics are listed. A weakness of the standard LM tests is that the LM test for λ responds to a non-zero ρ, and the LM test for ρ responds to a non-zero λ [35,36]. The robust tests are meant to correct for this, but are only meaningful when the standard tests are significant [36].

In contrast to the Moran's I statistic, the LM tests fail to reject the null hypothesis of no spatial dependence in either the error term or the dependent variable. To verify this result the SEM and SAR models were estimated for all three years, and Wald and LM tests run for $\lambda = 0$ and $\rho = 0$. The difference in results between the Moran's I statistic and the LM test for spatial dependence in the error term is an interesting result. It would seem that when one is significant the other should be as well. One possible explanation may be that the spatial clustering in the HPWD simply does not affect the error term, either because of the small size of the area or for some other unknown reason. Ultimately, however, the existence or lack thereof of spatial autocorrelation is of secondary importance. The goal of the analysis is to identify spatial spillovers. The results for these three counties make it impossible to reject the null hypothesis of no spatial dependence in the dependent variable; therefore the conclusion is that for these years there are no spillovers present.

While global Moran's I statistics showed no sign of spatial autocorrelation in the last four cross sections, OLS models were estimated to confirm these results, the results of which are listed in **Table 6**. For the years 1995 – 2005 the results are what would be expected based on the exploratory analysis; however, the Moran's I for spatial autocorrelation in the residuals in 2008 is significant at the 5% level. Estimating the SEM and SAR models, the results of which can be found in **Table 7**, produce similar findings as in earlier years, and no evidence of spatial dependence was found.

6. DISCUSSION AND CONCLUSIONS

This study measures the level of spatial dependence in the number of center pivots in existence in the HPWD area using spatial econometric methods. The results indicate that while center pivot technology is certainly clustered spatially, at least in the earlier years of the study, there is no evidence of spatial interdependence in either the dependent variable or the error term in any of the seven cross sections studied.

This curious result may be a consequence of the small size of the study area. With only a few counties to be neighbors with, it is likely that counties with high center pivot values can be clustered together and still be near neighbors to counties with low center pivot values simply because of

Table 3. Regression results for the 1986 cross section.

	OLS	SEM	SAR
Sat. Thickness	−3.821*	−4.122**	−3.754**
	(1.571)	(1.351)	(1.110)
Δ in S.T.	−1.863*	−1.935**	−1.819**
	(0.620)	(0.484)	(0.423)
% Irrigated Acres	3.113	3.296	3.129
	(3.415)	(2.711)	(3.967)
Dist. to Station	−0.014	−0.014	−0.016
	(0.022)	(0.017)	(0.017)
Aridity Index	−0.122	−0.141	−0.109
	(0.484)	(0.384)	(0.338)
constant	20.556**	21.814**	19.080**
	(6.067)	(5.253)	(3.216)
λ		−0.589	
		(1.422)	
ρ			0.242
			(0.516)
Observations	13	13	13
F-Statistic	4.00*		
R-squared	0.74		
Adj. R-squared	0.56		
Sq. Correlation		0.74	0.75
Moran's I (λ)	3.55**		
LM (λ)	0.07		
Robust LM (λ)	2.65		
LM (ρ)	0.07		
Robust LM (ρ)	2.65		
Wald Test ($\lambda = 0$)		0.17	
LM Test ($\lambda = 0$)		0.21	
Wald Test ($\rho = 0$)			0.22
Thiel's U	0.13	0.13	0.13

The SAR model uses robust Hubert-White standards errors; *significant at $p \leq 0.05$; **significant at $p \leq 0.01$.

Table 4. Regression results for the 1990 cross section.

	OLS	SEM	SAR
Sat. Thickness	−1.989	−1.936	−1.887
	(2.415)	(2.174)	(0.842)
Δ in S.T.	−0.226	−0.212	−0.163
	(0.709)	(0.156)	(0.081)
% Irrigated Acres	5.028	5.012	5.083
	(4.282)	(2.080)	(1.966)
Dist. to Station	0.012	0.012	0.011
	(0.027)	(0.016)	(0.015)
Aridity Index	−1.222	−1.203	−1.232
	(1.034)	(0.572)	(0.403)
constant	13.846	13.575*	11.506*
	(8.237)	(8.099)	(4.028)
λ		0.090	
		(2.482)	
ρ			0.395
			(0.379)
Observations	13	13	13
F-Statistic	1.63		
R-squared	0.54		
Adj. R-squared	0.21		
Sq. Correlation		0.54	0.55
Moran's I (λ)	3.92**		
LM (λ)	0.00		
Robust LM (λ)	4.59*		
LM (ρ)	0.16		
Robust LM (ρ)	4.75*		
Wald Test ($\lambda = 0$)		0.01	
LM Test ($\lambda = 0$)		0.01	
Wald Test ($\rho = 0$)			0.40
Thiel's U	0.15	0.15	0.15

The SAR model uses robust Hubert-White standards errors; *significant at $p \leq 0.05$; **significant at $p \leq 0.01$.

Table 5. Regression results for the 1993 cross section.

	OLS	SEM	SAR
Sat. Thickness	−4.317*	−4.712*	−4.281**
	(1.452)	(2.174)	(0.842)
Δ in S.T.	−0.370*	−0.396**	−0.370**
	(0.130)	(0.156)	(0.081)
% Irrigated Acres	1.216	1.239	1.281
	(2.701)	(2.080)	(1.966)
Dist. to Station	−0.015	−0.015	−0.016
	(0.020)	(0.016)	(0.015)
Aridity Index	0.312	0.386	0.288
	(0.549)	(0.572)	(0.403)
constant	22.843**	24.288**	19.340**
	(5.766)	(8.099)	(4.028)
λ		−0.534	
		(2.482)	
ρ			0.588
			(0.379)
Observations	13	13	13
F-Statistic	4.15*		
R-squared	0.75		
Adj. R-squared	0.57		
Sq. Correlation		0.75	0.79
Moran's I (λ)	3.06**		
LM (λ)	0.02		
Robust LM (λ)	9.85**		
LM (ρ)	0.93		
Robust LM (ρ)	10.76**		
Wald Test ($\lambda = 0$)		0.05	
LM Test ($\lambda = 0$)		0.09	
Wald Test ($\rho = 0$)			2.41
Thiel's U	0.09	0.09	0.08

The SAR model uses robust Hubert-White standards errors; *significant at $p \leq 0.05$; **significant at $p \leq 0.01$.

Table 6. OLS estimates for the 1995, 1998, 2005, and 2008 cross sections.

	1995	1998	2005	2008
Sat. Thickness	−0.252**	−0.105	0.022	−0.099
	(0.527)	(1.207)	(0.380)	(0.485)
Δ in S.T.	−0.216	2.651	4.582**	0.844
	(0.029)	(0.092)	(0.033)	(0.046)
% Irrigated Acres	−0.007	−0.004	0.009	−0.022
	(0.943)	(2.848)	(0.864)	(1.190)
Dist. to Station	−0.330*	−0.093	0.498	0.678
	(0.005)	(0.016)	(0.006)	(0.013)
Aridity Index	13.032**	11.615*	5.062**	7.162**
	(0.140)	(0.615)	(0.248)	(0.375)
constant	−1.445*	−1.559	−0.663	−0.560
	(1.692)	(4.296)	(1.456)	(1.825)
Observations	13	13	13	13
F-Statistic	32.18**	1.77	10.94*	6.43*
R-squared	0.96	0.56	0.89	0.82
Adj. R-squared	0.93	0.24	0.81	0.69
Moran's I (λ)	-2.17	1.56	0.38	2.33*
LM (λ)	1.99	0.31	0.73	0.44
Robust LM (λ)	3.76	5.48*	0.87	2.72
LM (ρ)	0.09	0.09	0.28	0.16
Robust LM (ρ)	1.85	5.26*	0.42	2.44

*significant at $p \leq 0.05$; **significant at $p \leq 0.01$.

Table 7. SEM and SAR results for the year 2008.

	SEM	SAR
Sat. Thickness	−0.784*	−0.519
	(0.353)	(0.406)
Δ in S.T.	−0.110**	−0.010*
	(0.032)	(0.041)
% Irrigated Acres	0.817	0.694
	(0.956)	(1.067)
Dist. to Station	−0.024**	−0.021*
	(0.009)	(0.009)
Aridity Index	0.686**	0.618*
	(0.269)	(0.311)
constant	8.135**	11.926
	(1.147)	(7.687)
λ	−2.07	
	(1.456)	
ρ		−0.693
		−1.092
Sq. Correlation	0.74	0.75
Wald Test ($\lambda = 0$)	2.02	
LM Test ($\lambda = 0$)	1.88	
Wald Test ($\rho = 0$)		0.4
Thiel's U	0.04	0.04

The SAR model uses robust Hubert-White standards errors; *significant at $p \leq 0.05$; **significant at $p \leq 0.01$.

aquifer characteristics. One way to deal with this might be use a different weighting matrix, such as a binary weights matrix that defines neighbors as only those observations within a critical distance, or a k nearest neighbor (knn) matrix that defines neighbors as only the k nearest observations.

Another explanation for the inability of the models to detect spatial dependence is the existence of an extreme outlier in Randall County. This would be especially true for the detection of spatial autocorrelation, which Moran's I statistics detected in 1986, 1990, and 1993. Again, though, the small sample size becomes a problem. While dropping Randall County out of the data set might remove an outlier, it also removes an observation.

With only thirteen observations the degrees of freedom in the model is very restricting. The problem is finding data for enough observations in the same time period. The ideal data set would provide information on the irrigation adoption decisions of individuals; however, if the goal is to model spatial dependence then collecting data for enough neighboring individuals for even one time period would be a daunting task.

Two options that might improve this analysis in the future would be to either analyze the study as a panel or to reduce the number of independent variables. Taking the data as a panel would increase the number of observations, but the semiannual center pivot counts make panel estimation difficult. Reducing the number of independent variables would free up some degrees of freedom, but in an aggregate level study that is already ignoring a number of micro level variables, what should be dropped?

It might be interesting to develop an index variable that takes into account all of the pertinent physical characteristics of a region, which would reduce the number of independent variables to one. Reducing the number of independent variables would also make the estimation of the Spatial Durbin Model (SDM) possible. The SDM was originally proposed as a means of addressing omitted variables in spatial studies. The SDM lags both the dependent variable and the independent variable, so how each observation is affected by the independent variables is different based on location. In this study, however, the SDM would have thirteen observations and eleven parameters. An index variable would reduce the number of parameters to three and make estimation more appealing.

Ultimately, the results of this study make it clear that while there is some sort of spatial relation, data limitations make understanding this relationship difficult. If patterns of spatial dependence in the adoption of irrigation technology on the Texas High Plains are important from either an academic or a policy making viewpoint, the analysis will need to make use of either different or more refined methods than those used here.

7. ACKNOWLEDGEMENTS

The authors would like to thank the Ogallala Water Project (United States Department of Agriculture, Agricultural Research Service) for their funding of this project.

We would also like to thank the staff of the High Plains Underground Water Conservation District no. 1 for their assistance with gathering data necessary for this paper.

Andrew Wright, Darren Hudson, Maria Mutuc Finally, A. Wright would like to thank Dr. Eduardo Segarra and Dr. Aaron Benson of Texas Tech University for their contributions to this research.

REFERENCES

1. M. Caswell and D. Zilberman, "The Effects of Well Depth and Land Quality on the Choice of Irrigation Technology," American Journal of Agricultural Economics, Vol. 68, No. 4, 1986, pp. 798-811.
2. R. Shrestha and C. Gopalakrishnan, "Adoption and Diffusion of Drip Irrigation Technology: An Econometric Analysis," Economic Development and Cultural Change, Vol. 41, No. 2, 1993, pp. 407-18.
3. A. Abdulai and W. Huffman, "The Diffusion of New Agricultural Technologies: The Case of Crossbred-Cow Technology in Tanzania," American Journal of Agricultural Economics, Vol. 87, No. 3, 2005, pp. 645-59.
4. P. Koundouri, C. Nauges and V. Tzouvelekas, "Technology Adoption Under Production Uncertainty: Theory and Application to Irrigation Technology," American Journal of Agricultural Economics, Vol. 88, No. 3, 2006, pp. 657-670.
5. G. Green, D. Sunding, D. Zilberman and D. Parker, "Explaining Irrigation Technology Choices: A Microparameter Approach," American Journal of Agricultural Economics, Vol. 78, No. 4, 1996, pp. 1064-1072.
6. E. Mansfield, "Technical Change and the Rate of Imitation," Econometrica, Vol. 29, No. 4, 1961, pp. 741-766.
7. F. Bass, "A New Product Growth for Model Consumer Durables," Management Science, Vol. 15, No. 5, 1969, pp. 215-227.
8. P. Geroski, "Models of Technology Diffusion," Research Policy, Vol. 29, No. 4-5, 2000, pp. 603-625.
9. B. Wejnert, "Integrating Models of Diffusion of Innovations: A Conceptual Framework," Annual Review of Sociology, Vol. 28, No. 1, 2002, pp. 297-326.
10. E. Rogers, "Diffusion of Innovations," 5th Edition, Free Press, New York, 2003.
11. N. McRoberts and A. Frank, "A Diffusion Model for the Adoption of Agricultural Innovations in Structured Adopting Populations," Working Paper No. 29, Land Economy Research Group, 2008. http://ageconsearch.umn.edu/handle/61117
12. G. Feder, R. Just and D. Zilberman, "Adoption of Agricultural Innovations in Developing Countries: A Survey," Economic Development and Cultural Change, Vol. 33, No. 2, 1985, pp. 255-298.
13. A. Foster and M. Rosenzweig, "Microeconomics of Technology Adoption," Center Discussion Paper No. 984, Economic Growth Center, Yale University, 2010. http://www.econ.yale.edu/growth_pdf/cdp984.pdf

14. K. Fuglie and C. Kascak, "Adoption and Diffusion of Natural Resource Conserving Technology," Review of Agricultural Economics, Vol. 23, No. 2, 2001, pp. 386-403.
15. S. Davies, "The Diffusion Process of Innovations," Cambridge University Press, Cambridge, 1979.
16. R. Just, D. Zilberman and G. Rausser, "A Putty Clay Approach to the Distributional Effects of New Technology under Risk," In: D. Yaron and C. Tapiero, Eds., Operations Research in Agriculture and Water Resources, North-Holland Publishing Co., Amsterdam, 1980, pp. 97- 121.
17. G. Feder and G. O'Mara, "Farm Size and the Adoption of Green Revolution Technology," Economic Development and Cultural Change, Vol. 30, No. 1, 1981, pp. 59-76.
18. F. Shah, D. Zilberman and U. Chakravorty, "Technology Adoption in the Presence of an Exhaustible Resource: The Case of Groundwater Extraction," American Journal of Agricultural Economics, Vol. 77, No. 2, 1995, pp. 291- 299.
19. W. Keller, "Are International R&D Spillovers Trade Related? Analyzing Spillovers among Randomly Matched Trade Partners," European Economic Review, Vol. 42, No. 8, 1998, pp. 1469-1481.
20. W. Keller, "The Geography and Channels of Diffusion at the World's Technology Frontier," NBER Working Papers No. 8150, National Bureau of Economic Research, 2001. http://www.nber.org/papers/w8150.pdf
21. M. Abreu, H. Groot and R. Florax, "Spatial Patterns of Technology Diffusion," Tinbergen Institute Discussion Paper, TI-2004-079/3, 2004. http://www. Tinbergen .nl/discussionpapers/04079.pdf
22. High Plains Water District No. 1, "The Cross Section," Various Issues, 1981-2005.
23. High Plains Water District No. 1, "Personal Communication, Personal Communication Regarding the Number of Center Pivots Systems in the Water District in Various Years," 2012.
24. Texas Tech University Center for Geospatial Technology, "Texas County Water Information," 2010. http://www.gis.ttu.edu/OgallalaAquiferMaps/
25. US Department of Agriculture, National Agricultural Statistics Service, "Quick Stats: US and All States County Data-Crops," 2010. http://www.nass. usda.gov /QuickStats/Create_County_All.jsp
26. S. Stadler, "Aridity Indices," In: J. E. Oliver, Ed., Encyclopedia of World Climatology, Spinger, Berlin, 2005. http://link.springer.com/referenceworkentry/10.1007/1-4020-3266-8_17/fulltext.html
27. Western Regional Climate Center, "Western US Climate Historical Summaries," 2012. http://www.wrcc.dri.edu/Climsum.html
28. Federal Reserve Board of Governors, "Selected Interest Rates," 2011. http://www.federalreserve.gov/releases/h15/data.htm
29. L. Anselin, "Under the Hood: Issues in the Specification and Interpretation of Spatial Regression Models," Agricultural Economics, Vol. 27, No. 3, 2002, pp. 247-267.
30. Texas State Historical Society, Various County WebPages, 2012. http://www. tshaonline.org/
31. P. Moran, "The Interpretation of Statistical Maps," Journal of the Royal Statistical Society, Series B (Methodological), Vol. 10, No. 2, 1948, pp. 243-251.
32. P. Jeanty, "Spmlreg: Stata Module to Estimate the Spatial Lag, the Spatial Error, the Spatial Durbin, and the General Spatial Models," 2010. http://ideas. repec.org/c /boc/bocode/s457135.html
33. P. Jeanty, "Splagvar: Stata Module to Generate Spatially Lagged Variables, construct the Moran Scatter Plot, and Calculate global Moran's I Statistics," 2010. http://ideas.repec.org/c/boc/bocode/s457112.html
34. P. Jeanty, "Spwmatrix: Stata Module to Create, Import, and Export Spatial Weights," 2010. http://ideas.repec.org/c/boc/bocode/s457111.html
35. M. Pisati, "Tools for Spatial Data Analysis," Stata Technical Bulletin, No. 60, 2001, pp. 21-37. http://www.stata.com/products/stb/journals/stb60.pdf

36. L. Anselin, "Exploring Spatial Data with GeoDa: A Workbook," 2005. https://geodacenter.asu.edu/system/files/geodaworkbook.pdf

CHAPTER 2

Irrigation and Drainage in Agriculture: A Salinity and Environmental Perspective

Sjoerd E.A.T.M. van der Zee[1, 3*], Sija F. Stofberg[1], Xiaomei Yang[1], Yu Liu[2], Md. Nazrul Islam[4] and Yin Fei Hu[2]

[1] *Soil Physics and Land Management, Environmental Sciences Group, WageningenUniversity, Wageningen, Netherlands*
[2] *Xinjiang Academy of Agriculture and Reclamation Science, Shihezi, China*
[3] *School of Chemistry, Monash University, Melbourne, Australia*
[4] *Soil Science Division, Bangladesh Rice Research Institute, Gazipur, Bangladesh*

ABSTRACT

Whereas irrigation and drainage are intended to address the shortage and surplus of soil water, respectively, an important aspect to address is also the management of salinity. Plants have a limited tolerance for soil water salinity, and despite significant gaps in our practical knowledge, an impression of acceptable salinities is available for many crops. To manage soil salinity, the Leaching Requirement is an old, yet useful, concept. In this chapter, we extend this concept for soils with shallow groundwater. Particularly if shallow groundwater is saline, management is needed to avoid capillary rise of this water into the root zone. One of the tools to do so is Climate Adaptive Drainage (CAD), for which many practical gaps in knowledge remain. Also, soil mulching, of which a special case is considered in more detail, i.e., using plastic covers, may be beneficial for many purposes, including improving the water and salt balances of the root zone. However, use of plastics may have significant adverse effects. Due to water shortage, also wastewater may be re-used for irrigation. For this reason, the hazard of sodicity due to elevated Na concentrations in domestic wastewater is highlighted.

Keywords: salinity tolerance, Leaching Requirement, groundwater, adaptive drainage, soil mulching, plastic mulching, wastewater re-use, sodicity, modeling

1. INTRODUCTION

Although arid and semi-arid regions have a potential for yielding multiple crops per year, due to the large incoming radiation, this potential is profoundly limited by the small annual rainfall. An impression of the distribution of dry areas is shown in **Figure 1**, which is based on the aridity index (AI), defined as annual precipitation divided by potential evapotranspiration.

$$AI = P/ET_p \tag{1}$$

where P is annual rainfall and ETp is potential evapotranspiration (both mm/y). Therefore, **Figure 1** shows the main regions, where an additional supply of water, besides rainfall, is needed or even crucial for agricultural production. However, also in the regions where water supply is sufficient, seasonal drought can be a farmer's risk. This is the case in a humid temperate country as the Netherlands.

Insufficient rain falls have to be compensated for, to ascertain sufficient agricultural yields. Irrigation using good-quality river water, groundwater and, sometimes, wastewater has been used for ages. Irrigation is often accompanied by drainage, which is the natural or artificial removal of surface and sub-surface water from a given area. Artificial drainage is always needed, except in cases with deep groundwater levels. In most river delta regions, drainage is a must to avoid water logging. This is not the only reason, because avoiding salinization is a recognized aim of drainage, as is illustrated in this chapter.

Research on irrigation and drainage has been considerable and this is not surprising as irrigated agriculture has been practiced during the past 5000 years, e.g,. in Mesopotamia. Due to long-term changes, water availability for irrigation could decline or salinity problems could develop. The underlying cause for agricultural yield depressions upon salinization is that most terrestrial plants tolerate only a limited salinity of soil water. The term salinity is usually associated with concentrations of sodium chloride (NaCl), but may be interpreted more broadly as the presence of ions in water. Rainfall usually contains few ions, but agricultural fields are often irrigated with groundwater or surface water, where this is different. The salts that may give problems in agriculture may be derived from different sources: (i)

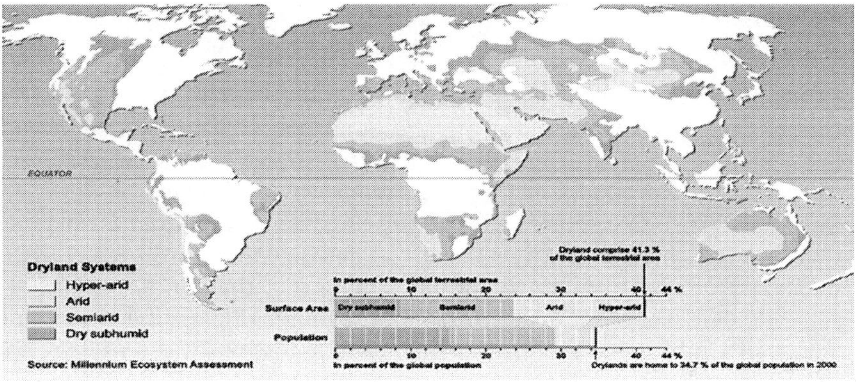

Figure 1. Distribution of dryland regions (from: Millenium Ecosystem Assessment, Chapter 22, 2005).

water in contact with soil material induces physical and chemical weathering, which is associated with release of ions, of which the concentrations may be increased due to evapotranspiration; (ii) groundwater may be brackish or saline due to different geohydrological causes such as sea water intrusion and the presence of old marine sediments; and (iii) use of river water with some level of salinity, originating from groundwater, sea, or industry. Salinity is often expressed as concentration (mass per volume of water), electrical conductivity (EC), or total dissolved solids (TDS).

The scope of this chapter is to address a number of environment-related aspects such as salinity and recent developments that may develop into contamination problems. Our motivation is that simple ways to predict developing environmental contamination are quite limited and sometimes not even recognized. After some background information on the vulnerability of plants for salinity, guidelines are provided for anticipating the salinity levels that may be expected in the root zone, which do not require complicated models. Some other general environmental hazards associated with irrigated agriculture are identified, as related with plastic mulching and re-use of wastewater in agriculture.

2. SALT TOLERANCE OF PLANT SPECIES

Efforts to limit the salinity of the root zone are, at the end, aimed at limiting adverse impacts of salinity on primary productivity of crops. In a somewhat broader sense, growth of both crops and botanic species, product quality, and other measures related to plant well-being are important in the assessment of salt-induced adverse effects.

Plants exposed to elevated salinity may experience different forms of stress. Due to the high osmotic value of saline solutions, soil water may become less available for plants to accommodate their transpiration, which directly affects primary production [1, 2] in a similar way as drought. However, it is also well known that salts (e.g. involving Na^+, Cl^-) may be toxic for plants or that toxic components such as boron (B) and selenium (Se) become better bio-available under saline conditions. For boron, the concentrations that separate deficiency and toxicity are close together [3], making an optimum availability difficult to accomplish. In addition, induced nutrient deficiency has been well documented, e.g. for iron and nitrate [4, 5]. In many dose-effect types of exposure studies of salinity, permanent stress is considered. However, in many field situations, salt stress occurs only in limited periods or one season. Typically in temperate conditions saline stress is periodic, e.g. due to tidal or seasonal fluctuations, and is followed by salt leaching conditions. However, research on salt stress duration and plant response is quite limited.

Salt tolerance has been investigated much for agricultural crops, both in field and greenhouse conditions and, particularly, for the case where salts enter the root zone. Different plant species have different salt tolerance and strategies to deal with salinity [6, 7]. It has been well established that also different genotypes of the same species may differ according to their salt tolerance ability [8]. The accumulated experimental evidence of salt tolerance functions (MH) for crops has been summarized [2], which describe the salt concentration range in the root zone where transpiration (hence primary production) is unaffected, and in the larger salt concentrations where transpiration is reduced. Since 1977, such 'MH'-functions have been adjusted and developed for other crops or for different environmental conditions [9]. MH-functions make it is possible to compare different plant species with regard to their vulnerability to salinity.

The MH-functions describe how soil or soil water salt concentrations affect primary production. They seem to differ for different soil types and climate and weather conditions. MH-functions, apart for comparing species for salt vulnerability, are also important for predicting crop yields, because most modelling uses the MH-concept [10–13]. The aim of such modelling is often to really predict yields, e.g. if climate changes, or to generalize experimental observations to other soils or geohydrological situations. MH-functions are often used to measure water availability and salinity of the root zone throughout the growing season, which are relatively constant. However, in the field, this is generally not the case. Particularly for distributed modelling, additional assumptions on how plants spatiotemporally deal with variability are needed [14]. For instance, plant roots may compensate for dry or saline layers in soil by taking up water from other layers, but do they do so based on the water content, water potential, or a mixture of these properties? At the moment, the experimental evidence is predominantly based on water contents and salt concentrations that are as homogeneous as possible. Compensation strategies may differ according to plant species or even genotype.

In agricultural modelling, emphasis has been on salinity effects on the osmotic potential of soil water and its availability for plants. Ion specific effects as toxicity or induced deficiency are important, but generally disregarded in modelling. The reason is that crops are often well fertilized, but also that for modelling ion-specific effects, a multicomponent approach is needed as in HP2, UNSATCHEM, and SWAP-ORCHESTRA [15]. Not only is multicomponent modelling much more demanding with regard to parameterization, data, and computational efforts, but also data are needed for different plant species. This is not a trivial effort.

As much as crops have been investigated for salt tolerance, this not the case for landscape/decorative plants. Native vegetation has been investigated much from the perspective of plant ecology [16] or mapping [17]. Spatiotemporal variability of abiotic factors is often ignored, although salt tolerance of species and systems is estimated by investigating their spatial distribution and plant traits. Out of thousands of papers, only about 50 dealt with deciduous trees [18, 19]. Nevertheless, also for these categories of plants lists have been produced for their salt tolerance [20, 21]. In some cases, such listings give a quantitative grading, but a relative classification may be more appropriate [22] because the experimental techniques, quality of data, and used metrics to describe the damage may differ profoundly among papers. In view of the limited data for non-crop plants, at this moment modelling of climate change on the distribution of botanic species may be hard or impossible.

An important salt exposure pathway is that of salt spray and sprinkler irrigation. Although the importance of exposure of the leaves has been demonstrated [21, 23–25], this pathway is generally ignored in modelling. Mostly, it is mentioned that uptake of salts (notably Na and Cl) by the leaves and toxicity lead to necrosis, but osmotic effects are seldom mentioned. Timing of sprinkling plays a role, because sprinkling at night, or at frequencies that limit the duration that leaves are wet, appear to affect the degree of adverse effects: the impact is smaller as leaves are wet shorter, or transpiration is smaller. The direct (leaf exposure) pathway was also recognized [26] to enhance the impact of exposure through the root zone.

Browsing through the literature, it is clear that salt effects depend on plant species, genotype, salt composition, soil type, climate and weather conditions, and geohydrology. These factors all modify the plant's response to salinity, but some are ignored completely (ion specific effects; leaf exposure) in crop and soil models. The large complexity and significant gaps reveal that data can be fitted and described by models, but true prediction may be feasible only for broad features and large uncertainty bands.

3. IRRIGATION AND SALINITY

All sources of soil water, except for artificially desalinated (deionized) water, have one thing in common: they all contain dissolved salts (together with non-ionic compounds dissolved compounds are comprehensively called solutes) to some degree. In case of rain water, little salt is dissolved but in case of brackish or saline water, salt may be present in such quantities, that the density of the liquid is increased. Thus, sea water contains 35 g/l of salt (mainly NaCl) and has a density of 1.025 g/l or 1025 kg/m^3. These density differences may have a significant effect on water flow, as will be considered later in this chapter. Groundwater and river water often contain solutes that originate from the geological formations that the water has flown through, as well as from biological or human activities. These ions include calcium (Ca^{2+}), magnesium (Mg^{2+}), carbonates (HCO_3^-), sulphates (SO_4^{2-}), nitrates (NO_3^-) and phosphates (PO_4^{3-}).

When water evaporates from the soil or is transpired by plants, most salts are left behind in the soil. Therefore, it is necessary to be alert for the hazard that too large quantities accumulate in the soil solution. This understanding was communicated already in the first half of the previous century, and described in the famous Handbook 60 [27]. For illustrative purposes, the Leaching Requirement (LR) is of use. It is based on annual water and salt balances for the root zone or 'plough layer' (here the term root zone is used). In words, both balances tell how large the (water or salt) accumulation rate is in dependence of the inflow of irrigation and rainfall water and the outflow of drained water. On the longer (multi annual) term, the water balance is zero, i.e., the net water accumulation or depletion is zero. If this were not the case, the soil would become completely water saturated, or water deficient (note: at the scale of one or several years, the water balance may be positive or negative, but on the long run, this cannot be the case as otherwise we deal with an inundated soil or completely dry one). Hence,

$$D_{iw} = D_{ET} + D_{dw} \tag{2}$$

where D stands for a 'depth of water layer', as related to a flux through a designated area (in $m^3\,m^{-2}y^{-1}$ or in $mm\,y^{-1}$), *iw* stands for irrigation water, *ET* for evapotranspiration, and *dw* for drained water. For the salt balance, it is illogical that the balance is zero: salts may accumulate or be depleted until no salt is present. Accordingly, the salt balance would be

$$\frac{\Delta\theta C z_r}{\Delta t} = D_{iw}C_{iw} - D_{ET}C_{ET} - D_{dw}C_{dw} \tag{3}$$

where θ is the volumetric water fraction, C the salt concentration, *zr* the root zone thickness, and t is time.

Usually, water taken up by plants for evapotranspiration contains some salts (and nutrients), but far less than in irrigation water for those cases where salinity is an issue. Therefore, the term for ET can be neglected and Eq. (3) can be modelled for if more salt is coming in than is in equilibrium with initially present salt. The salt concentration increases, giving a behaviour as is shown in **Figure 2**.

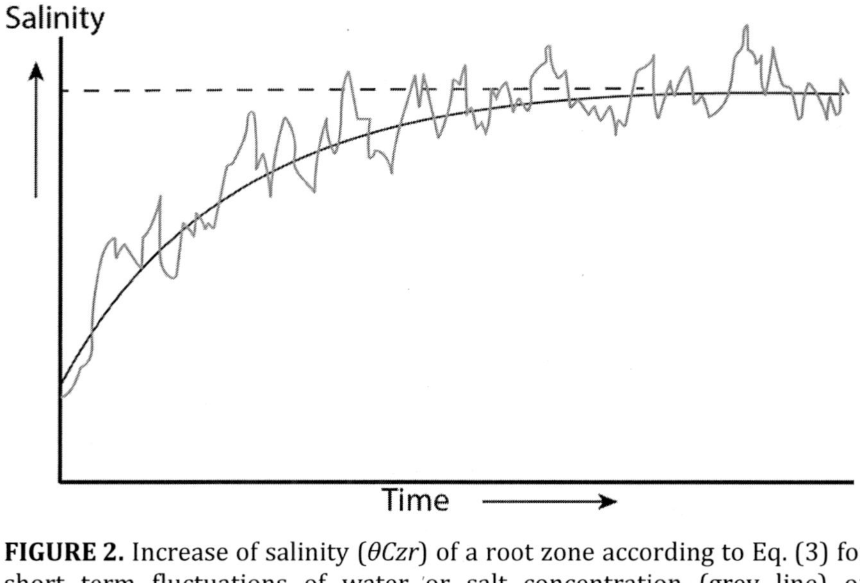

FIGURE 2. Increase of salinity (θCzr) of a root zone according to Eq. (3) for short term fluctuations of water or salt concentration (grey line) or constant irrigation water concentrations and rainfall, ET and drainage rates (black line).

As **Figure 2** reveals, short term variability can lead to an erratic pattern of root zone salinity. Although this variability may be important, for instance if it is a seasonal pattern, particularly the long term level at which salinity stabilizes is of main interest in a sustainability assessment: will the adopted management (involving how much is irrigated, how much is leached, in view of the quality of irrigation water) lead to unacceptable salinization or not?

On the long term, the salinity acquires the level of the dashed line in **Figure 2** and does not change anymore: the term to the left of the =– sign of Eq. (3) becomes zero. Then, Eq. (3) can be recast as follows:

$$D_{iw}C_{iw} = D_{dw}C_{dw} \tag{4}$$

where we remember that *Diw* = *DET* + *Ddw*.

Already in Handbook 60 [27], it was realized, that the concentration in water that drains out of the root zone is equal to the concentration in the root zone, if this zone is at field capacity. Hence, $C_{dw} = C_{soil,FC}$. Inserting this in Eq. (4) and re-arranging leads to

$$\frac{D_{iw}}{D_{dw}} = \frac{C_{soil,FC}}{C_{iw}} \tag{5}$$

For illustration, this is a very useful result. Usually, the quality of irrigation water (here represented by its salt concentration) is known, as it is easily measured. Then, it is easy to determine with Eq. (5) how much irrigation water must be applied (in view of the amount of evapotranspiration, represented by *DET* = *Diw* – *Ddw*) to drain enough to keep the soil salinity (*Csoil,FC*) below the level that a certain crop demands.

Practically convincing is it, to show what happens if one does not drain at all. In that case, *Ddw* = 0, and the division by zero implies that *Csoil,FC* becomes infinitely large. The compelling message of this simple illustration is:

1. If regular leaching of salts out of the root zone does not occur, salinity will increase unbounded: concentrations continue to increase.

2. This increase of salinity may take time, particularly if irrigation water is of good quality, but it is still inevitable. On the long term, salinity must increase if salts are not leached, and agricultural soil will become too saline for agriculture, as old civilizations in Mesopotamia testify [28]!

3. Really unlimited increase of the concentration will not occur, of course, in reality. This is due to chemical precipitation of salts as soon as the solubility product is exceeded. Where this solubility product is salt as well as environment dependent, it defines an upper limit to salt concentrations in solution. The environment dependency is related with e.g. temperature, but also the solution composition. For instance, gypsum is known to be better soluble under saline (or more general: higher electrolyte concentration) than low salt conditions.

4. IMPACT OF GROUNDWATER

To determine the need for leaching salts without consideration of groundwater is often inappropriate. In many regions, and particularly in lowlands as coastal plains and delta areas, groundwater is shallow. For the salt-affected Hortobágy area of Hungary, the guideline was developed, that if groundwater is shallower than say 3 m below the soil surface, the upward capillary flux may be significant enough to be important for the water and salt balances [29]. Simulations support this guideline [30].

The relationship of the impacts on the root zone wetness between natural precipitation (mostly rainfall) and groundwater (capillary rise) can be complex. Whereas rainfall is an independent forcing, capillary rise of water depends on the root zone water content pressure, soil hydraulic properties, and groundwater depth: root zone water affects capillary rise and is affected by it. The same is true for infiltration and runoff [31]. The complexity was one of the reasons of a new school of Ecohydrology to analyze low-dimensional flow models [11].

In the simplest form, that approach of modelling considers a root zone of which the volumetric water fraction θ depends on infiltration, evapotranspiration, and drainage. At the same time, θ is the variable that determines these fluxes (plus runoff). Vervoort and Van Der Zee [32] extended such models with capillary rise of ground water. These water balance models represent the layer between root zone and groundwater level with a functional dependency of θ and these fluxes.

The impact of seasonality on the water dynamics can be profound, and many countries are characterized with seasonality. This is also the case for root zone salinity, because dissolved salts (of alkali and earth alkali cations) respond quickly on the water fluxes. For this reason, the salt concentration of the root zone often shows a distinct short term fluctuation as in **Figure 2**. Using the ecohydrological framework with minimalist (parsimonious) models [11, 32], salt accumulation was investigated [12]. The latter took capillary rise of brackish groundwater into account. Their root zone water balance, was therefore given by

$$\varphi Z_r \frac{ds}{dt} = P - R(s) - ET(s) + U(s) - L(s) \tag{6}$$

where φ is porosity, *Zr* is root zone thickness, *s* is water saturation (about equal to θ/φ), *t* is time, *P* is rainfall or precipitation rate, *R* is runoff rate, *ET* is evapotranspiration rate, *U* is capillary upward flow rate, and L is leaching or drainage rate.

The complexity of Eq. (6) is that the rainfall is erratic: rain occurs with irregular intervals and each shower has a different intensity and duration. Therefore, water saturation s and salt concentration *c* vary irregularly with time.

The salt mass balance is given by

$$\frac{dM}{dt} = \varphi Z_r \frac{dsC}{dt} = U(s).C_Z - L(s)C \tag{7}$$

with *M* the total mass of salt, *CZ* is the salt concentration in the groundwater (fixed) and *C* is the salt concentration in the root zone. In Eq. (7), it is assumed that the other water fluxes (*P*, *R*, *ET*) do not transport appreciable quantities of salt, which is often reasonable. Solving Eqs. (6) and (7) shows patterns similar as **Figure 2**. If saline soil is washed with good-quality water, the mean concentration decreases but short term variability still occurs. If seasonality is profound, intra-annual periods of drought will often coincide with periods of root zone salinity, both having a tendency to reduce crop yields.

Taking into account the often erratic variability of weather [33], recently the salt accumulation in the root zone was investigated, considering that root zone salinity levels respond fast to the water quality of incoming fluxes (rainfall, irrigation, capillary rise). In general, say in 10 years, the root zone concentration attains a dynamic equilibrium with rain, irrigation water, and groundwater quality.

For water managers, it is important to be able to assess whether intended water and salt management is sustainable on the longer term: will the use of irrigation water with a known salinity lead to too much salinity? And if it does, to what degree: will crops that are planned to be part of the rotation scheme show reduced yield? If for such questions only the longer term level of the concentration in the root zone is important (i.e., the plateau in **Figure 2**), a simpler assessment may be sufficient, than analyzing Eqs. (6) and (7).

Comparing simulated salt concentrations using a root zone model [12], and analytical approximations as Eq. (8) that are as simple as the Leaching Requirement approach, a good agreement was obtained, as shown in **Figure 3**, with ratios close to one. Apparently, the longer term mean concentration levels can be quantified well if short term variability of weather is ignored using

$$\langle C \rangle = \frac{\langle D_{cr} \rangle C_Z + \langle D_{irr} \rangle C_{irr}}{\langle D_{dw} \rangle} \quad (8)$$

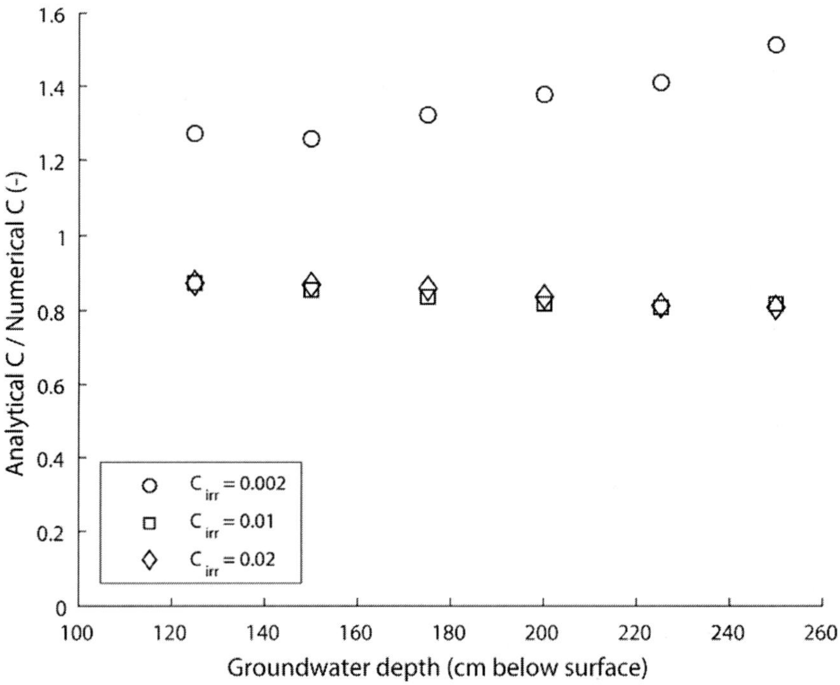

Figure 3. Ratio of calculated root zone salt concentrations with Eq. (8) and as modelled numerically, as a function of groundwater level (*Zf* in cm below surface) for a temperate climate as in The Netherlands. Irrigation water salinity given by *Cirr* in mol l^{-1}, and groundwater salt concentration of 0.02 mol l^{-1}. Adapted with permission from Ref. [33].

where brackets <.> refer to time averaging, subscripts *cr*, *Z*, *irr*, and *dw* stand for capillary rise, groundwater level, irrigation water, and drainage water, and D is the respective flux in mm or m water layer for the averaging time (usually one year, so D in mm/y). Eq. (8) holds if both groundwater (concentration *CZ*) and irrigation water (concentration *Cirr*) carry significant salt quantities, and implies that the mean soil concentration, and the mean leaching concentration, is given by a weighted average concentration of the water fluxes.

Irrigation can be modelled similar as rainfall, except that it is a managed source of water (timing is a managing decision) and that it may contain more dissolved salts than is common for rainwater. To predict salinization using simple means, for cases not considered by the Leaching Requirement (e.g. if capillary rise is important, and possibly it contains appreciable dissolved salts, if both rain water and irrigation water infiltrate, with different salinities) is useful and Eq. (8) is an example of such a simple management tool.

With numerical models and with approximations as LR and Eq. (8), a judgement can be made whether root zone salinity is expected to become too high under the current practice, and how much and how good quality irrigation water will be needed to limit salinity to desirable levels. These tools, however, are generic and for a good desalinization of soil, the local conditions need to be known well. Besides the management decision to install a drainage system, various measures can be taken to homogenize fields (e.g. surface level) and improve the efficiency of salt leaching. Tools as discussed in this chapter may fine-tune the leached salt concentrations to agree with the farmer's needs and the salinity in discharged drainage water that are acceptable downstream.

In most model approximations for the long-term salinity of the root zone, it is assumed that level and quality of groundwater are unaffected by the root zone water and salt balance. This is a simplification. Root zone drainage water may affect groundwater, and cause fresh water bodies to develop in coastal dunes [34]. The thickness of the freshwater bodies floating on top of saline water as follows from Archimedes' law. Various approximations have been developed since the nineteenth century for the thickness of fresh water lenses, assuming a sharp interface between fresh and saline water and different assumptions on the outflow at the edge of the lens and movement of the saline water [35] (**Figure 4**).

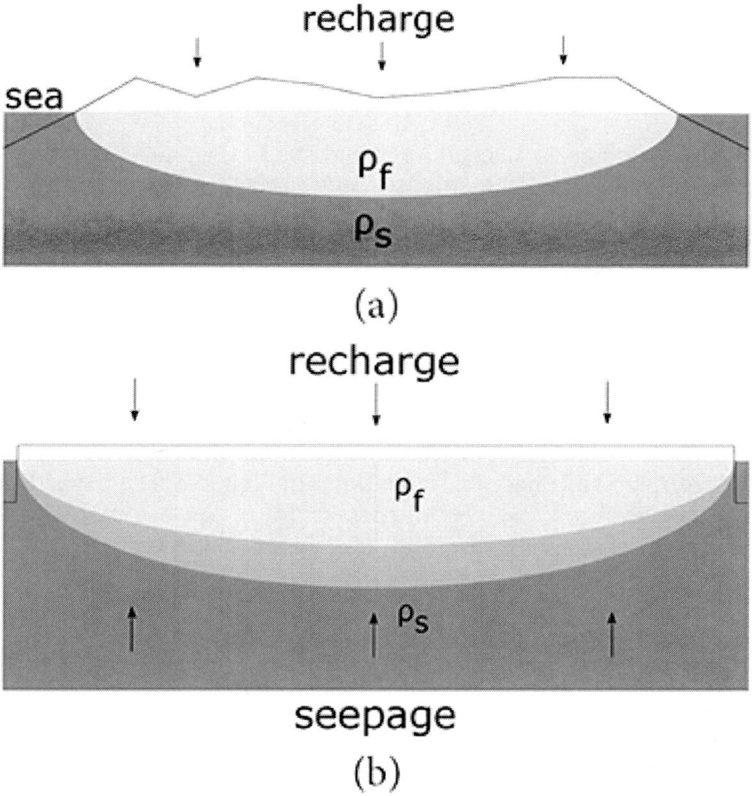

Figure 4. The freshwater lens (a) in a dune (the classical Ghyben-Herzberg lens) and (b) in an agricultural lowland field as considered by Refs. [35, 36].

Mostly, fresh water lenses outside the dune areas have been studied for situations where the soil surface is close to the sea level, and open water is at or even below this level. Then, groundwater seeps upward into the open water, and thin fresh water lenses form, that may respond quickly to changes in rainfall and evapotranspiration. They are found not only in coastal regions [36], but also in central Australia due to incidental flooding, to form an important source of fresh water to indigenous vegetation [37, 38] and 'inland' parts of PR China.

Because small fresh water lenses are vulnerable to disappear during extended periods of drought, Stofberg et al. [33] assessed how this vulnerability depends on a number of conditions. To avoid this depletion of fresh water, active management of water storage in wet periods is needed and one of the tools may be Climate Adaptive Drainage (CAD).

5. DRAINAGE DIMENSIONING AND ADVANCED OPTIONS

Drainage of water is needed to avoid water logging, poor mechanical behaviour, and increasing salinity. This recognition has made drainage technology as well as theory to become topics of high importance in soil, water, and agricultural science. Due to the good economic situation, high population density, and usually shallow groundwater table, drainage has particularly been investigated in detail in The Netherlands. Basic theory has been developed by, e.g. Hooghoudt and others.

Dimensioning of drainage for (often flat) fields in agricultural use, is commonly aimed at management of groundwater levels to avoid water logging, poor tractability, structure degradation due to grazing cattle, and water saturated and anaerobic conditions. In their simplest form, the steady state water level between two ditches or drains can be obtained from the combination of Darcy's law and the water balance, with the Dupuit assumption. For a steady-state rainfall rate of P, we obtain

$$P = q = \frac{4K\left(H^2 - D^2\right)}{L^2} \tag{9}$$

where P is recharge (net infiltration) rate (m/day), q is the discharge into the drain or ditch (m/day), K is the hydraulic conductivity of soil (m/day), H is the water level halfway two drains measured from the impervious basis of the phreatic aquifer (m), D is the drain level measured from the same basis (m), and L is drain spacing (m). With Eq. (9), it is possible to determine the spacing needed to limit D to a value sufficiently below the soil surface.

As was already mentioned, in many climate zones it may be important to save water, for instance because precipitation may occur in a season that is unfavourable for crop production, and primary production in general. One of the tools to enable temporary storage is Climate Adaptive Drainage (CAD). In CAD, the level of drainage can be dynamically varied to improve flexibility with regard to water stored in the soil profile. The basis of this technique is quite simple: drainage, e.g. using tile drains, is no longer passive but active by using a collector between the drain and the receiving ditch. The collector (**Figure 5**) has an overlet shown in black of which the height can be varied. If this height above the base of the collector's overlet becomes larger, the groundwater becomes more shallow, closer to the soil surface and the root zone. Hence, adjusting the overlet height is the same as manipulating groundwater level.

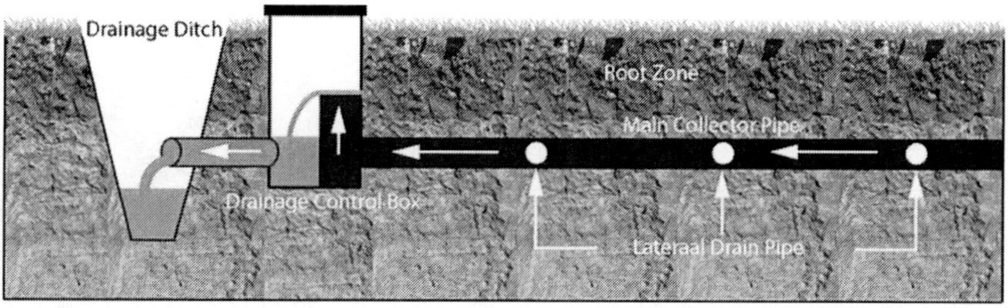

FIGURE 5. Climate Adaptive Drainage (CAD), with drains merging into a collector pipe, that discharges into a control box. The level in this box can be adjusted (manually or remote control), to anticipate on weather conditions in the next weeks or months.

CAD is beneficial to dynamically store water for the dry season. In regions with saline groundwater, and particular in areas where such saline groundwater seeps upward, as in deltas, dynamic drainage can be very useful to diminish the upward saline fluxes. With a shallow groundwater level enabled by CAD in the wet season, the upward saline seepage can be diminished. In the growing season, fresh water is consumed for evapotranspiration, and CAD can lower the saline water table, or perhaps be used to introduce better quality into soil.

6. PLASTIC MULCHING

Irrigation is often combined with other techniques that improve the water use efficiency (WUE), including several cropping and tillage techniques [39, 40]. Mulching involves putting a barrier between soil and atmosphere, using different materials such as crop residue, such as straw [41], leaves, paper, old carpet, plastic (**Figure 6**) or gravel. It may serve purposes other than water use efficiency as well, such as weed growth reduction.

In China, plastics have been used for many years after importing from Japan in 1970s. The technology has been transformed and developed and become important in agronomy, especially in NW China, where agriculture became much dependent on plastic mulching. In China, it is mainly used for cash crops, such as cotton, tobacco, medical herbs, bean nuts, loofah, and crops such as maize, bean, and potato.

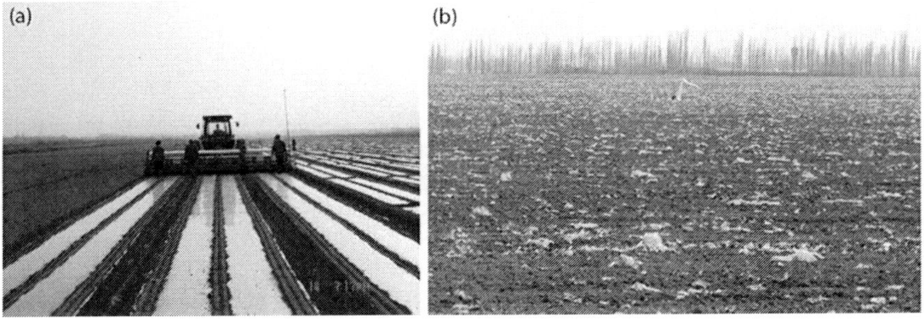

FIGURE 6. (a) Installation of plastic mulch and (b) postharvest field contaminated with plastics in Xinjiang, China.

Mulching serves various purposes.

1. Increasing soil temperature. Plastic mulching decreases heat exchange between soil and atmosphere, and leads to steadily increasing soil temperatures. For NW China, topsoil temperatures were found to increase 3–6°C in the top layer, and some heat is transferred into deeper layers. The sowing or planting time can therefore be earlier in spring (10–20 days) increasing the growing season. It is reported that crop yields can increase 20–30% compared with situations without plastic mulching and also crop quality can improve.

2. Reducing water evaporation and water conservation. The plastic blocks water vapor exchange between soil and open air and reduces soil evaporation. However, infiltration of rain water is not affected. Hence, soil moisture in 0–20 cm layer may increase 2% by plastic covering and in semi-arid and arid areas, around 1950 m^3 ha^{-1} water can be saved during the growing season.

3. Enhanced fertilizer efficiency. This beneficial effect is attributed to the reduction of nutrient losses via leaching, runoff and erosion. There are indications that soil available nitrogen, phosphorus and potassium might increase 30–50%, 20–30%, and 10–20%, respectively.

4. Improved solar light irradiation efficiency. Transparent plastic film and the water droplets which adhere onto this film improve soil absorption of light, benefitting photosynthetic intensity and postponing the senescence time of lower leaves of plant. It is reported that lower leaves can capture 12–14% extra catoptric light with plastic mulching, and tomato can capture 3–4 times more light, photosynthesis intensity may enhance 13.5–46.8%, and chlorophyll content in tomato may increase 5%.

5. Improving soil physical properties. Due to mulching, soil temperature rises, vapour pressure effects have been claimed to enlarge the pore sizes within soil particles, increase soil porosity, reduce soil bulk density and increase soil aggregate stability. This affects physical properties as well as soil fertilizer, water, atmosphere and heat balances, besides protecting the soil surface, avoid raindrop detachment and erosion of soil. Weed control may be less frequent and reduce soil compaction.
6. Improving soil microbial activity. By positive impacts on temperature and soil moisture, mulching is beneficial for soil microbial activity, microbial reproduction and growth, and organic matter decomposition and mineralization in the root zone, and improves nutrient availability. Also plastic use might inhibit diseases and pests.
7. Avoiding soil salinity. In saline regions, salts can be transferred by water evapotranspiration to the soil surface. By its water conserving effect, this transport is reduced.

Great changes have occurred since plastic mulching became use in China. Crop yields and quality improved. However, at the same time, plastic film residues have become a pollutant in the plough layer of agricultural soils. About 313.3×10^4 ha farmland is covered by plastic mulching, or 75% of the entire farming land of Xinjiang province, with an annual weight of plastics of 18.5×10^4 tons. In the main land for cotton planting, the plastic film residue is 25.32 g m^{-2} and has become a big problem for soil quality and crop growing. Recycling and natural degradation have therefore become important.

In China, plastic film residue recycling is mainly a human labour and machine-supported effort. About 60% of the plastic film residue is recycled with machinery and the main difficulty is the low intensity of residue and plastic pieces. To strengthen film intensity, China government has already issued a regulation since 1990s, entitled GB13735-92, for standard thicknesses, that changed from 0.008 ± 0.003 mm to 0.01–0.02 mm. However, the efficiency of machine recycling is limited. The paradox is that if the thickness increases, the cost will also be increased which farmers cannot afford. But if plastic film residue is recycled manually, the cost will become higher, especially in the regions with a labour shortage and efficiency of collecting plastic is low. Hence, machine supported recycling needs further development.

One of the concerns with all this use of plastics is that they may be gradually transformed into microplastics: particles smaller than 5 mm. Particularly for marine ecosystems, microplastics and their effects on various organisms have been investigated [42]. Also the plastics may interact with other pollutants accumulation in organisms [43], and possibly affect their impacts. For soil organisms, the study of microplastics is still in its initial stages as far as accumulation and effects is concerned, and one of the few papers addressing this issue is very recent [44]. Therefore, it is quite well possible, that bio-availability of pollutants and of microplastics, and their adverse environmentally and ecological effects may be linked, and this deserves closer investigation.

An alternative to deal with plastic residues in soil is to use biodegradable plastic film. However, such types of plastic film tends to be unstable for the range of natural conditions encountered, and therefore, these plastic films have not been applied widely yet. The use of organic mulch, such as crop residues, may be more sustainable from a chemical and biological perspective, as it would not lead to pollution. At the same time, it may have benefits regarding organic matter content, nutrient recycling and soil biology [45]. Although many studies are available regarding the effect of plastic cover on WUE, there are few studies available that compare the use of organic mulch to plastic covering. The results of these studies do not seem very promising for using organic mulches as an alternative for plastic, with lower water use efficiencies [46]. The use of plastic mulch was compared with the use of traditional practice and crop residue mulching [47]. It appeared that using plastic mulch not only led to higher yields in two of the three studied years, but also lead to significant soil water depletion in deeper soil layers, which may be explained by increased evapotranspiration [48]. However, in warmer climates, straw seems to conserve more soil water than plastic cover, although the effects on yield may differ [49, 50]. One study not only reported positive effects of straw mulching on salinity, but also mentioned that many uncertainties still remain [51].

7. RE-USE OF WASTEWATER AND SOIL SODICITY

To make optimal use of resources, also marginal water, such as sewage or domestic wastewater is used for irrigation, in rural and peri-urban agriculture. However, this 'wastewater' suffers often of poor quality.

Compounds in wastewater that may be beneficial are nutrient elements, such as C, N, P, and K, that make wastewater irrigation similar to fertigation. Nutrient supply via wastewater is generally unbalanced, compared with the needs of crops. This may lead to accumulation in soil, or excessive leaching to groundwater and surface water.

Several other elements that may cause adverse (toxic) effects, for plants, soil fauna, animals and humans are boron, selenium, and (heavy) metals. It is worthwhile to mention the presence of Na in most domestic wastewaters. Sodium is often present due to the high salt concentration in the human diet. The hazard of Na-fluxes into agricultural root zones irrigated with wastewater cannot be judged well by only measuring the aqueous Na-concentration. As all cations, Na may sorb onto suspended materials such as organic matter (or organic carbon) and the load of Na that reaches soil with organic carriers might be appreciable. Moreover, also other cations (Ca, Mg) sorb, preferentially, onto suspended and dissolved (DOC) organic carbon. This may make the effect of Na larger, by suppressing the beneficial effect of Ca and Mg. Hence, the adverse effects of adding Na via domestic wastewater needs to be anticipated on the concentrations of Na, Ca, and Mg, as well as Biological Oxygen Demand (BOD), as a measure of suspended and dissolved organic matter.

The adverse effect of sodium is related to it being monovalent: Na^+. Accordingly, it may react with mineral and organic surfaces and surface groups, but different from multivalent cations as Ca^{2+}, Mg^{2+}, Fe^{3+} and Al^{3+}, it is less equipped to form charged bridges between different reactive surface groups. Therefore, why cementing agents as organic matter are known to detach from the sodic soil's solid phase and disperse and accumulate at the soil surface.

For similar reasons, sodic soils may be susceptible to structure deterioration. Being monovalent, Na-ions are less able to electrostatically neutralize the solid surface negative charge of soil clay colloids. Under saline conditions, this may not be so much of a problem, in view of the high concentration of ions that neutralize the electrostatic field of the negative clay colloids. If the salinity drops, diffuse double layers at the clay colloid surface expand (say an order of magnitude, of e.g. 40–400 nm) due to enormous swelling pressures [3]. Swelling causes very small pores (tens of nm) to swell at the expense of macropores, which are important for conveying water and (soil) air. Soil is then dominated by small pore sizes, and becomes almost impermeable to water. An impression of the reduction of the hydraulic conductivity due to decreasing salinity of a sodic soil is given in **Figure 7** [30]. In **Figure 7**, the sodicity of soil is expressed with the common metric of Exchangeable Sodium Percentage (ESP): the percentage of the cation exchange capacity that is occupied by sodium. Usually, a soil is called sodic if ESP exceeds 15%. As shown, the hydraulic conductivity decreases rapidly in case salinity decreases. In reality, the effect of sodicity on hydraulic properties is even more complex, as also the retention function is significantly affected.

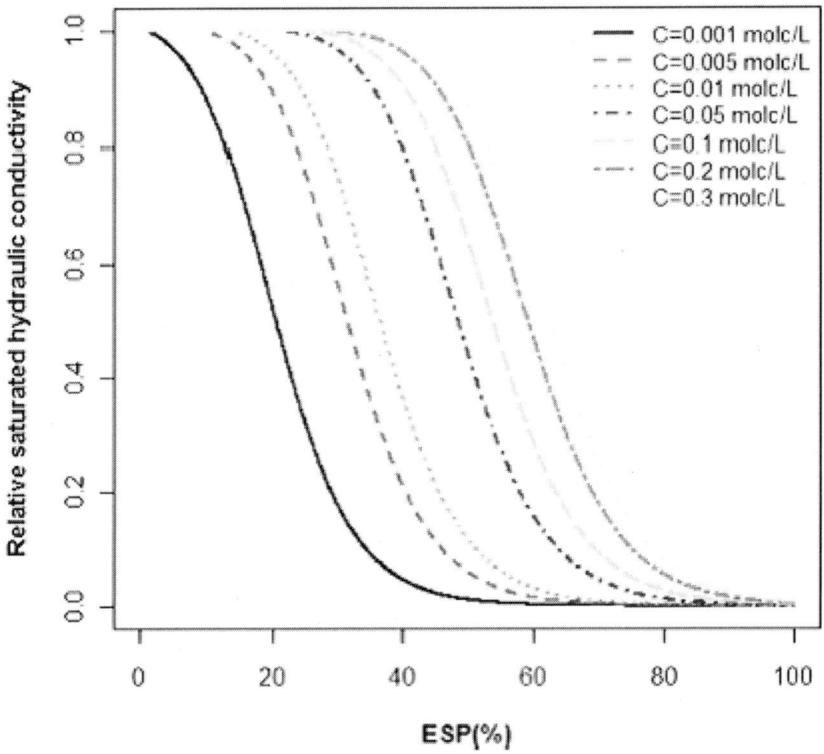

Figure 7. Relative saturated hydraulic conductivity as a function of ESP for different total salt concentrations in the soil solution (after Ref. [30]).

Sodic soils can be found in the Central Plain of Hungary, in the Hortobágy region, developed after the regulation of the main rivers, one and a half century ago. These physically degraded soils, with distinct B_2t horizons at say 20–40 cm depth that virtually impede vertical water transport, are perhaps only useful for extensive agriculture. Due to the small thickness of the A-horizon, that supplies water for the crops (capillary rise from the ground water being impossible due to the B_2t horizon), the water supply is very small. Remediation during the past century proved quite challenging and a large part of the Hungarian central plain is still suffering from small yields. This indicates that sodicity is a process to avoid.

8. CONCLUSIONS

Water scarcity can be countered by improving water use efficiency, but as inevitably, irrigation leads to salinity under conditions of scarce fresh water, salts should be leached and removed. The Leaching Requirement (LR) shows how to irrigate and avoid salinity, for deep groundwater. Also for shallow groundwater, minimalist modelling as with LR may lead to very useful approximations for the estimation of the long term salinity of the root zone.

In many agriculturally important regions worldwide, such as deltas, groundwater is sufficiently shallow to be taken into account in assessments of salinization risks. Accordingly, drainage and irrigation dimensioning necessitate the consideration of groundwater levels, which are strongly related with capillary rise replenishment of root zone water. Simple guidelines, as proposed in this chapter, may become very relevant to avoid salinity. Furthermore, advanced dynamic approaches to drainage, such as Climate Adaptive Drainage (CAD), may be very useful for saving water till the dry season, suppressing saline upward seepage and capillary rise, and removing salts from the root zone more effectively. In the Dutch WaterNexus program (with the lead author), CAD is experimentally and theoretically investigated for this purpose.

Besides salinity, plastic mulching may have environmental risks in view of scarcely known impacts on environment such as soil ecology, and ultimately, human health. Likewise, sodicity, which is often a salinity-related issue, requires urgent attention: it is a stealthily developing condition, that cannot well be observed by the land user without advanced chemical tools, its adverse effects may occur quite suddenly, and remediation may require enormous efforts and significant resources. For these reasons, sodicity needs to be avoided as much and as early as possible. That requires expertise, excellent communication, and great awareness.

This chapter has been aimed at providing awareness and illustrating the value of theoretical understanding (e.g. the Leaching Requirement and further developments based on it). Such understanding is important to anticipate when currently common practice may be in conflict with long term sustainability.

9. ACKNOWLEDGEMENTS

This contribution was made possible through the project WaterNexus that is funded by the Dutch Technology Foundation STW and the Dutch Ministry of Infrastructure and Environment.

REFERENCES

1. De Wit CT. Transpiration and crop yields. Versl. Landbouwkd. Onderz. 64, 88pp. 1958.
2. Maas EV, Hoffman GJ. Crop salt tolerance—current assessment. ASCE Journal of the Irrigation and Drainage Division. 103 (2), 115–134,1977.
3. Bresler, E, MacNeal BL, Carter DL. Saline and Sodic Soils: Principles-Dynamics-Modeling. Springer-Verlag, New York. 236 pp., 1982.
4. Schinas S, Rowell DL. Lime-induced chlorosis. Journal of Soil Science. 28, 351–368, 1977.
5. Grattan SR, Grieve CM. Mineral element acquisition and growth response of plants grown in saline environments. Agriculture, Ecosystems and Environment. 38, 275–300, 1992.
6. Parida AK, Das AB. Salt tolerance and salinity effects on plants: a review. Ecotoxicology and Environmental Safety. 60, 324–349, 2005,
7. Munns R, Tester M. Mechanisms of salinity tolerance. Annual Review of Plant Biology. 59, 651–681, 2008,
8. FLowers TJ, Gaur PM, Gowda CLL, Krishnamurthy L, Samineni S, Siddique KHM, Turner NC, Vadez V, Varshney RK, Colmer TD. Salt sensitivity in chickpea. Plant, Cell & Environment. 33, 490–509, 2010,
9. Shannon MC, Grieve CM. Tolerance of vegetable crops to salinity. Scientia Horticulturae. 78, 5–38, 1999.
10. Homaee M, Feddes RA, Dirksen C. A macroscopic water extraction model for nonuniform transient salinity and water stress. Soil Science Society of America Journal. 66, 1764–1772, 2002.
11. Rodriguez-Iturbe I, Porporato A. Ecohydrology of Water-controlled Ecosystems: Soil Moisture and Plant Dynamics. Cambridge University Press, Cambridge, UK, 2004.
12. Shah SHH, Vervoort RW,Suweis S, Guswa AJ, Rinaldo A, Van Der Zee SEATM.Stochastic modeling of salt accumulation in the root zone due to capillary flux from brackish groundwater. Water Resources Research. 47, W09506, 2010,
13. Stöckle CO, Donatelli M, Nelson R. CropSyst, a cropping systems simulation model. European Journal of Agronomy. 18(3–4), 289–307, 2003,
14. Kuhlmann A, Neuweiler I, Van Der Zee SEATM, Helmig R. Influence of soil structure and root water uptake strategy on unsaturated flow in heterogeneous media. Water Resources Research. 48, W02534, 2012.
15. Meeussen J. ORCHESTRA: an object-oriented framework for implementing chemical equilibrium models. Environmental Science and Technology. 37(6), 1175–1182, 2003.
16. Niu G, Cabrera RI. Growth and physiological responses of landscape plants to saline water irrigation: a review. HortScience. 45(11), 1605-1609, 2010.
17. Tóth T, Csillag F, Biehl LL, Michéli EE. Characterization of semivegetated salt-affected soils by means of field remote sensing. Remote Sensing of the Environment. 37, 167-180, 1991.

18. Francois LE, Maas EV. Plant Response to Salinity: An Indexed Bibliography. USDA-ARS, ARM-W-6, Washington, D.C., 1978.
19. Francois LE, Maas EV. Plant Response to Salinity: A Supplement to An Indexed Bibliography. USDA, ARS, ARS-24, Washington, D.C., 1985.
20. Ehlig CF, Bernstein L. Foliar absorption of sodium and chloride as a factor in sprinkler irrigation. Journal of the American Society for Horticultural Science. 74, 661-670, 1959.
21. Bernstein L, Francois LE. Effects of frequency of sprinkling with saline waters compared with daily drip irrigation. Agronomy Journal. 67, 185-190, 1975.
22. Hop MECM. Salt tolerance of plants. Dendroflora. 47, 43-73, 2010 (in Dutch).
23. Benes SE, Aragues R, Grattan SR, Austin RB, Foliar and root absorption of Na and Cl in maize and barley: implications for salt tolerance screening and the use of saline sprinkler irrigation. Plant Soil. 180, 75-86, 1996.
24. Maas, EV, Grattan SR,Ogata G. Foliar salt accumulation and injury in crops sprinkled with saline water. Irrigation Science. 3, 157-168, 1982.
25. Busch CD, Turner Jr F. Sprinkler irrigation with high salt-content water. Transactions of the ASAE. 10, 494-496, 1976.
26. Jordan LA, Devitt DA, Morris RL,Neuman DS. Foliar damage to ornamental trees sprinkler-irrigated with reuse water. Irrigation Science. 21, 17-25, 2001,
27. Richards LA, Allison LE,Bernstein L, Brown CA, Fireman JW, Hatcher M, Hayward JT, Pearson HE, Reeve GA, Wilcox RC. Diagnosis and improvement of saline and alkali soils. Agric. Handbook. vol. 60, 160 pp, USDA, Washington, D.C., 1954.
28. El-Ashry MT, Van Schilfgaarde J, Schiffman S. Salinity pollution from irrigated agriculture. Journal of Soil Water Conservation. 40(1), 48-52, 1985.
29. Szabolcs I. Salt Affected Soils. 274 pp., CRC Press, Boca Raton, FL, 1989.
30. Van Der Zee SEATM, Shah SHH, Vervoort RW. Root zone salinity and sodicity under seasonal rainfall due to feedback of decreasing hydraulic conductivity. Water Resources Research. 9431-9446, 2014,
31. Appels WM, Bogaart PW, van der Zee SEATM. Surface runoff in flat terrain: How field topography and runoff generating processes control hydrological connectivity. Journal of Hydrology. 534, 493-504, 2016.
32. Vervoort RW, Van Der Zee SEATM. Simulating the effect of capillary flux on the soil water balance in a stochastic ecohydrological framework. Water Resources Research. 44, W08425, 2008,
33. Stofberg SF, Oude Essink GHP, Pauw PS, De Louw PGB, Leijnse A, Van Der Zee SEATM. Fresh water lens persistence and root zone salinization hazard under temperate climate. Water Resources Management. 2016,
34. Raats PAC. Salinity management in the coastal region of the Netherlands: a historical perspective. Agricultural Water Management. 157, 12-30, 2014,
35. Maas K. Influence of climate change and sea level rise on a Ghijben Herzberg Lens. Journal of Hydrology. 347, 223–228, 2007.
36. Eeman S, Leijnse A, Raats PAC, Van Der Zee SEATM. Analysis of the thickness of a fresh water lens and of the transition zone between this lens and upwelling saline water. Advances in Water Resources. 34(2), 291–302, 2011,
37. Cendón DI, Larsen JR, Jones BG, Nanson GC, Rickleman D, Hankin SI, Pueyo JJ, Maroulis J.Freshwater recharge into a shallow saline groundwater system, Couper Creek floodplain, Queensland, Australia. Journal of Hydrology. 392, 150-163, 2010.
38. Cartwright I, Weaver TR, Simmons CT, Fifield LK, Lawrence CR, Chisari R, Varley S. Physical hydrogeology and environmental isotopes to constrain the age, origins and stability of a low-salinity groundwater lens formed by periodic river recharge: Murray Basin, Australia. Journal of Hydrology. 380, 203-221, 2010.
39. Hatfield JL, Sauer TJ, Prueger JH. Managing soils to achieve greater water use efficiency: a review managing soils to achieve greater water use efficiency: a review. Agronomy Journal. 93, 271–280, 2001.
40. Howell TA. 561bb7ef08ae6d17308b03eb.tif. Enhancing water use efficiency in irrigated agriculture. Agronomy Journal. 93, 281–289, 2001.

41. Huang Y, Chen L, Fu B, Huang Z, Gong J. The wheat yields and water-use efficiency in the Loess Plateau: straw mulch and irrigation effects. Agricultural Water Management. 72, 209–222, 2005,
42. Wright SL, Rowe D, Thompson RC, Galloway TS. Microplastic ingestion decreases energy reserves in marine worms. Current Biology. 23(23), R1031–R1033, 2013.
43. Koelmans AA, Besseling E, Wegner A, Foekema EM. Plastic as a carrier of POPs to aquatic organisms. A model analysis. Environmental Science and Technology. 47 (14), 7812–7820, 2013.
44. Huerta Lwanga E, Gertsen H, Gooren H, Peters P, Salanki T, Van Der Ploeg M, Besseling E, Koelmans AA, Geissen V. Microplastics in the terrestrial ecosystem: implications for Lumbricusterrestris (Oligochaeta, Lumbricidae). Environmental Science and Technology. 50, 2685–2691, 2016,
45. Hobbs P, Sayre K, Gupta R. The role of conservation agriculture in sustainable agriculture. Philosophical Transactions of the Royal Society B: Biological Sciences. 363, 543–555, 2008,
46. Li SX, Wang ZH, Li SQ, Gao YJ, Tian XH. Effect of plastic sheet mulch, wheat straw mulch, and maize growth on water loss by evaporation in dryland areas of China. Agricultural Water Management. 116, 39–49, 2013,
47. Zhang S, Li P, Yang X, Wang Z, Chen X. Effects of tillage and plastic mulch on soil water, growth and yield of spring-sown maize. Soil and Tillage Research. 112, 92–97, 2011,
48. Li S, Kang S, Li F, Zhang L. Evapotranspiration and crop coefficient of spring maize with plastic mulch using eddy covariance in northwest China. Agricultural Water Management. 95, 1214–1222, 2008,
49. Ghosh PK, Dayal D, Bandyopadhyay KK, Mohanty M. Evaluation of straw and polythene mulch for enhancing productivity of irrigated summer groundnut. Field Crops Research. 99, 76–86, 2006,
50. Yang Y, Liu X, Li W, Li C. Effect of different mulch materials on winter wheat production in desalinized soil in Heilonggang region of North China. Journal of Zhejiang University Science. B7, 858–867, 2006,
51. Pang HC, Li YY, Yang JS, Liang YS. Effect of brackish water irrigation and straw mulching on soil salinity and crop yields under monsoonal climatic conditions. Agricultural Water Management. 97, 1971–1977, 2010,

CHAPTER 3

Estimation of Crop Coefficient and Water Requirement of Dutch Roses (Rosa hybrida) under Greenhouse and Open Field Conditions

Vikas Kumar Singh*, KN Tiwari and Santosh DT

Department of Agricultural and Food Engineering, Indian Institute of Technology, Kharagpur, India

ABSTRACT

Precise estimation of reference evapotranspiration (ETo) and crop evapotranspiration (ETc) on a daily basis is important to apply water through drip system for crops grown in the greenhouse. Crop coefficients and crop water requirement were determined for the Dutch roses cultivated in the greenhouse and open field for the sub-humid climatic conditions of Kharagpur, India. Reference evapotranspiration (ETo) was estimated using the method suggested by FAO-56. The crop ET was determined using soil water balance approach. The soil moisture data was collected using TDR moisture meter at three depths. The maximum daily values of Crop ET were 4.99 and 5.28 mm day-1 for greenhouse and open field conditions respectively. Maximum values of crop coefficient were found during conjunction period of mid-season and late season stage of the crop growth period. During different growth stages of rose, crop coefficient values were found in the range from 0.48 to 0.96 and 0.59 to 1.01 for greenhouse and open field conditions. The Dutch rose planted in 200 μ diffused poly film cladded greenhouse and given 100% water requirement through drip irrigation resulted in maximum number of flowers/m2 (212.3) annually. Total annual water requirement of rose plant was 999.51 mm and 1210.94 mm for greenhouse and open field condition respectively.

Keywords: Dutch rose; Crop coefficient; Crop evapotranspiration; Greenhouse

INTRODUCTION

Rose is a leading cut flower grown commercially all over the world. It ranks first in global cut flower trade. This flower has a worldwide consumption of more than 40 billion. The various purposes for rose cultivation includes garden flowers, aesthetic value, decoration and preparation of various products such as rose oil, rose water, gulkhand rose attar, garland etc. The heavy demand of rose cut flowers in the European **MARKETS** is mainly from November to March due to the shortage of local production because of severe winter. Fortunately, this is the most congenial condition for successful production of most of the cut flowers, including roses in India. It is pointed out that buyer at international market prefers a very high quality rose cut flowers. As it is difficult to obtain good quality cut flowers under open conditions throughout the year, the crops should be cultivated under the greenhouse to get good quality produce.

Greenhouse cultivation or protected cultivation is a kind of farming systems used to maintain a controlled or partially controlled environment suitable for maximum crop production. This includes creating an environment suitable for working efficiency as well as for better crop growth. The presence of a cover, characteristics of greenhouses, causes a change in the climatic conditions compares to those in open field. The property of cladding materials alters the magnitude and quality of **SOLAR RADIATION** incidence in the greenhouse, thereby affecting the micro climate of greenhouse. UV stabilized diffused film does not allow the shadow formation of top canopy on the lower leaves. Diffused radiation penetrates deeper into plants canopy in comparison to direct radiation; thus, it is desirable to use diffused film. At high irradiation, diffused film greenhouse cover leads to better light distribution, lower plant temperature, decreased transpiration and increased photosynthesis and growth [1].

Drip irrigation has been proved as an effective water-saving irrigation method and it is important component of greenhouse cultivation system. Drip irrigation research studies carried out at Precision Farming Development Centre, IIT Kharagpur, India on several vegetable and fruit crops showed increase in yield, saving in water, higher water use efficiency and net increase in profit.

Crop evapotranspiration (ET) is required for determination of crop water requirements, irrigation scheduling and water productivity. The water balance method is a simple and easy to use as compared to several methods reported in the literature. ET is estimated using water balance method considering change in total soil water content between sampling dates plus rainfall minus any known drainage or surface runoff occurred during the period [2]. Time Domain Reflectometry (TDR) probes and capacitance-based **SENSORS** are popular devices for measuring inside soil water content directly. Advantages of these electronic sensors are in-situ measurements with continuous recording [3].

Crop coefficient is a significant parameter for estimation of crop evapotranspiration, because it accounts biological characteristics of crops, crop condition, soil texture, soil tillage conditions, crop growing environment etc. [4]. The crop coefficient Kc, is ratio of crop ET to reference ET, is needed to estimate crop evapotranspiration for irrigation planning at a regional scale. The crop coefficient value represents crop specific water use and is required for accurate estimation of irrigation requirement of single or more than one crop grown under different climatic conditions [5].

Several studies on estimation of water requirement and crop coefficient have been conducted for open field crops like Cotton (Abdelhadi et al. [6] Fisher [7]), Maize (Akinmutimi, [8] Chuanyan and Zhongren [9]), Onion and Spinach, Okra (Odofin et al. [10]), Capsicum (Miranda et al. [11]), Cabbage, Paddy (Kuo et al. [12]), Wheat and Sorghum, Teff (Araya et al. [13]), Sapota, Jujube (Hu et al. [14]), Pomegranate and greenhouse crops like Tomato (Maldonado et al. [15] Gómez et al. [16] Harmanto et al. [17]), Eggplant, Melon, Green beans, Watermelon and Pepper (Orgaz et al. [18]), Gladiolus (Bastug et al. [19]), Cucumber (Zhang et al. [14] Blanco and Folegatti [20]) However, limited studies are found on cultivation of rose crop under greenhouse conditions (Ehret et al. [21] Katsoulas et al. [22]) and no literature is reported for the estimation of water requirement and crop coefficient of Dutch roses under greenhouse conditions. Therefore, the present study have been undertaken to determine the crop coefficient and water requirement of **DUTCH ROSE** under greenhouse and open field conditions.

MATERIALS AND METHODS

Description Of The Experimental Area

Experiment was conducted at the Field Water Management Laboratory of Agricultural and Food Engineering Department, Indian Institute of Technology, Kharagpur, India. The experimental site is located on flat land at 22°18.5' N latitude, 87°19' E longitude and 48 m altitude above mean sea level. The local climate is sub-humid subtropical with an average annual rainfall of 1390 mm, of which about 80% is received during June to October. The mean monthly minimum temperature is 12°C in January, whereas the mean monthly maximum temperature is 41.5°C in May. The mean monthly **RELATIVE HUMIDITY** varies from 35% in February to 96% in July-August. The site was deep and well drained sandy loam soil. The average values of sand, silt and clay was 63%, 28% and 9% respectively in the soil depth of 1 m.

Experimental Greenhouses

Two experimental sawtooth shape greenhouses (N-S oriented) one cladded with 200μ diffused (PAR transmissivity as 90% and 42% diffusivity) film and another with 200μ clear UV stabilized film installed at the Field Water Management Laboratory of Agricultural and Food Engineering Department, Indian Institute of Technology Kharagpur, India. The total ground area of each greenhouse was 84 m^2 with central height of 4.5 m and gutter height of 4 m **(FIGURE 1)**. Greenhouses had provision to vary ventilation area maximum upto 60% (60% of the floor area) to allow hot air to escape during peak summer. Ventilation was provided at ridge and both sides of the greenhouses. These ventilated openings were covered with an insect-proof net of 60 mesh size to prevent the entry of insects. These greenhouses were also equipped with fogging system (16 L h^{-1} discharge **CAPACITY**) and shade net (75% shading intensity) beneath the greenhouse roof. Two exhaust fans with a capacity of 1100 m^3/min with 2.2 kW power were installed to replace hot air with ambient air in the greenhouse. Natural ventilation was maintained in both greenhouses, however fans were operated only when the air temperature in the greenhouse was high during peak summer.

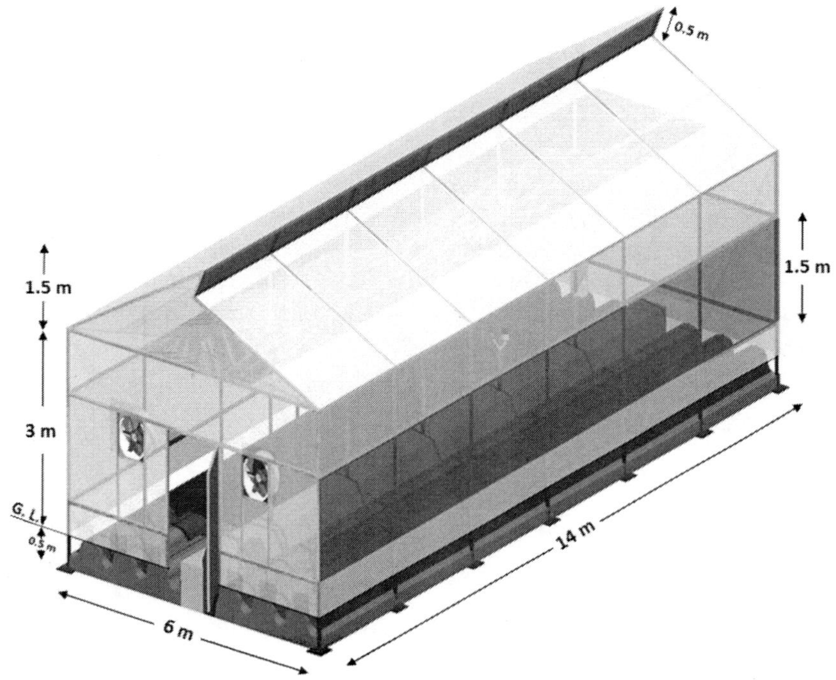

FIGURE 1: Isometric view of the experimental greenhouse.

The diffused film used for cladding of one of the greenhouses has inbuilt property to **SCATTER** the radiation and does not form shadow on the top of the canopy of lower leaves.

MONITORING OF GREENHOUSE MICROCLIMATE: Automatic weather station of M/S Campbell Scientific, Canada comprising a datalogger (model CR1000) and sensors were installed in the greenhouse to monitor soil temperatures (models 107 BL and CS616 L), air temperatures and relative humidity (model HMP 45 C), global radiation and **PHOTOSYNTHETICALLY** Active Radiation (SPLITE and PARLITE of Kipp and Zonen). Outside air and soil temperatures, relative humidity and solar radiation were measured manually at 8:30 AM, 12:30 PM and 4:00 PM in a day.

REFERENCE EVAPOTRANSPIRATION, ET_O

Reference evapotranspiration (ET_O) was estimated using the method suggested by FAO-56 [2]. The equation used to estimate ET_O is described below:

$$ET_o = \frac{0.408\Delta(R_n - G) + \gamma \frac{900}{T+273} u_2(e_s - e_a)}{\Delta + \gamma(1 + 0.34u_2)} \quad (1)$$

Where,

ET_0 - Reference evapotranspiration [mm day^{-1}], Rn - Net radiation at the crop surface [MJ m^{-2} day^{-1}], G - Soil heat flux density [MJ m^{-2} day^{-1}], T - Mean daily air temperature at 2 m height [°C], u_2 - Wind speed at 2 m height [m s^{-1}], e_s - Saturation vapour pressure [kPa], e_a - Actual vapour pressure [kPa], e_s-e_a - Saturation vapour pressure deficit [kPa], Δ - Slope vapour pressure curve [kPa °C^{-1}], γ - Psychrometric constant [kPa °C-1].

In a greenhouse, wind speed approximately equals to zero; therefore, the Eq. (3) can be simplified written as:

$$ET_o = \frac{0.408\Delta(R_n - G)}{\Delta + \gamma} \quad (2)$$

CROP DETAILS

Dutch rose plants (Variety First Red) were transplanted at a spacing of 0.5 m between rows and 0.3 m between plants in a row in both the greenhouses and also in open field on October, 2009. Rose nursery plants were planted on raised bed of 1 m wide with two rows per bed. A trench of 30 cm wide and 50 cm deep was dug in between two beds for removal of excess moisture from root zone and separation of one treatment from other. To improve the soil organic matter and nutrient (NPK) contents, manures and fertilizers were applied 1 month prior to the planting. The soils of beds were incorporated with 15 kg m^{-2} of Farm Yard Manure, 5 kg m^{-2} of vermi compost, 0.5 kg m^{-2} Neem cake and 0.2 kg m^{-2} of single superphosphate. The crop was drip-irrigated using one lateral per bed of two rows and one emitter of 2 L h^{-1} capacity, which met the water requirement of 4 plants of both rows. The drip irrigation system's emission uniformity was determined and maintained above 90%. Recommended dose of soluble fertilizers were applied twice in a week using a venturi injector. After 180 days of planting, plants were pruned to a height of 0.3 m above the ground. There after pruning was done once in a year to increase the number of branches for good flowering.

MEASUREMENT OF CROP EVAPOTRANSPIRATION, ET_C

The actual crop evapotranspiration was estimated with the soil water balance method. The soil moisture changes in 0-90 cm soil layer during the study period were measured and used to estimate actual crop evapotranspiration with equation (1) [2]. Three circular shape drums having closed bottom, each having 0.5 m^2 cross sectional area, 1 m deep and 5 mm thick wall **(FIGURE 2)**were installed in the greenhouse. Three drums of the same size and specifications were installed outside in open field. Considering the rose plants spacing and the drum dimensions, one rose plant was planted in the each drum installed inside and outside the greenhouse. The packages of practices for the plant inside and outside the drum were maintained identical. Rose plant planted in the drum **LYSIMETER** installed outside the greenhouse is mulched with plastic film to prevent the entry of rainwater in the lysimeter during rainy season. Soil water content was measured daily using TDR moisture meter at 0-30, 30-60 and 60-90 cm depth. The root zone depth of rose plant was maximum up to 45-50 cm. The soil moisture measured at 60-90 cm was considered as vertical drainage and deducted from amount of irrigation water applied to estimate crop evapotranspiration. Appropriate amount of irrigation water was applied manually to the plant in lysimeters to replenish depleted soil moisture at field capacity. Soil-water depleted was determined by the TDR moisture meter readings and replenished by a factor of 1.1 on the same day, to meet the crop evapotranspiration in the next following day [11]. Amount of irrigation water varied according to crop water use in different growth stages. These measurements and irrigation water supply were done for the full rose crop season. Crop evapotranspiration was calculated using the change in soil moisture content and measurement of other water balance parameters from following water balance equation

$$ET_C = P + I - R - D \pm \Delta W \tag{3}$$

Where, ET_C: Crop evapotranspiration (mm); P: Precipitation (mm); I: Irrigation water depth (mm); R is the surface runoff (mm); D: Amount of water drained from the root zone (mm); ΔW: change in soil water storage (mm). Contribution of water due to rain in the greenhouses was zero, hence, P = 0. The contribution of lateral flow and surface runoff during irrigation events was also zero, hence R = 0.

Thus Eq. (1) can be simplified as follows

$$ET_C = I - D \pm \Delta W \tag{4}$$

The change in the amount of soil-water into the drum (ΔW) is computed by the difference between soil moisture observations of two consecutive days.

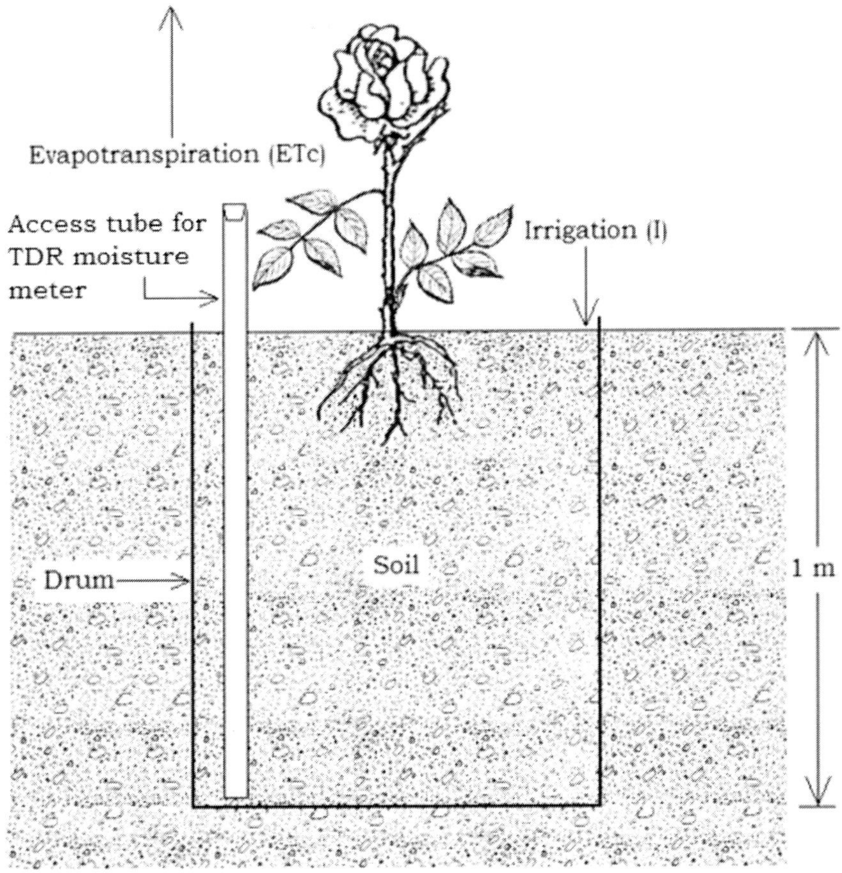

Figure 2: Cross sectional view of drum lysimeter used for determination of ET_C of rose.

ESTABLISHMENT OF CROP COEFFICIENT (K_C) OF DUTCH ROSES FOR GREENHOUSE AND OPEN FIELD

The **CROP COEFFICIENT** values for Dutch roses were determined for the crop grown both in greenhouse and open field conditions. The average crop coefficients for Dutch roses are estimated with an assumption that one annual cycle of crop is started after planting of crop in October and completed in next September month when pruning is done. The next annual cycle is started just after pruning and completed in next September month when pruning is done. Crop coefficient was computed on a daily basis from transplanting to pruning and pruning to pruning in next year using following equation:

$$Kc = \frac{ET_c}{ET_o} \tag{5}$$

Average crop coefficients were calculated for the initial ($Kc_{initial}$), development (Kc_{dev}), mid-season (Kc_{mid}) and late-season (Kc_{late}) growth stages. The lengths of the growth stages were determined based on the crop phenological phase and the Kc curve. $Kc_{initial}$ begins from transplanting to plant covers 10% of the plant area. The Kc_{Dev} begins after $Kc_{initial}$ (i.e., 10% canopy to 60% of canopy area). The Kc_{mid} begins after Kc_{Dev} ends (i.e., canopy attains 60% and ends in the later part of the season when the crop begins to enter in the end of the season) and the Kc_{late} starts when crops has more than 60% canopy and end when crop attains dormant stage.

WATER REQUIREMENT OF ROSE PLANT UNDER GREENHOUSE AND OPEN FIELD CONDITIONS

To optimize the water requirement of rose plants grown under two saw tooth shape greenhouses cladded with UV stabilized clear and diffused film of 200 μ thickness, an experiment was conducted for three years (January 2012 to December 2014) with four irrigation levels of 100%, 80%, 60% and 40% of estimated **CROP EVAPOTRANSPIRATION** (ETc) application. The experiment was designed to assess the effect of irrigation levels and greenhouse cladding materials on biometric growth and yield (No. of flowers) of rose plant. Eight treatments replicated three times were planned for the greenhouses and one treatment for open field as a control to compare the results. Statistical analysis was done for analysis of variance (ANOVA) using the Least Significant Differences (LSD) test at the 5% probability level ($P = 0.05$).

Treatment details are stated below:

Drip irrigation was installed to irrigate plants inside the **GREENHOUSE** and outside (open field). The time of operation of drip system was computed using daily water requirement of four plants and emitter discharge for each irrigation. In drip system, irrigation was scheduled on alternate days and the quantity of total water applied was calculated on the basis of daily crop water requirement. The lateral lines of 16 mm diameter were laid parallel to plant rows. Online emitters of 2 L h^{-1} capacity were fitted with lateral and each emitter served the water requirement of four plants. Irrigation duration for delivery of water for different treatments was controlled with the help of gate valve provided at the inlet of each laterals. The operating pressure of about 1 kg/cm^2 was maintained to obtain design **DRIPPER** discharge. The layout of the experiment and the division of beds along with the laterals are shown in **FIGURE 3**.

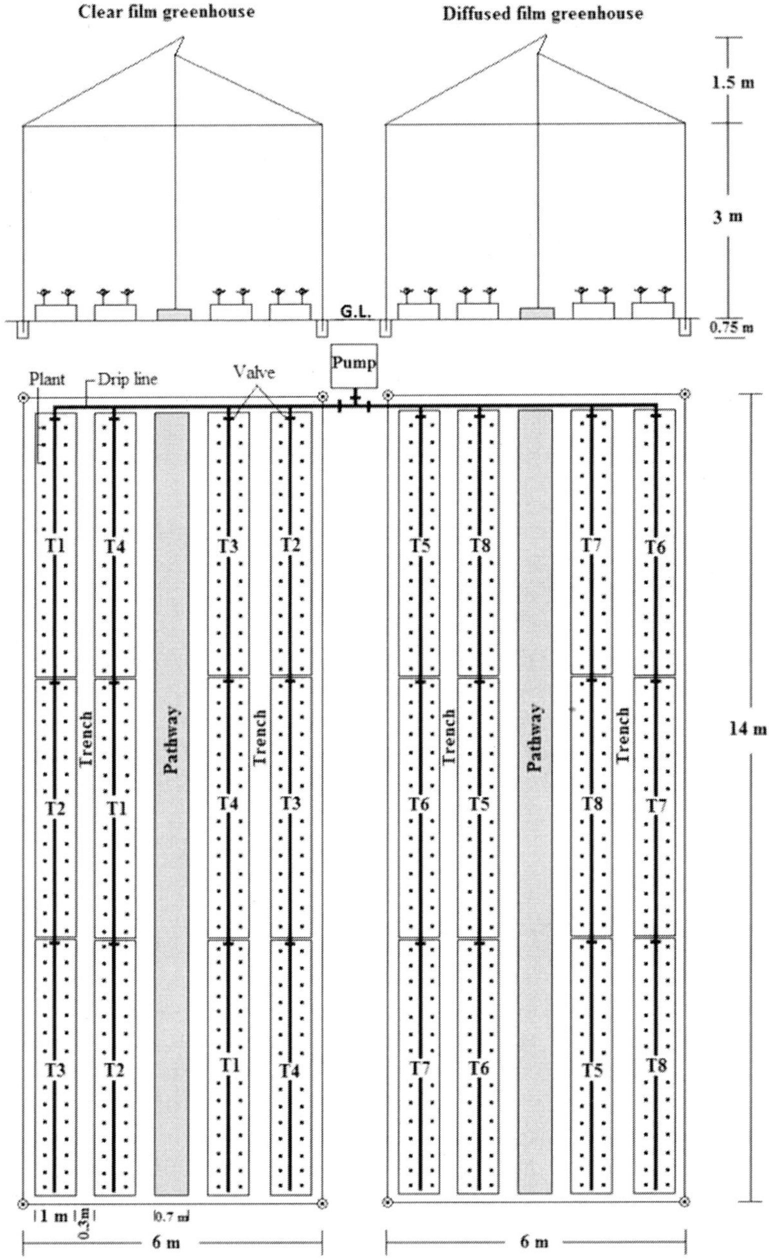

Figure 3: Relationship between ETo and ETc in greenhouse and outside conditions.

BIO-METRIC OBSERVATIONS AND WATER PRODUCTIVITY: Bio-metric observations viz. plant height, number of shoots per plant, shoot length, number of flowers/m^2/year, flower diameter and number of petals per flower were recorded at 15 days interval to evaluate the effect of irrigation levels and cladding materials on yield and yield attributes of Dutch roses inside the greenhouse and in open field (outside).

The annual water productivity (number of flowers/m^2/mm) for greenhouse and open field treatments were calculated using the following equation [2]:

$$AWP = \frac{Y}{AWU} \tag{6}$$

Where,

Y is total flower yield (number of flowers/m2) and AWU is annual water use (mm)

RESULTS AND DISCUSSION

ESTIMATION OF ETO FOR INSIDE THE GREENHOUSE AND OPEN FIELD

The meteorological observations collected from automatic weather stations were substituted in FAO Penman-Monteith equation to calculate daily ETo for fifteen years (January 2000 to December 2014) for inside the greenhouse and outside (open field), and the results are shown in **TABLE 1** and **FIGURE 3**. It can be seen from **FIGURE 3** that average ETo value inside the greenhouse and outside (open field) are relatively of greater amount from late March to October months as compared to January to early March and November – December. In April, the average daily ETo value of month is below 4 mm except a few high values occurred at the end of this month. With the increase of sunshine hours and the intensity of radiation, ETo value gradually increases. Peak value of the ETo (i.e., 5.39 mm) was observed in the month of May. June months onward average daily ETo values was reduced due to incidence of rainfall, low solar radiation, and high humidity especially during monsoon months (June to September). Further due to lowering of temperature from October months onward till March, daily ETo values reduced and reached to lowest in the month of January. ETo values get affected due to cloud cover and rains, hence fluctuating ETo values were found in the same months.

TABLE 1: Estimated water requirement and crop coefficient of Dutch rose crop and time of operation of drip system for different months.

Month	Inside/ outside condition	Reference evapo-transpiration, ETo (mm day^{-1})	Kc	Actual evapo-transpiration*, ETc (mm day^{-1})	Area occupied by four plants (4×0.5×0.3) (m^2)	Required discharge through single emitter (L day^{-1})	Time of operation of drip system per day with 2 L h^{-1} emitters (min)
		A	B	C=A×B	D	E=C×D	
Jan	Greenhouse	1.67	0.86	1.43	0.6	0.86	25.9
	Open field	2.40	0.89	2.14	0.6	1.28	38.5
Feb	Greenhouse	2.05	0.87	1.78	0.6	1.07	32.1
	Open field	3.12	0.90	2.81	0.6	1.69	50.6
Mar	Greenhouse	2.81	0.89	2.51	0.6	1.50	45.0
	Open field	4.18	0.93	3.89	0.6	2.33	70.0
Apr	Greenhouse	3.88	0.92	3.57	0.6	2.14	64.3
	Open field	4.99	0.96	4.79	0.6	2.87	86.2
May	Greenhouse	5.20	0.96	4.99	0.6	3.00	89.9
	Open field	5.39	0.98	5.28	0.6	3.11	93.4
Jun	Greenhouse	4.41	0.96	4.23	0.6	2.54	76.2
	Open field	4.59	1.01	4.63	0.6	2.78	83.4
Jul	Greenhouse	3.84	0.89	3.42	0.6	2.05	61.5
	Open field	4.05	0.97	3.93	0.6	2.36	70.7
Aug	Greenhouse	3.89	0.82	3.19	0.6	1.91	57.4
	Open field	4.11	0.84	3.45	0.6	2.07	62.1
Sept	Greenhouse	3.74	0.76	2.84	0.6	1.71	51.2
	Open field	4.02	0.78	3.14	0.6	1.88	56.4
Oct	Greenhouse	3.67	0.48	1.76	0.6	1.06	31.7
	Open field	3.96	0.52	2.06	0.6	1.24	37.1
Nov	Greenhouse	2.70	0.60	1.62	0.6	0.97	29.2
	Open field	2.96	0.63	1.87	0.6	1.12	33.6
Dec	Greenhouse	1.93	0.76	1.47	0.6	0.88	26.4
	Open field	2.34	0.81	1.90	0.6	1.14	34.1

* This is 100% of crop evapotranspiration estimated for Dutch rose crop with drum lysimeter in this study (treatments T1 & T5). ETc for treatment T2 & T6 is 80%, T3 & T7 is 60% and treatment T4 & T8 is 40% of this estimated value of ETc.

FAO Penman-Monteith equation contains two parts: one radiation and other aerodynamic components and they represent the impact of solar radiation and convection, turbulence and degree of drying above the evaporating surface respectively. Both of these two components are dynamic processes vary with time with influenced by location and climatic conditions. Analysis of data shows that the part of solar radiation component is more than 60% of total evapotranspiration during study period, and it is maximum in May. From May month to August month, the solar radiation decreases from the 23.5 MJ/m^2/day in May to 18.5 MJ/m^2/day in August, during which the value of each month are 23.5, 19.0, 17.4 and 18.5 MJ/m^2/day during May, June, July and August respectively. This is due to decrease in solar intensity and day length during monsoon months.

Comparison of monthly ETO values of greenhouse and open field conditions shows that ETO of greenhouse is always lower and this is due to the reduced evaporative demand inside the greenhouse **(FIGURE 3)**. The evaporative demand is lower inside the greenhouse than outside (open field) due to the decrease in solar radiation (20% on average) and the wind speed is nearly zero. The evaporative demand inside the greenhouse is 60 percent of that of outside and the same finding is reported by Möller and Assouline [23] Fernández et al. [18]. They used lime suspension on greenhouse cover surface to cut down the entry of solar radiation to reduce air temperature that caused ETo reduction. During May to August there is very less difference between ETo of inside and outside of greenhouse, is because of high humidity with minor difference in dry bulb and wet bulb temperature of prevailing micro climate.

SOIL WATER CONTENT AT DIFFERENT DEPTH IN DRUM LYSIMETER

Soil moisture content at three depths (0-30, 30-60 and 60-90 cm) of drum lysimeters kept inside and outside of greenhouse was measured daily with the help of TDR moisture meter. Analysis of the soil moisture data collected from TDR moisture meter shows that moisture content in the middle layer (30-60 cm) was found maximum (15.4-9.1%) in comparison to top layer (14.6*18%) and bottom layer (14-16.1%). The top layer has high evaporation due to greater soil temperature hence lesser soil moisture content than middle layer. The top soil (up to 50 cm) of lysimeters had rich organic matter which might have increased in water holding capacity of soil, hence moisture content at bottom layer was found to retain lesser than top and middle layers **(FIGURE 4)**. Similar trend of soil moisture content was observed in all the three layers of soils of lysimeters kept outside the greenhouse (open field). Within the whole growth period, the change in values of soil water content at different layers reflects crop evapotranspiration demand. In the drip irrigation water supply, the evaporation of soil water is lesser than the surface irrigation and changes of soil water content indirectly reflect characteristics of the crop water consumption [14].

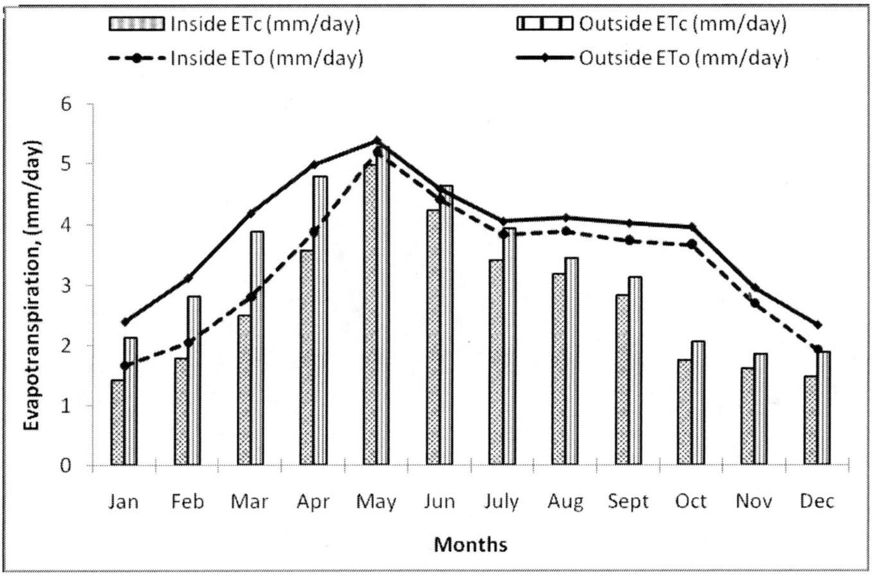

Figure 4: Soil moisture variations at different depth of lysimeter.

CROP EVAPOTRANSPIRATION (ETC) AND CROP COEFFICIENT (KC)

The evapotranspiration (ETc) of the rose crop was determined daily from the drum lysimeters kept inside and outside the greenhouse for the full crop season is presented in **TABLE 1**. Daily Rose plant ETC values varied from 1.43 to 4.99 mm day^{-1} and 1.86 to 5.28 mm day^{-1} for greenhouse and outside conditions respectively. Evapotranspiration (ET_c) requirement of the plant in drum lysimeters in greenhouse is lower than the plant of drum lysimeters kept outside (open field) due to the reduced evaporative demand inside the greenhouse [14,23]. Maximum ETC values occurred in the month of May when the crop is fully matured with greater level of ground cover [11]. ETc values reduced considerably and reached to 3.5 mm day^{-1} which coincided during the flower harvesting period. Then, during the second flowering peak, ETc values increased again rapidly, but did not reach the same level as observed during the first flowering peak.

Daily crop coefficient value was estimated for three years (October 2009 to October 2012) of experimental period and averaged on monthly basis. The daily average value of crop coefficient for one complete cycle under greenhouse and open field conditions are presented in **TABLE 1** and **FIGURE 5**. The crop coefficient for greenhouse conditions is 0.48 to 0.6 during initial stage (K_{CIn}), 0.6 to 0.86 during development stage (K_{CDev}), 0.87 to 0.96 during middle stage (K_{CMid}) and 0.96 to 0.76 during late season ($K_{C\ Late}$). In the same way the crop coefficient for open field conditions was estimated and found to vary from 0.50 to 0.63 during initial stage ($K_{C\ In}$), 0.63 to 0.89 during development stage ($K_{C\ Dev}$), 0.90 to 1.01 during middle stage ($K_{C\ Mid}$) and 0.97 to 0.78 during late season ($K_{C\ Late}$).

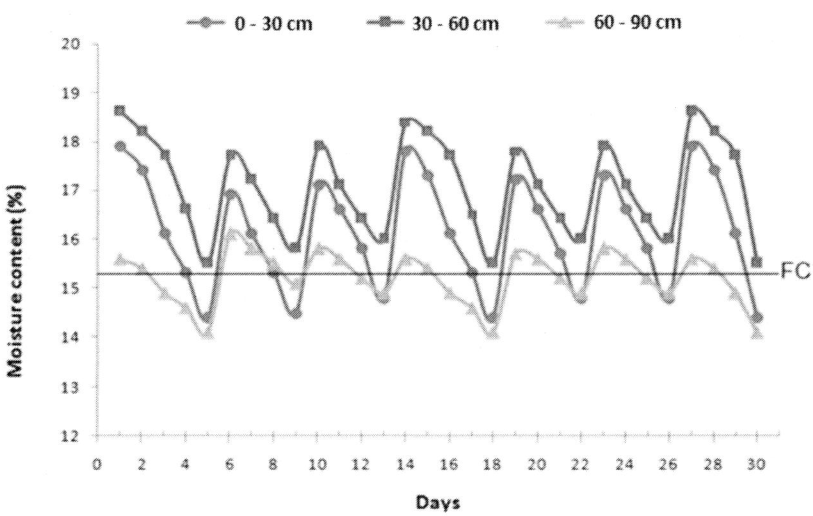

Figure 5: Experimental layout of different treatments.

During different growth stages of rose, crop coefficient value varied from 0.48 to 0.96 and 0.59 to 1.01 for the plant under greenhouse and open field conditions. During the initial stage (30 days) of the plant crop coefficient is found lower value because of lesser ground cover and low crop water requirement whereas during the development stage (90 days) it is more than earlier [20]. During the mid-season (120 days) stage crop coefficient is found maximum for both greenhouse and open field conditions.

Crop coefficient values decreased steadily after the first harvest, reaching minimum values close to 0.75. After the pruning is done, a new cycle of vegetative growth and flowering began, that increased Kc values again. Finally, during the second harvest period Kc values decreased again. The similar trend of crop coefficient under both the conditions is found during all three years of experimental period. The behaviour of the crop coefficient curve throughout the Dutch rose growth period, with a short period of low Kc values kept between two phases of high Kc values, matched with the crop phonological behaviour observed in the lysimeter. Although the Dutch rose produced incessantly after the first harvest, it presented basically two phases when flowering was more intense.

The rose plant present more than one harvest cycle, it is important to adjust the Kc value accordingly, increasing the Kc during periods of full ground cover and decreasing the Kc when pruning is done. Since there are no published Kc values specific for Dutch rose for both the conditions, it was assumed that for irrigation management purposes one would use Kc values recommended in this study.

MICRO CLIMATE BEHAVIOR OF GREENHOUSE CLADDED WITH UV STABILIZED CLEAR AND DIFFUSED FILM

During summer season (May to September), the average daily variation of temperatures was found to vary between 35°C to 44°C and 34°C to 42°C in the UV stabilized clear film and diffused film respectively and during winter season (November to February), average daily air temperature in the UV stabilized clear and diffused films at 12:30 PM varied from 25°C to 31°C and 23°C to 29°C respectively against open field condition (22 to 28°C). The maximum temperature reduction in winter at 12:30 PM in the diffused film was 3°C as compared to the clear film.

The study of temperature variation in the clear and diffused covers indicated that the UV stabilized clear film maintained 3°C higher temperature than diffused film during winter and 2°C higher temperature than diffused film during summer months. Temperature reduction in the UV stabilized diffused film is more prominent than in the UV stabilized clear film due to the diffusive property. UV stabilized clear film maintained 10-

15% lesser relative humidity than the diffused film whereas outside relative humidity is even lesser than diffused film. Weekly variation of solar radiation levels in the greenhouse at 12:30 PM during winter season varied from 32 to 56 Wm^{-2} and 38 to 67 W m^{-2} in the **CLEAR AND DIFFUSED FILM** respectively against the ambient solar radiation of 74 to129 W m^{-2}. Whereas in summer months, it varied from 34 to 61 W m^{-2} and 41 to 82 W m^{-2} in the clear and diffused covers, respectively against the ambient solar radiation of 86 to 170 Wm^{-2}. UV stabilized diffused film passes 20-34% more solar radiation than clear film. This solar radiation is directly related to the Photo synthetically Active Radiation (PAR) which is used in the photosynthesis of crop, hence better crop growth is found under diffused film cladded greenhouse. It is clear from other studies that Ultraviolet (UV), Photosynthetic Active Radiation (PAR) and Near-Infrared (NIR) is the part of solar radiation and which transmit through the greenhouse covering material. The PAR is essential for photosynthesis which is favourable to plants growth. Thus, greenhouse cladding films which diffuse the incoming solar radiation can offer several advantages and provide better micro climate (Chou and Lee [24], Kondratyev [25], Lamnatou and Chemisana [26], Espí et al. [27]). Soil temperature in the greenhouses cladded with clear and diffused film at 25 cm depth during winter season ranged from 19 to 21°C and 18 to 21°C respectively. During summer season, it varied from 25 to 29°C and 24 to 28°C in clear and diffused film cladded greenhouse. It was observed that no significant difference in the soil temperature was found in the greenhouses cladded with clear and diffused film.

OPTIMUM WATER REQUIREMENT FOR DUTCH ROSES UNDER TWO GREENHOUSES OF DIFFERENT CLADDING MATERIALS

To optimize the water requirement of the Rose crop under two different types of cladded film greenhouse (clear and diffused UV stabilized film of 200 μ thickness), the treatment wise biometric observations were recorded and analyzed for different irrigation levels.

The analysis of biometric data revealed that Dutch rose cultivation with 100% water requirement for the greenhouse cladded with diffused film (T5) resulted in best performance than control plots in terms of plant height (68.8 cm), shoot length (46.2 cm), flower diameter (6.97 cm), number of petals per flower (25.1), number of shoots per plant (29.3) and number of flowers/m^2/year (212.3) **(FIGURE 6)**.

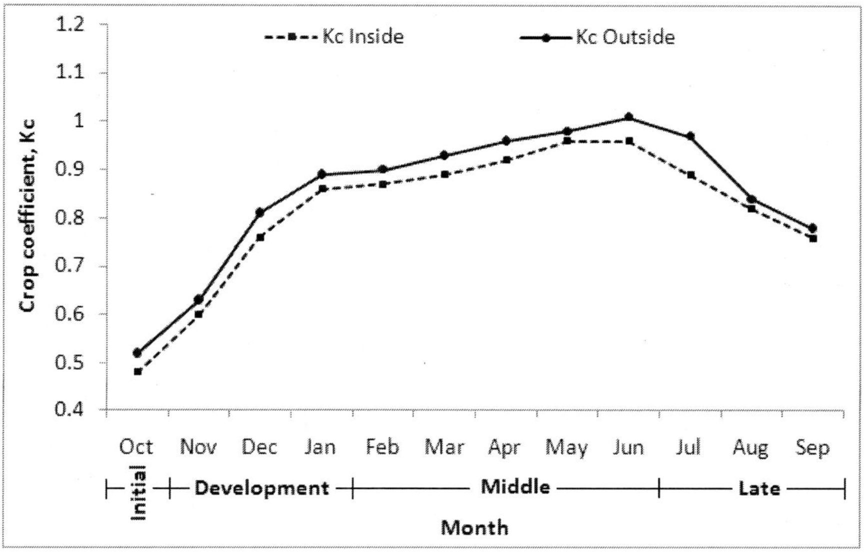

FIGURE 6: Crop coefficient for Dutch roses under greenhouse and open field conditions.

Time of operation of the drip irrigation system with 2 Lh^{-1} emitters for different months of growth period of rose crop planted inside and outside of greenhouse was also estimated and results are presented in **TABLE 2.** The maximum time of operation of drip system to irrigate rose plants was 89.9 minutes and 93.4 minutes required in May month under greenhouse cladded with diffused film and open field conditions respectively however, minimum time of operation is needed in January month i.e., 25.9 minutes and 38.5 minutes for inside and open field conditions respectively.

TABLE 2: Amount of water applied and yield response of Dutch roses for different amount of water applications under different treatments.

Treatment	T_1	T_2	T_3	T_4	T_5	T_6	T_7	T_8	T_9
Water applied (mm)	999.5	799.6	599.7	399.8	999.5	799.6	599.7	399.8	1210.9
Yield (No. of flowers/m²/year)	197	192	169	148	212	208	176	158	136
Annual water productivity (No. of flowers/m²/mm)	0.20	0.24	0.28	0.37	0.21	0.26	0.29	0.40	0.11

EFFECT OF DIFFERENT LEVELS OF IRRIGATION ON BIOMETRIC RESPONSE OF DUTCH ROSES

The effect of different levels of irrigation on biometric parameters such as plant height, number of shoots per plant, shoot length, number of flowers/m²/year, flower diameter and number of petals per flower were analyzed statistically and compared with that of open field experimental treatment (T9). The experimental results of these biometric observations for all the three years are presented in **FIGURE 7**. The analysis of variance results showed that variation among the replications for all the treatments was found to be statistically insignificant at 5% level of significance.

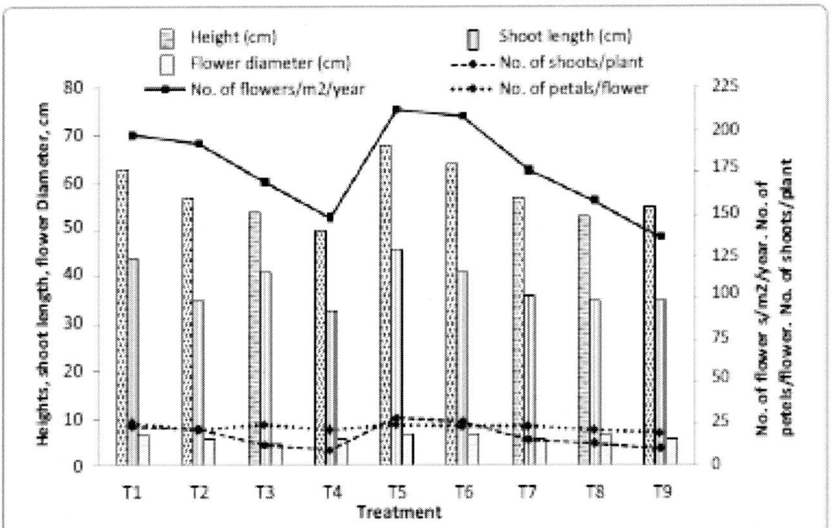

Figure 7: Biometric response of Dutch roses due to different treatments.

FIGURE 7: Biometric response of Dutch roses due to different treatments.

The plant height responded significantly with different irrigation levels of drip irrigation, however, plant height at 100% irrigation level (T_1 and T_5) was statistically at par with 80% irrigation level (T_2 and T_6). Similar trend was found for all the 3 years. The plant height was found to be highest in the year 2012 as compared to other 2 years data. Irrigation levels had no significant influence on flower diameter and number of petals per flower.

Significant increase in the number of shoots per plant was obtained due to drip irrigation treatment at all the levels of irrigation. Among various levels of water supply 100% irrigation requirement met through drip resulted in best over 60% and 40% irrigation levels in terms of number of shoots per plant and shoot length. However, biometric response at 100% and 80% of water requirement supply through drip was statistically at par. The maximum shoot length and number of flowers/m²/year were recorded for the year 2012 in comparison to other two years (2011 and 2013). It can be seen that the 100% irrigation supply has significant influence on number of flowers/m2/year over two other irrigation levels that is 60% and 40% and 100% irrigation supply in open field treatment.

In all the three years the number of shoots per plant and number of flowers/m^2/year at 100%, 80% and 60% irrigation levels was significantly greater than that of open field conditions. Based on the average yield (number of flowers/m^2/year) of three years, all the irrigation levels at 100%, 80% and 60% (i.e., T_5, T_6 and T_7 respectively) resulted in 55.8%, 52.9% and 29.4% higher yield respectively as compared to open field treatment (T_9). Hence, it showed that even by 60% deficit water supply through drip irrigation resulted in 29.4% higher flower yield of Dutch roses as compared to open field cultivation (T_9). The flower yield (presented in **TABLE 2**) was found to decrease as the amount of irrigation water supply was reduced from 100 % to 40 % of irrigation water requirement.

EFFECT OF CLADDING MATERIALS ON BIOMETRIC RESPONSE OF DUTCH ROSES

The plant growth and flower yield were greater in the greenhouse cladded with diffused film as compared to clear film cladded greenhouse. Among the various treatments under both the greenhouses treatment T5 i.e., 100% irrigation water level in greenhouse cladded with diffused film responded the highest plant height in all the years considered in this study. There was significant influence of irrigation levels and cladding material on plant height, number of shoots per plant, shoot length and number of flowers/m^2/year. However, there is no significant influence of irrigation level and cladding material on flower diameter and number of petals per flower. Plant height and shoot length were found to be highest in case of greenhouse cladded with diffused film as compared to greenhouse cladded with clear film at the same level of irrigation water supply.

Analysis of biometric observations revealed that flower yield of Dutch roses (number of flowers/m^2/year) was recorded greater in all the 3 years for the crop grown in the greenhouse cladded with diffused film at all the levels of irrigation. Maximum number of flowers/m^2/ year was found as 212 in treatment T_5 (i.e., 100% irrigation water level under greenhouse cladded with diffused film). With the same level of irrigation water application in both the cladded materials, the flower yield was always greater for greenhouse cladded with diffused film. This may be due to fact that the diffused film has inbuilt property to disperse solar radiation which penetrate into the crop canopy that increases photosynthesis process of the crop. Diffused film does not allow the radiations to interact directly to the plant leaves hence protect plats from scorching. Moreover, the plants in the greenhouse covered with diffused film were healthy compared to the clear film greenhouse during summer season. Lamnatou and Chemisana also observed the same results that diffused light have the ability to penetrate deeper into a plant canopy in comparison to the direct light [28]. The seasonal variation of solar radiation, affect variations in the crop response. The temperature in diffused film greenhouse was 2-3°C lower than clear film cladded greenhouse. Higher temperature prevailing in Kharagpur **SUB-HUMID** climatic condition under greenhouse cladded with clear film resulted in inferior biometric response and reduced flower yield.

WATER PRODUCTIVITY

The annual water productivity (Number of flowers/m^2/mm) values (AWP) ranged from 0.2 to 0.37 and 0.21 to 0.40 for greenhouse cladded with clear and diffused film, however it is 0.11 for open field conditions **(TABLE 2)**. The AWP values are always greater for the same amount of water application under diffused film cladded greenhouse. The AWP value was found to increase as the amount of irrigation water supply was reduced from 100% to 40% of crop evapotranspiration [29,30]. From the analysis of AWP values, it is clear that for water scarce region Dutch roses can be grown under diffused film cladded greenhouse at 40% of crop evapotranspiration with the compromise in 25.5% reduction in yield.

CONCLUSION

The evapotranspiration requirement (ETc) of Dutch roses established under greenhouse and open field conditions is useful for the determination of the irrigation water requirement. The crop coefficient (Kc) values established on a monthly basis for both the conditions from the beginning to the end of each annual cropping period are the basic information for estimation of ETc. Four Kc values are the irreducible minimum for describing and constructing the Kc curve, namely Kc values at the initial stage ($K_{c\ Initial}$), development stage ($K_{c\ Dev}$), mid-season stage ($K_{c\ mid}$) and the end of the late season stage ($K_{c\ end}$). The optimum water requirement of Dutch rose crop estimated under both the greenhouses can be successfully used by the rose growers and irrigation planners in other parts of the sub-humid climatic conditions.

There is significant influence of cladding materials on the Dutch rose plant height, number of shoots per plant, shoot length and flower yield during winter and summer seasons. The number of shoots per plant in the clear and diffused film is almost same up to 30 days after transplanting of the crop, and thereafter plant response gradually differs up to the flowering stage. The plant height is better under the diffused film greenhouse than that in the clear film greenhouse from vegetative stage to the flowering stage in winter and summer seasons due to inbuilt property of diffused film. The values of Crop ET are more for open field conditions than greenhouse. Total annual water requirement of rose plant is 999.51 mm and 1210.94 mm for greenhouse and open field condition.

ACKNOWLEDGEMENT

Authors are thankful to the National Committee on Plasticulture Applications in Horticulture (NCPAH), Department of Agriculture and Cooperation, Ministry of Agriculture, Government of India for providing necessary funds and Department of Agricultural and Food Engineering, IIT Kharagpur, India for making available necessary facilities to conduct this research studies.

REFERENCES

1. Hemming S, Mohammadkhani V, Dueck T (2008) Diffuse greenhouse covering materials-material technology, measurements and evaluation of optical properties. ActaHorticulturae 797: 469-475.
2. Allen RG, Pereira LS, Howell TA, Jensen ME (2011) Evapotranspiration information reporting: I.Factors governing measurement accuracy. Agricultural Water Management 98: 899-920.
3. Evett SR, Tolk JA, Howell TA (2006) Soil profile water content determination: sensor accuracy, axial response, calibration, temperature dependence, and precision. Vadose Zone Journal5: 894-907.
4. Ma HY, Jiao XY (2006) Research progress of the crop water demand calculation. Water Science Engineering Technology 16: 5-7.
5. Doorenbos J, Pruitt WO (1975) Guidelines for predicting crop water requirements. Irrigation and Drainage. Italy.
6. Abdelhadi AW, Hata T, Tanakamaru H, Tada A, Tariq MA (2000) Estimation of crop water requirements in arid region using Penman-Monteithequation with derived crop coefficients: a case study on Acala cotton in Sudan Gezira irrigated scheme. Agricultural Water Management 45: 203-214.
7. Fisher DK (2012) Simple weighing lysimeters for measuring evapotranspiration and developing crop coefficients. Int J Agric&BiolEng 5: 35-43.
8. Akinmutimi AL (2015) Estimation of water requirements of early and late season maize in Umudikesoutheastern Nigeria, using Penman's equation of soil science and environmental management. J Soil Sci Environ Manage6: 24-28.
9. Chuanyan Z, Zhongren N (2007) Estimating water needs of maize (Zea mays L.) using the dual crop coefficientmethod in the arid region of northwestern China. Afr J Agric Res 2: 325-333.
10. Odofin AJ, OladiranJA, Oladipo JA, Wuya EP (2011) Determination of evapotranspiration and crop coefficients for bush okra (Corchorusolitorius) in a sub-humid area of Nigeria. Afr J Agric Res 6: 3949-3953.

11. MirandaFR, Gondim RS, Costa CAG (2006) Evapotranspiration and crop coefficients for tabasco pepper (Capsicum frutescens L.). Agricultural Water Management 82: 237-246.
12. Kuo SF, Ho SS, Liu CW (2009) Estimation of irrigation water requirements with derived crop coefficients for upland paddy crops in ChiaNan Irrigation Association, Taiwan. Agricultural Water Management 82: 433-451.
13. Arayaa A, Stroosnijderb L, Girmayc G, Keesstrab SD (2011) Crop coefficient, yield response to water stress and water productivity of teff (EragrostistefZucc.). Agricultural Water Management 98: 775-783.
14. Hu Y, Li Y, Zhang Y (2012) A study oncrop coefficients of jujube under drip-irrigation in Loess Plateau of China. Afr J Agric Res 7: 2971-2977.
15. Maldonado AJ, Mendoza AB, Alba-RomenusKd, Morales-Díaz AB (2014) Estimation of the water requirements of greenhouse tomato crop using multiple regression models. Emir J Food Agriculture 26: 885-897.
16. Gómez HV, Farías SO, Argote M (2009) Evaluation of water requirements for a greenhouse Tomato cropusing the priestley-taylor method. Chilean Journal of Agricultural Research 69: 3-11.
17. Harmantoa, Salokhea VM, Babelb MS, Tantau HJ (2005) Water requirement of drip irrigated tomatoes grown in greenhouse in tropical environment. Agricultural Water Management 71: 225-242.
18. Fernández MD, Bonachela S, Orgaz F, Thompson RB, López JC, et al. (2010) Measurement and Estimation of plastic greenhouse reference evapotranspiration in a Mediterranean climate. IrrigSci 28: 497-509.
19. Bastug R, Karaguzel O, Aydinsakir K, Buyuktas D (2006) The effects of drip irrigation on flowering and flower quality of glasshouse gladiolus plant. Agricultural Water Management 81: 132-144.
20. Blanco FF, Folegatti MV (2003) Evapotranspiration and crop coefficient of cucumber in greenhouse. RevistaBrasileira de EngenhariaAgrícola e Ambiental 7: 285-291.
21. Ehret DL, Menzies JG, Helmer T (2005) Production and quality of greenhouse roses in recirculating nutrientsystems. ScientiaHorticulturae 106: 103-113.
22. Katsoulas N, Kittas C, Dimokas G, LykasCh (2006) Effect of irrigation frequency on rose flower production and quality. Biosystems Engineering 93: 237-244.
23. Möller M, Assouline S (2007) Effects of a shading screen on microclimate and crop water requirements. Irrigation Science 25: 171-181.
24. Chou MD, Lee KT (1996) Parameterizations for the absorption of solar radiation by water vapor and ozone. American Meteorolological Society 53: 1203-1208.
25. Kondratyev KY (1969) Radiation in the atmosphere. New York.
26. Lamnatou C, Chemisana D (2013) Solar radiation manipulations and their role in greenhouse claddings: Fresnel lenses, NIR and UV-blocking materials. Renewable and Sustainable Energy Reviews18: 271-287.
27. Espí E, Salmerón A, Fontecha A, García Y, Real AI (2006) Plastic films for agricultural applications. J Plast Film Sheeting 22: 85-102.
28. Lamnatou C, Chemisana D (2013) Solar radiation manipulations and their role in greenhouse claddings: Fluorescent solar concentrators, photoselective and other materials. Renewable and Sustainable Energy Reviews 27: 175-190.
29. Manohar RK, Khan MM, Shyamalamma, Kariyanna, Biradar MS (1996) Gerbera under low cost greenhouse. Published by Plasticulture Development Centre. Bangalore.
30. Orgaz F, Ferna´ndezb MD, Bonachelac S, Gallardoc M, FereresaE (2005) Evapotranspiration of horticultural crops in an unheated plastic greenhouse. Agricultural Water Management 72: 81-96.

CHAPTER 4

Wastewater Reuse for Irrigation — Practices, Safe Reuse and Perspectives

I. Navarro[1], A. Chavez[1], J.A. Barrios[1], C. Maya[1], E. Becerril[1], S. Lucario[1] and B. Jimenez[1]

[1] Institute of Engineering, UNAM, Mexico

1. INTRODUCTION

1.1. Wastewater for Irrigation and its Role in Sanitation and Human Health

Human impact on water bodies has become relevant since water withdrawal, especially for agriculture, has resulted in overexploitation of rivers, lakes and aquifers. In most countries agriculture represents by far the largest use of water and worldwide, it represents about 70% of total withdrawal and 90% of water consumption [1]. To address this problem wastewater reuse has proven to be an alternative to reduce anthropogenic impacts [2]. In addition, raw wastewater reuse in agriculture is a valuable tool available to developing countries to control pollution and tackle the challenge of increasing food production in water scarce areas. The benefits of reusing water in agriculture are many and beyond doubt: it saves considerable amounts of first-use water that may be assigned to critical uses; it provides nutrients that may substitute chemical fertilizers, increasing soil fertility and crop yield, and reducing production costs; it makes it possible to expand agricultural land in arid areas; it is a relatively cheap disposal method for raw wastewater; and it may avoid pollution of surface water. In addition, it has been demonstrated that the health risks of reusing water in agriculture are minimal as long as its biological quality meets established criteria [3, 4].

However, it should be considered that there are potential negative impacts that may arise, such as soil salinization as well as soil and groundwater pollution with metals and organic compounds. In addition, the use of raw wastewater for crop production poses health risks due to its microbial content, especially bacteria, viruses and parasites, which produce a wide range of diseases since many of them may survive on the environment for long periods of time.

The main issue associated to wastewater reuse is related to public health and infection risks, either real (produces a disease) or potential (transmits the infection but the disease does not develop). Infection rates may be high, low or minimal, depending on the type of pathogen, the infective dose, and the susceptibility of the affected person (host). According to epidemiological studies over the last 20 years, when untreated wastewater is applied to land for crop production, there exist real infection risks caused by pathogens. To reduce such risks, control actions must be implemented, such as treating wastewater to comply with regulation limits; developing fast, cheap, easy, and efficient detection techniques; breaking the disease-infection cycle with medical treatment; and developing education campaigns for the population [4-6].

2. INTERNATIONAL AND REGIONAL GUIDELINES AND COUNTRY REGULATIONS FOR TREATED WASTEWATER REUSE IN IRRIGATION

Globally, wastewater reuse has become significant and this has encouraged many countries to develop local regulations to control water quality for reuse with the aim of reducing health and environmental risks. Due to different geographic, economic, and social characteristics, development of such regulations has been gradual and dissimilar among countries. Developed countries have worked on these regulations for several years and among them, the United States of America, and in particular the state of California, applied the first regulations on agricultural reuse in 1918 [7]. As a result, California has one of the most strict and complete regulatory frameworks. In addition, the United States Environmental Protection Agency (US EPA) has developed regulations and criteria for water reuse, which are used as a reference by many countries. Moreover, the creation of international organizations has led to the publication of general recommendations. The World Health Organization (WHO) published a series of four volumes that include information about agricultural irrigation and reuse in aquaculture [8]. At the same time, the Food and Agriculture Organization of the United Nations (FAO) has its own guidelines that are similar to those of the WHO [9]. WHO proposes limits for indicator bacteria (total or faecal coliforms) and helminth eggs when wastewater is reused for agriculture, considering that helminth eggs are highly resistant to treatment process and common in the environment.

In Europe, Mediterranean countries have detailed legislation about this topic, while countries with high water availability, such as Germany or the United Kingdom, do not regulate reuse as it is seldom practiced. Spain sets limits on faecal coliforms and nematodes based on their 1985 Water Law.

In the Americas, countries like Brazil, Costa Rica, Chile, and Mexico have made progress in developing regulations focused mainly on restricting water reuse based on microbial content, including indicator bacteria, helminth eggs and some metals, allowing organic matter and nutrients that are beneficial for agriculture to be used on land.

In Africa, despite the fact that several countries face a water crisis, many of them have lax or non-existent regulations and thus wastewater reuse is practiced uncontrolled. However, countries like Tunisia have detailed guidelines that include physicochemical and biological parameters, as well as heavy metals, an approach shared by some Mediterranean countries like Saudi Arabia, Israel, and Jordan [9].

In contrast, countries like Palestine, Libya, and Afghanistan, which have low water availability, have not been able to develop their own standards due to political conflicts and generally use either FAO or WHO guidelines. Finally, Indian regulation is outdated (1974) and limits just a few parameters, even though it has the world's second largest population.

Usually, most regulations for reuse in irrigation establish limits on one or more of the following parameters: indicator bacteria (total or faecal coliforms), helminth eggs (intestinal worms), nematodes (a subgroup of helminthes), organic matter (as biochemical oxygen demand, BOD), dissolved and suspended solids, and heavy metals.

3. SUCCESSFUL CASES OF WASTEWATER REUSE FOR IRRIGATION

Demographic growth and economic development of emerging American countries has promoted the implementation of several agricultural reuse projects, some of which are summarized below.

In Mendoza, Argentina, an area known as Campo Espejo used to be irrigated with raw wastewater (2,000 ha) but currently, 129,600 m^3/d originating from stabilization ponds are supplied for the irrigation of 1900 ha [10].

Chile has several successful reuse projects such as the Maipo and Maipocho regions where 130,000 ha are irrigated [11]; Antofagasta where about 20,000 m^3/d of treated water are produced and 65 ha are irrigated; and Santiago de Chile with 110,000 ha that use reclaimed water mixed with first-use water [10].

In Mexico, the central and northern part of the country, where 80% of the population live, is considered arid or semi-arid, and cities like Ciudad Juarez irrigate 26,000 ha with approximately 400,000 m^3/d of reclaimed water. Additionally, the Mezquital Valley, covering more than 90,000 ha, is one of the largest areas in the world where agricultural reuse is practiced [12] and where a large wastewater treatment plant (35 m^3/s) is under construction to improve water quality for irrigation.

In South America, Peru encompasses a number of different biomes that range from Amazon rainforest (more than 50% of the country) to the west coast, an area with low precipitation where most of the population lives, and where small irrigation projects like San Agustin (535 ha) and Tacna (738 ha), have been developed [10].

4. CASE STUDY: TULA VALLEY, MEXICO

The Tula Valley (also known as Mezquital Valley) is one of the largest areas irrigated with untreated wastewater in the world. The fact that it has been receiving wastewater since the late XIX century, and specifically for irrigation since the early XX century, makes this a unique site with regard to wastewater reuse. A large number of studies have described several processes and phenomena that emerge out of this practice, including incidental aquifer recharge, increase in crop yield, and treatment of the wastewater by the soil (in the same way as a soil-aquifer treatment system, SAT).

Concurrently, other studies have found an increase in health risks after using untreated wastewater in agriculture due to the presence of pathogens, heavy metals, and organic compounds in soil and even in groundwater. This section summarizes the history and current situation of the Tula Valley to share some lessons learned for more than 100 years of water reuse.

4.1. Metropolitan Area of The Valley of Mexico and Tula Valley

The Metropolitan Area of the Valley of Mexico (MAVM) has a population of 21.2 million and covers the Federal District (Mexico City) and 60 surrounding Municipalities (59 located in the State of Mexico and 1 corresponding to the State of Hidalgo). Water use is estimated to be 82 m^3/s which includes 91% first use water and 9% reclaimed water. The rate of wastewater generation (including collected precipitation) ranges from 52 to 300 m^3/s depending on the season, and it is conveyed to the Tula Valley. The reason for this practice is that Mexico City is located in an endorheic (closed) basin from which wastewater and excess precipitation needs to be transported to avoid flooding. On average, 60 m^3/s are sent by gravity or pumping to the Tula Valley via four artificial exits (Figure 1): the Tajo de Nochistongo (deep cutting through Nochistongo Hill; 1607-1789); the Tequixquiac Tunnel (1900); the New Tequixquiac Tunnel (1955) and the Central Emitter (1975). Currently, a fifth exit (East Emitter) with a capacity of 150 m^3/s is being constructed along 62 km and will be operating by 2016.

The Tula Valley is located 100 km north of Mexico City, with an elevation that ranges from 1,700 mamsl in the north, to 2,100 mamsl in the south. The climate is semiarid with an annual precipitation of 550 mm (national average is 790 mm) mainly between May and October [10] which contrasts with evapotranspiration that reaches 1,524 mm/y.

Originally, soils were low in organic matter and nutrients, and with such low precipitation, productivity was also low. By the end of the XIX Century, the Tula Valley started receiving raw wastewater from the MAVM. The first documented use of wastewater for irrigation in areas closed to the Salado River was reported around 1896. However, officially, reuse initiated in 1889 when wastewater was used to produce energy at the Juandhó and La Cañada hydroelectric plants [13], and in 1912 for irrigation [14].

As wastewater generation gradually increased, the irrigated area grew from 10,000 ha in 1920, to 80,888 ha today (Figure 2). This means that about 1,350 Mm^3/y are transported a distance of 98 km to the Tula Valley. Once in the Valley, wastewater is distributed through a complex hydraulic system that includes six storage dams (combined capacity: 347 Mm^3), 323 km of main distribution canals, 264 km of lateral distribution canals, and 101 km of agricultural drains. Year-round availability of wastewater, as well as its organic matter and nutrient content which act as fertilizers, allowed the development of three important Irrigation Districts (ID 003, 100, and 112) that today cover more than 85 thousand hectares [15].

Figure 1. Wastewater transport and use in the Tula Valley.

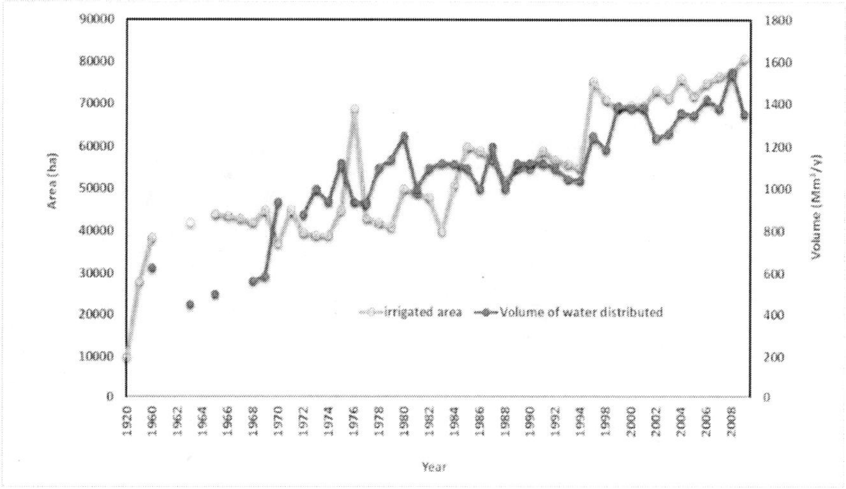

Figure 2. Growth of area irrigated and volume of water distributed since 1920 (adapted from [16], with data from [15]).

4.2. Water Reuse in The Tula Valley

The gradual growth of the productive area in the Tula Valley allowed the differentiation of regions with various irrigation ages (Figure 3). Irrigation was first performed in the south-central area of the Valley [14]. Subsequently, when the City expanded and had to dispose of a larger amount of combined wastewater, the irrigation area grew towards the north and the east following the Tula River. Subsequent expansions of the Irrigation Districts carried wastewater to the east and southeast (ID 100), where reuse had started by 1970 [16].

Recently, wastewater was taken to the far east of the Valley, close to the Ajacuba Municipality (ID 112). As a result, the quality of irrigation water exhibits spatial and temporal variations. For example, in the southern part of the Valley, approximately 10,000 ha are irrigated with raw wastewater, while about 35,000 ha in the central and eastern regions receive diluted wastewater (80% wastewater and 20% river water/precipitation from the Taxhimay and Requena dams) with a different composition.

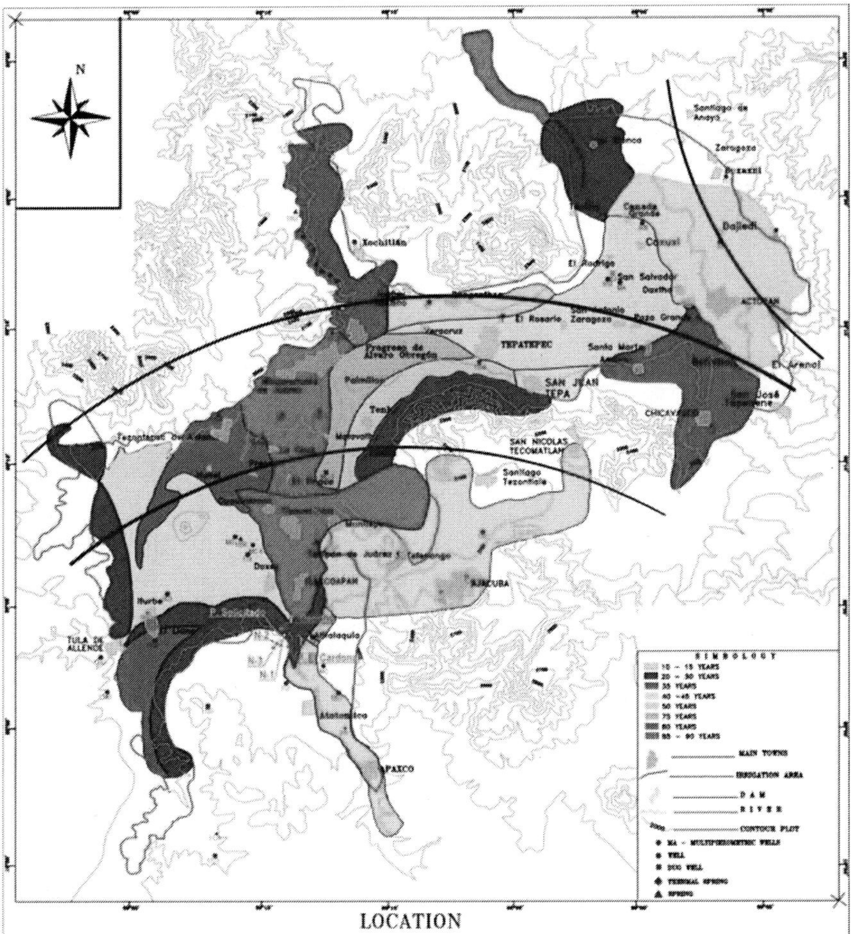

Figure 3. Areas under irrigation for different lengths of time (adapted from [16]).

On the other hand, 25,000 ha located in the far west region of the Valley are irrigated with partially treated wastewater after being stored in the Endho dam which acts as a large settling tank with a hydraulic retention time of up to three months. Finally, the northern areas use well water or return flows and are mainly utilized for growing vegetables and fruits [17].

4.3. Water Quality

4.3.1. Microbiological indicators

The potential risk of bacterial, viral and parasitic diseases that can be transmitted through the human-water-soil-crop-human cycle constitutes the greatest problem associated with the use of wastewater in agriculture, for human consumption or for other uses. Thus the study of their removal is especially relevant in developing countries where they represent the higher risk of disease.

The presence of microorganisms can be a serious cause for concern. For example bacteriophages (viruses) are resistant in the environment, have been shown to have the capacity to penetrate and reach confined aquifers, and have low infective doses [7]. Cysts of *Giardia lamblia* can survive and remain active for months or even years, are resistant to chlorination, have a low infective dose and can have serious consequences for vulnerable individuals [18]. Helminth eggs also survive for long periods of time in the environment, have a low infective dose and their incidence in Mexico is very high [19]. Faecal coliforms are a universally accepted indicator of faecal contamination and have been found to migrate through soil [20].

Wastewater from the MAVM reaches the distribution system at the Tula Valley without any treatment and thus it contains a high concentration of faecal coliforms (between 10^5 to 10^8 colony forming units, CFU/100 mL), *Streptococcus faecalis* (10^2-10^6 CFU/100 mL), *Clostridium perfringens* (10^3 to 10^6CFU/100 mL), somatic bacteriophages (10^2 to 10^6 plaque forming units, PFU/mL), *Giardia* spp. (450 to 10,000 cysts/L), and helminth eggs (1.8 to 23 helminth eggs/L) (Table 1). This quality has shown little temporal variation but the concentrations registered pose a significant health risk in areas under irrigation, according to WHO [4].

To evaluate temporal and spatial variability along the Valley, samples from different regions have been taken for microbial analyses during the wet and dry seasons. Zone 1 (south) uses raw wastewater for irrigation and has shown higher concentrations of biological indicators than Zones 2 or 3 which correspond to areas with some sedimentation and dilution (rainfall and groundwater). This variability was due to the nature of the wastewater but there was a large decrease in the *Giardia* spp. content in all zones during the dry season. Helminth eggs showed little variability within the three zones and there was no obvious difference between the wet and dry seasons. However, both *Giardia* spp. and helminth eggs were relatively high in zone 3 at the farther reaches of the distribution system, which suggests a possible direct contribution from local discharges, since population in this zone is relatively high (Figure 4). The concentrations of microorganisms reported are similar to those previously described for untreated wastewater in Mexico [19] and are similar to untreated wastewater from other countries [21, 22], but much greater than concentrations seen in effluents of treatment plants [19].

TABLE 1. Microbiological quality of raw wastewater and characteristics of selected indicators (values in parenthesis indicate range of concentration).

Parameter	Concentration	Description
Faecal Coliforms, Log(CFU/100 mL)	6.53 ± 0.58 (5.15-7.84)	Traditional indicator of faecal contamination; may indicate the presence of pathogens; behaves similarly to other bacteria under different environmental conditions.
Streptococcus faecalis, Log(CFU/100 mL)	5.15 ± 1.0 (2.04-6.20)	Intestinal bacteria found in the faeces of all warm-blooded mammals; useful for indicating the quality of water for recreational use and also for reuse.
Clostridium perfringens spores, Log(CFU/100 mL)	4.70 ± 0.81 (3.00-5.97)	Anaerobic bacteria that have been recently used as an indicator of faecal contamination; they form a resistant spore commonly found in faeces which is more resistant to disinfection and adverse environmental conditions that many pathogens; the presence of vegetative cells indicates recent contamination while spores imply prior contamination.
Somatic bacteriophages, Log (PFU/mL)	3.63 ± 0.77 (2.41-6.41)	Viruses that infect bacteria; used as indicators based on ease of detection by analytical laboratories; the coliphage group has been the best model for enteroviruses, given their similar physical structure and their greater resistance to treatment processes (such as chlorination); it is always present and relatively abundant in wastewater; easily detected over a short time period (24 h).
Giardia spp., Cysts/L	2,231 ± 231 (450-10,000)	Pathogenic protozoa; single celled organisms that develop in two ways: as trophozoites, and as cysts; Infection results from the consumption of a mature cyst that is resistant to the gastric juices; the trophozoites can become cysts again in a process that is apparently aided by unideal luminal conditions, and they are then expelled in the faeces of persons with symptoms of the illness or with no apparent symptoms. Cysts can survive and remain active for weeks, months or even for periods of up to 7 years; capable of forming cysts (resistant structures) under adverse conditions, this organism represents a serious health risk.
Helminth eggs, eggs/L	13.1±6.2 (1.8-23.0)	Group of parasitic and free-living worms of various sizes and shapes; they cause mechanical deterioration, tissue damage, toxic effects, and blood loss; intestinal parasites cause anaemic malnutrition and/or delayed growth; they present high resistance to chemical and physical conditions and thus are considered the most resistant form of parasites; they have the ability to survive long periods of time in biosolids and soil (up to 6 years after their initial application).

Although it was expected that concentrations of microorganisms would be lower in the wet season due to rainfall dilution, this happens intermittently as the wet season is characterized by irregular intense storms. An additional issue is that the wastewater is mixed with stored rainwater and wastewater depending on the amount of water required by farmers and thus its quality exhibits some variability.

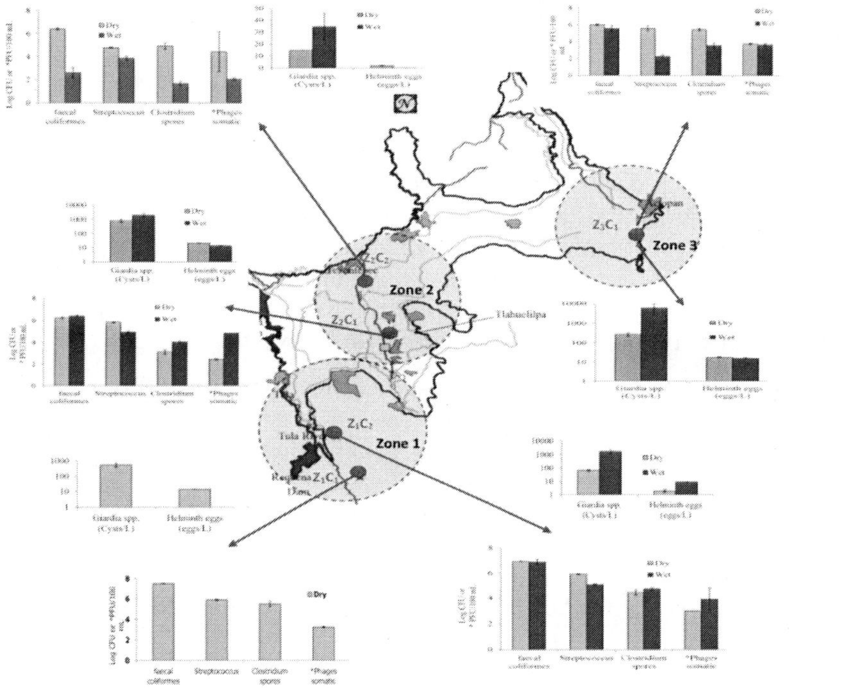

Figure 4. Concentration of selected microorganisms in different zones of the Tula Valley (Adapted from [**10**] and [**23**])

4.3.2. Emerging pollutants

Emerging pollutants are defined as those unregulated pollutants that may be controlled by future regulations depending on their potential effects on health and ecosystems [24]. Advances in analytical techniques for their detection and an increased understanding of their effects on public health and the environment, has increased the need to establish discharge limits worldwide. During the last two decades, the study of emerging pollutants (such as pharmaceuticals, personal care products, and endocrine disruptors included as additives of gasoline, plastics, or detergents) in the environment has gained scientific attention. Many of these studies have been performed in aquatic ecosystems in the United States of America and Europe [24]. However, for most of the emerging pollutants there are still insufficient data to indicate their risk and ecotoxicity and thus their effects on human health or aquatic organisms remains unknown. Many of these compounds reach the environment through wastewater and they may be removed in wastewater facilities, however, some of them remain in treated effluents and enter the environment, eventually reaching groundwater [25, 26].

Initial reports on pharmaceuticals in wastewater were published in the United States in the 1970's [27], however, even though their presence has been related to fish toxicity, little attention has been given to them [28, 29]. Considering that developed countries have detected emerging pollutants in water and wastewater, even after treatment, their presence in wastewater from the Metropolitan Area of the Valley of Mexico was clearly expected. Furthermore, the fact that many prescription drugs in developed countries are considered over-the-counter medications in Mexico, there is a higher probability of finding their active ingredients in wastewater.

On the other hand, agricultural irrigation favors the transport of pollutants, such as pesticides, from soil to groundwater. Several factors influence their incidence in groundwater, including soil permeability, the depth of the unsaturated zone, geological composition, as well as the solubility, partition ($K_{o/w}$), and dissociation (pKa) values of the pollutant. These conditions may promote their adsorption to soil particles or their leaching to groundwater. It has been reported that wastewater irrigation, treated or untreated, increases the concentration of pesticides in groundwater [30].

As mentioned previously, the presence of emerging pollutants was expected in wastewater from the MAVM, and thus at the Tula Valley, and was confirmed by different studies. Some of them [31] analyzed wastewater samples and found a variety of compounds with the exception of clofibric acid. Concentrations found were similar to those reported for wastewater samples taken at treatment plants (Table 2), but higher than those reported for their effluents [32, 33]. The concentrations of naproxen, a widely use pharmaceutical in Mexico, as well as salicylic acid were higher than those reported in other countries, while ibuprofen, diclofenac, are variable in comparison. Other compounds like 4-nonylphenol and diethylhexyl phthalate were the predominant endocrine disruptors in wastewater. Levels of Triclosan, bisphenol-A, butylbencylphtalate, estrone, and 17β-estradiol were similar to those reported elsewhere [33].

It was observed that dilution occurred during the rainy season when reported concentrations of endocrine disruptors were about half of those during the dry season. It should be noted that wastewater at the Tula Valley includes precipitation from the MAVM. Nonetheless, acid pharmaceuticals remain similar in both seasons which may suggest another phenomenon.

Siemens *et al.* [39] concluded that six acid and five basic pharmaceuticals measured in raw wastewater, irrigation water, and spring water, were reduced in concentration along the wastewater distribution canals. However, the authors suggest that acid compounds may permeate clay soils and exhibit poor removal compared to basic or neutral chemicals. Nevertheless, none of the reported compounds were found in spring water.

Table 2. Concentration of pharmaceutical compounds in wastewater.

Compound	Germany[a,c] ng/L	Spain[d] ng/L	Finland[e] ng/L	Brazil[b,c] ng/L	Mexico[f] ng/L
Ibuprofen	3,400	-	13,100	3,300	2,500
Diclofenac	2,000	2,600–5,700	-	800	1,607
Ketoprofen	300	-	2,000	500	447
Salicylic acid	340-5,400	-	-	-	29,867
Naproxen	440	1,800–4,600	4,900	600	13,620

[a][34, 35] [b] [36]; [c][25]; [d][37]; [e][38]; [f][31].

4.4. Natural Attenuation of Basic Parameters, Organic Pollutants and Pathogens in Soil

In the Tula Valley, known as the largest case of indirect wastewater reuse for human consumption in the world, natural soil aquifer treatment (SAT) has been taken place for more than 100 years, recharging the aquifer and acting as a barrier to prevent contaminants from entering it. The local aquifer is being recharged at a rate of 25 m^3/s due to the infiltration of untreated wastewater from unlined irrigation channels, storage dams, and excess water used for irrigation (flood irrigation practice; [12]). During percolation, natural soil infiltration occurs through unsaturated soil; as the effluent moves through the soil and the aquifer, it can undergo significant quality improvements through physical, chemical and biological processes.

Non-intentional natural SAT has treatment benefits in the unsaturated zone in the Tula Valley. It acts as a natural filter, and produces groundwater of acceptable quality due to the characteristics of the soil-aquifer structure, the residence time, and the history of the complex geohydrological system. Table 3 showed the historical data of wastewater and aquifer water quality variability obtained from studies performed at the site. It indicates that parameters related with salinity behavior are present in groundwater, in addition to parameters such as nitrate, and solids in dissolved phase. In contrast, the percentage removal of pathogenic organisms and emerging pollutants through unsaturated soil is greater than 50%.

There is evidence that during infiltration the vast majority of microorganisms are retained in the first few centimeters of soil [40], however their potential fate is influenced by size ranges of microorganisms (20-80 μm for helminth eggs, 1 μm for bacteria, 4-12 μm for protozoa), type of soil, and even soil organic matter content for bacteriophages (20 to 200 nm), and therefore for viruses; thus the migration of microorganisms through soil does not always allow complete attenuation.

Table 3. Water quality variability for wastewater and the Tula Valley aquifer and natural removal percentages (SAT)

Parameters	Units	Wastewater (Min-Max)	Aquifer (Min-Max)	% removal (mean)
		Basic analysis		
Total Dissolved Solids, TDS	mg/L	409-1123	546 -1586	-55.5
Alkalinity	mg/L	324-600	256-748	-7.8
Nitrate (as N)	mg/L	0.02-0.47	0.0-32	-8014
		Salinity related analysis		
Electrical conductivity	μS/cm	1734-3000	1755 -4187	-18.0
Redox potential	mV	-51 - -37	-79.1 - -11.0	-38.7
Sodium	mg/L	56-215	103 -361	-38.3
Potassium	mg/L	19-13190	10.5 -107	98.0
Hardness Ca	mgCaCO$_3$/L	90-131	70-787	-218.4
Hardness Mg	mgCaCO$_3$/L	12-51	17 -301	-65.7
Total hardness		122-172	114-1006	-181.8
Bicarbonate	mg/L	270-504	212-652	-12.8
Sulfate	mg/L	53-2492	190 -3025	-15.0
		Microbiological analysis		
Total Bacteria count (37 ºC)	Log CFU/100 mL	6.4-7.7	1.3-4.2	57.6
Faecal coliforms	Log CFU/100 mL	5.6-7.5	0.0-2.9	70.8
Enterococci	Log CFU/100 mL	2.3-5.9	0.0-1.9	79.3
Clostridium spores	Log CFU/100 mL	3.2-5.6	0.0-2.6	61.1
Somatic Bacteriophages	Log PFU/100 mL	2.5-4.9	0.0-2.8	52.6
Giardia spp.	Cysts/L	70.3-3233	0.0-600	89.4
Helminth egg	Ova/L	1.9-21.9	0.0-2.9	96.0
		Emerging pollutants		

Clofibric acid	ng/L	< LOD	< LOD-0.39	--
Ibuprofen	ng/L	1325-4700	0.05-1.46	99.99
Salicylic acid	ng/L	13580-72979	0.02-27.7	99.98
2,4-D	ng/L	295-2641	< LOD-0.48	99.99
Gemfibrozil	ng/L	640-750	< LOD-0.1	99.99
Naproxen	ng/L	5861-16336	0.04-3.266	100.00
Ketoprofen	ng/L	82-500	0.02-0.83	99.89
Diclofenac	ng/L	1240-3424	< LOD-3.75	99.98
4-nonylphenol	ng/L	6970-38130	0.81-67.6	99.93
Pentachlorophenol	ng/L	40-110	< LOD-0.33	99.81
Triclosan	ng/L	360-2880	< LOD-22.6	99.83
Bisphenol-A	ng/L	700-6230	0.02-153	99.54
Butil-bencilphthalate	ng/L	125-2959	0.03-308	97.78
Bis-2 ethyl(hexyl)phthalate	ng/L	4664-70200	3.07-933	99.97
Estrone	ng/L	14-100	< LOD-0.24	99.92
17β-estradiol	ng/L	6.8-22	< LOD-0.06	99.90
17α-etinilestradiol	ng/L	<LOD	< LOD-0.05	--
Carbamazepine	ng/L	200-275	0.14-193	85.89

Adapted from [23, 31, and 39] (ND: Not detected LOD: Lower limit of detection).

Data recorded over many years regarding the attenuation of emerging pollutants through soil show a mean reduction of 86 percent of the chemical load in the wastewater used for irrigation (Table 3). The main attenuation processes such as adsorption and biodegradation have been confirmed in experimental studies with local soil samples, and enriched samples [23]. The most important result of this research is the observation that large amounts of suspended and dissolved organic matter in soils [41], and even in raw wastewater, improve the adsorption of emerging pollutants in soils, and therefore their final low content in groundwater; this verifies previously published results with regard to the role of natural SAT in removing pollutants [26, 42]. In fact, the sorption behavior of three pharmaceuticals (naproxen, carbamazepine, and triclosan), a plasticizer Bis-2 ethyl(hexyl)phthalate, and the surfactant metabolite 4-nonylphenol in wastewater irrigated soil, was analyzed for different soil depths sampled from the Tula Valley, showing that the potential migration of these compounds to the aquifer depends on the physical and chemical characteristics of soil, such as organic matter content, and clay/sand percentages [43, 44].

Other studies have shown that heavy metals have accumulated in soil in the Tula Valley (for example cadmium, nickel, and lead) and their retention is associated with the length of time that wastewater has been used for irrigation [45].

It should be mentioned that a large wastewater treatment plant is currently under construction to treat the MAVM's wastewater and reduce pollutant load to the Tula Valley. This plant, Atotonilco, is designed to treat 35 m^3/s with a combined process (23 m^3/s with a biological process and 12 m^3/s with a physicochemical process) that may partially remove organic matter and the bulk of pathogens. However, the effect on soil attenuation must be evaluated as Gibson *et al.* [41] suggest that a reduction in the amount of organic matter that gets into irrigated soils may affect sorption processes, which could have an impact on the removal of emerging pollutants, pathogens, and heavy metals.

4.5. Impacts of Wastewater Irrigation in The Mezquital Valley

Reuse in the Tula Valley has functioned as a discharge route for the closed basin of the MAVM at a relatively low cost. The characteristics of the water, in terms of organic matter and nutrients have significantly increased crop productivity in the Valley and have allowed farmers to harvest up to five crops per year of alfalfa, fodder oats, tomato, barley, and maize. Resulting yields in the Tula Valley are 71 to 150% higher than those obtained with rain-fed agriculture [19]. Thus, the contribution of nutrients represents estimated annual savings of $180 to $200 million USD. To exemplify this, agricultural production for 2011-2012 in the Irrigation Districts of the Tula Valley reached $418 million USD [46].

Since 81% of the main distribution canals as well as 52% of lateral distribution canals are unlined, 80.2 Mm^3/y of conveyed wastewater infiltrates to the aquifer [47]. It is with this infiltration that wastewater quality is improved before reaching the groundwater through processes like adsorption or biodegradation, which depend on contact time and filtering distance. As an example, travel times from the irrigated fields to the springs are estimated to be 3 to 5 days with groundwater velocities of 0.02 to 6.0 m/d [16].

This infiltration has also incidentally recharged the Tula Valley aquifer for more than 100 years, at a recharge rate estimated at 25 m^3/s and equivalent to 13 times the natural recharge [16]. As a consequence, the water table has risen from a depth of 60 meters in the 1950's to 4 meters in the southern part of the Valley and several springs with flows between 40 and 600 L/s have appeared [16]. The amount of available water with a relatively good quality allows some areas to be used to grow vegetables like tomato, lettuce, cabbage, beetroot, cilantro, radish, carrot, spinach, and parsley [16]. At the same time, surface and groundwater produced by wastewater infiltration provide drinking water to approximately 500,000 people after only treatment with chlorine for disinfection [12].

In contrast, wastewater reuse in the Tula Valley has some negative impacts on the local population and the environment. Due to the pathogens and parasites contained in wastewater, the incidence of gastrointestinal diseases has increased by more than 16 times in children living in the irrigation area, compared to children unexposed to wastewater [48]. In addition, microbial and organic pollutants have been detected in the soil matrix of the irrigation areas at various depths [41, 44]. Heavy metals have accumulated in the upper soil layer [49], and Siebe [45] suggests that eventually the retention capacity of the soil might be exceeded with the risk of groundwater pollution. Moreover, the gradual salinization of local soils has caused the loss of more than 2,000 ha of cropland [17].

4.6. Conventional Wastewater Treatment Processes, Control Of Micropollutants, and Hazardous Substances In Water

Even though the soil filters a large amount of pollutants from wastewater with removals above 90% [12], some studies have demonstrated that in some areas the filtered water still contains bacteria (coliforms and enterococci) and viruses [40, 50]. Filtered water presents a high concentration of salts (indicated by sodium, sulfates, carbonates, bicarbonates, chloride, calcium hardness, electrical conductivity, and potassium), and furthermore, total dissolved solids and nitrates exceed the limits set by the Mexican regulatory authorities (1,000 and 10 mg/L, respectively).

At the same time, several studies [23, 31, 40, 41, 43, 44] report the presence of emerging pollutants in springs and wells, which poses a health risk to local population (approximately 500,000 inhabitants) that use the water for human consumption (Figure 5). These pollutants exhibit low removal by conventional treatment processes and thus alternative technologies, such as membrane filtration (nanofiltration or reverse osmosis), have been proposed [51]. Different studies have demonstrated that nanofiltration removes emerging pollutants, including those with low molecular weight, in particular pesticides [52], pharmaceuticals [51], and endocrine disruptors [52].

As a result, to improve the quality of spring water in the Tula Valley, a pilot plant with a nanofiltration membrane was installed in the Cerro Colorado

Spring (Figure 5) [53]; The pilot plant (11.4 m^3/d) operated for 800 hours and included a pumping system, a prefilter, a softening unit for reducing potential scaling, and a nanofiltration module (Figure 6). The nanofiltration (NF) membrane was selected after laboratory trials on a pilot cell. The pilot plant had the instrumentation required to measure flow, pH, TDS, as well as pressure along the system. Recovery (permeate) was maintained at more than 66 % of the influent.

Figure 5. Cerro Colorado Spring at the Tula Valley.

Nanofiltration demonstrated its effectiveness for removing organic matter (92% of total organic carbon), salinity (60% of total dissolved solids and 75% of electrical conductivity) and 100% for selected microorganisms (faecal coliforms, faecal streptococci, *Salmonella* spp., *Clostridium perfringens*, *Giardia* spp., and bacteriophages). With respect to emerging pollutants, membrane selectivity varied from 5 to 6% for salicylic acid and nonylphenol up to more than 75% for gemfibrozil, butylbenzyl phthalate, carbamazepine, and diclofenac. Treated water (permeate) met the limits for drinking water and thus it could be considered suitable for human consumption.

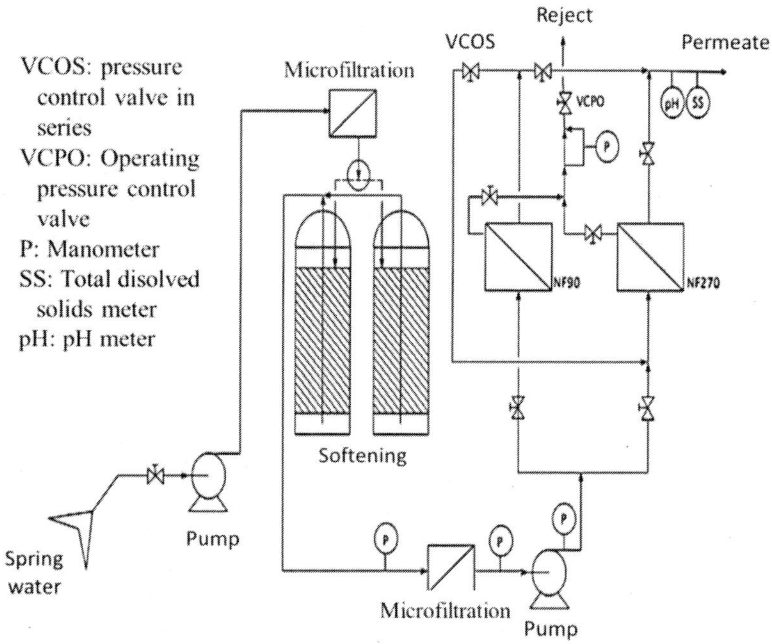

Figure 6. Flow diagram of the nanofiltration pilot plant at the Cerro Colorado Spring, Tula Valley.

Based on such studies, filtered water from the Tula Valley is being considered as a potential source for supplying drinking water for Mexico City and studies have determined that 6.5 m^3/s may be sourced from local groundwater [54].

5. FORTHCOMING EXPECTATIONS AND RECOMMENDATIONS

Findings confirmed by 15 years of research into the complex Tula Valley hydrological and hydraulic system, are:

- Wastewater reuse for irrigation has proven to be an alternative in the water-scarce region of Tula Valley.
- Agricultural irrigation in the Tula Valley using raw wastewater from the Metropolitan Area of the Valley of Mexico has been performed for

more than 100 years and several studies have demonstrated that complex phenomena occur.

- One of these is the natural SAT in the saturated soil zone, which even occurs today after a century of irrigation, mainly in terms of the retention of typical pathogenic organisms and emerging contaminants in wastewater of urban origin.
- The observed level of natural removal through filtration by soil exceeds the expected removal levels from conventional wastewater treatment, not only for emerging contaminants, but also for microorganisms that are difficult to remove by chlorine disinfection.
- Non-intentional aquifer recharge with water purification by natural infiltration, achieved through agricultural reuse of untreated wastewater, has provided new water sources for the local population with the potential to supply water for human consumption to the large city from which comes the reused wastewater originates.
- Advanced treatment studies conducted by our research group have shown that it is possible to treat water from aquifer for safe water supply.

There is no doubt about the benefits of wastewater reuse for irrigation, but some potential impacts should be considered:

- Even though the soil matrix retains pollutants, the limit of the soil retention capacity remains uncertain, largely for metals and emerging compounds.
- The content of organic matter in soil has been a key factor in the natural attenuation of water pollution, however, the impact of future changes in the quality of water for irrigation purposes in Tula Valley, may affect removal capacity.
- Soon, the Atotonilco WWTP will supply wastewater to the irrigation canals with a lower content of organic matter and nutrients. These are factors which have previously favored the natural process of the SAT and the development of agricultural activity in the area. These reductions may affect complex physicochemical processes in the future, and may have impacts on soil and groundwater quality that cannot be predicted.
- The use of treated wastewater for irrigation, will allow the growing of crops normally eaten raw, which up till now has been forbidden by national legislation. Other benefits will be the reduction of direct exposure of farmers to pollutants in untreated water.

- Finally, with regard to national and international legislation for irrigation, it is evident that local studies are required for establishing permissible limits. This is because existing guidelines require wastewater treatment levels which are economically onerous, mainly for developing countries, and moreover, do not include limits for other pollutants present in urban wastewater (pharmaceuticals and personal care use products). Wastewater will become more and more attractive for agricultural reuse, given the current and future problems of water scarcity.

REFERENCES

1. FAO. Monitoring agricultural water use at country level. Experiences of a pilot project in Benin and Ethiopia. Land and Water Discussion Paper 9. ISSN 1729-0554. 2011.
2. Hamilton A.J., Mebalds M., Aldaoud R., Heath, M. A survey of physical, agrochemical and microbial characteristics of waste-water from the carrot washing process: implications for re-use and environmental discharge. Journal of Vegetable Science. 2005. 11, 57–72.
3. Peasey A., Blumenthal U., Mara D., Ruiz-Palacios G.A. Review of Policy and Standards for Wastewater Reuse in Agriculture: A Latin American Perspective. London School of Hygiene & Tropical Medicine. 2000.
4. World Health Organization. Guidelines for the Safe Use of Wastewater, Excreta, and Greywater. Vol. 2, Wastewater Use in agriculture. Geneva, Switzerland. 2006.
5. Blumenthal, U., Peasey, A. Critical Review of Epidemiological Evidence of the Health Effects of Wastewater and Excreta Use in Agriculture. London School of Hygiene and Tropical Medicine. 2002.
6. Page, D., Dillon, P., Toze, S., Bixio, D., Genthe, B., Jiménez Cisneros, B.E., Wintgens, T. Valuing the subsurface pathogen treatment barrier in water recycling via aquifers for drinking supplies. Water Research. 2010;44(6) 1841-1852.
7. Asano, T., editor. Wastewater Reclamation and Reuse. Water Quality Management Library Vol. 10. Technomic Publishing Inc., Lancaster, Pennsylvania, USA. 1998.
8. FAO (1992) Wastewater treatment and use in agriculture-FAO irrigation and drainage paper 47 http://www.fao.org/docrep/T0551E/t0551e00.htm
9. Medaware. Development of Tools and Guidelines for the Promotion of the Sustainable Urban Wastewater Treatment and Reuse in the Agricultural Production in the Mediterranean Countries. European Commission Euro-Mediterranean-Partnership. Technical Guidelines on Wastewater Utilization. ME8/AIDCO/2001/0151/59341-P033. 2005.
10. Jimenez, B. and Asano, T. Water reuse: an international survey of current practices, issues and needs. IWA Publishing. ISBN: 1843390892, 650 pp. 2008.
11. Jiménez, B. Treatment technology and standards for agricultural wastewater reuse: a Case Study in Mexico. Irrigation and Drainage. 2005;54(1) 23-25.
12. Jiménez, B., Chávez, A. Quality assessment of an aquifer recharged with wastewater for its potential use as drinking source: "El Mezquital Valley" case. Water Science and Technology. 2004;50(2) 269-276.
13. Domínguez, R. El sistema principal de drenaje del Área Metropolitana del Valle de México. Series Instituto de Ingeniería. CI-22, 34 p. In Spanish. 2001.
14. Cruz Campa, S. Rehabilitación Integral del Distrito de Riego 03, Tula, Hidalgo. Thesis, Escuela Nacional de Agricultura, Chapingo, México. In Spanish. 1965.

15. CONAGUA. Comisión Nacional del Agua. Estadísticas del agua en México, 2010. Secretaría del Medio Ambiente y Recursos Naturales. México D.F. In Spanish. 2010.
16. British Geological Survey-Comisión Nacional del Agua (1998). London School of hygiene and tropical Medicine and University of Birmingham. Impact of wastewater reuse on groundwater in the Mezquital Valley, Hidalgo State, Mexico. Department for International Development, DFID Final Report. Technical Report WC/98/42. 1998.
17. Siebe C. and Cifuentes E. Environmental impact of wastewater irrigation in central Mexico: An overview. International Journal of Environmental Health Research. 1995; 5: 161-173.
18. Brusseau, M., Oleen, J., Santamaria, J., Cheng, L., Orosz-Coghlan, P., Chetochine, A., Blandford, W., Rykwalder, P., Gerba, C. Transport of *Microsporidium Encephalitozoon* Intestinales Spores in Sand Porous Media. Water Research. 2005; 39: 3636-3642.
19. Jiménez, B., Chávez, A., Hernández, C. Alternative Wastewater Treatment Intended for Agricultural Use. Water Science and Technology. 1999; 40(4-5) 355-362.
20. Dunsmore, B., Bass, C., Lappin-Scott, H. A Novel Approach to Investigate Biofilm Accumulation and Bacterial Transport in Porous Matrices. Environmental Microbiology. 2004; 6(2) 183-187.
21. Blumenthal, U., Mara, D., Ayres, R., Cifuentes, E., Peasey, A., Scott, R., Lee, D., Ruiz-Palacios, G. Evaluation of the WHO nematode egg guidelines for restricted and unrestricted irrigation. Water Science and Technology. 1996; 33:277-283.
22. Modella, P., Blanco, I., Ansola, G., de Luis, E. Removal of Wastewater Pathogen Indicators in a Constructed Wetland in Leon, Spain. Ecological Engineering. 2008; 33(3-4) 252-257.
23. Jiménez, B., Chavez, A., Gibson, R., Maya, C. Unplanned aquifer recharge in El Mezquital/Tula Valley, México. In: Kazner, C., Wintgens, T., Dillon, P. (eds.) Water reclamation technologies for safe managed aquifer recharge. IWA Publishing. London, 2012. 141-143.
24. Barceló, D. Emerging pollutants in water analysis. Trends in Analytical Chemistry. 2003; 22(10): 685-696.
25. Daughton, C.G. and Ternes, T.A. Pharmaceuticals and Personal Care Products in the Environment: Agents of Subtle Change? Health Perspective. 1999; 107: 907-928.
26. Kümmerer, K. (ed). Pharmaceuticals in the Environment. Source, Fate, Effects and Risks. 2nd edition. Springer. 2004.
27. Tabak, H.H. and Bunch, R.L. Steroid hormones as water pollutants. I. Metabolism of natural and synthetic ovulation inhibiting hormones by microorganisms of activated sludge and primary settled sewage. Developments in Industrial Microbiology. (1970). 11, 367–376
28. Kramer, V.J., Miles-Richardson, S., Pierens, S.L., Giesy, J.P. Reproductive impairment and induction of alkaline-labile phosphate, a biomarker of estrogen exposure, in fathead minnows (*Pimephales promelas*) exposed to waterborne 17β-estradiol. Aquatic Toxicology. 1998; 40(4) 335–360.
29. Snyder, S.A., Villeneuve, D.L., Snyder, E.M., Giesy, J.P. Identification and quantification of estrogen receptor agonists in wastewater effluents. Environmental Science and Technology. 2001; 35(18) 3620–3625.
30. Bixio, D., Thoeye, C., De Koning, J., Joksimovic, D., Savic, D., Wintgens, T., Melin, T. Wastewater reuse in Europe. Desalination. 2006; 187: 89-101.
31. Gibson, R., Becerril-Bravo, E., Silva-Castro, V., Jiménez-Cisneros, B. Determination of acidic pharmaceuticals and potential endocrine disrupting compounds in wastewaters and spring waters by selective elution and analysis by gas chromatography–mass spectrometry. Journal of Chromatography A. 2007; 1169: 31–39.
32. Verenitch, S.S., Lowe, C.J., Mazumder, A. Determination of acidic drugs and caffeine in municipal wastewaters and receiving waters by gas chromatography–ion trap tandem mass spectrometry. Journal of Chromatography A. 2006; 1116: 193-203.

33. Lee, H.B., Peart, T.E., Svoboda, M.L. Determination of endocrine-disrupting phenols, acidic pharmaceuticals, and personal-care products in sewage by solid-phase extraction and gas chromatography-mass spectrometry. Journal of Chromatography A. 2005; 1094: 122-129.
34. Heberer, T., Fuhrmann, B., Schmidt-Baumler, K., Tsipi, D., Koutsouba, V., and Hiskia, A. Occurrence of pharmaceutical residues in sewage, river, ground and drinking water in Greece and Berlin (Germany). In: Daughton, CGTJ-L, (ed.) Pharmaceuticals and Personal Care Products in the Environment: Scientific and Regulatory Issues. American Chemical Society. Washington, DC. 70-83. 2001.
35. Farré, M., Ferrer, E., Ginebreda, A., Figueras, M., Olivella, L., Tirapu, L., Villanova, M. and Barcelo, D. Determination of drugs in surface water and wastewater samples by liquid chromatogrphy-mass spectrometry: methods and preliminary results including toxicity studies with *Vibrio fishery*. Journal of Chromatography A. 2001; 938: 187-197.
36. Stumpf, M., Ternes, T.A., Wilken, R.D., Rodrigues, S.V., Baumann, W. Polar drugs residues in sewage and natural waters in the state of Río de Janeiro, Brazil. Science of the Total Environment. 1999; 225(1-2) 135-141.
37. Carballa, M., Omil, F., Lema, J.M., Llompart, M., García-Jares, C., Rodríguez, I., Gómez, M., Ternes, T. Behavior of pharmaceuticals, cosmetics and hormones in a sewage treatment plant. Water Research. 2004; 38(12) 2918-2926.
38. Linqvist, N., Tuhkanen, T., Kronberg, L. Occurrence of acidic pharmaceuticals in raw and treated sewage and in receiving waters. Water Research. 2005; 39(11) 2219-2228.
39. Siemens, J., Huschek, G., Siebe, C., Kaupenjohann, M. Concentrations and mobility of human pharmaceuticals in the world's largest wastewater irrigation system, Mexico City–Mezquital Valley. Water Research. 2008; 42(8-9) 2124 –2134.
40. Chávez, A., Maya, C., Gibson, R., Jiménez, B. The removal of microorganisms and organic micropollutants from wastewater during infiltration to aquifers after irrigation of farmland in the Tula Valley, Mexico. Environmental Pollution. 2011; 159(5) 1354-1362.
41. Gibson, R., Durán-Álvarez, J.C., León, K., Chávez, A., Jiménez, B. Accumulation and leaching potential of some pharmaceuticals and potential endocrine disruptors in soils irrigated with wastewater in the Tula Valley, Mexico. Chemosphere. 2010; 81(1) 1437–1445.
42. Birkett, J.W.; Llester, J.N. Endocrine disrupters in wastewater and sludge treatment processes. CRC Press. 2002.
43. Murillo, R., Durán-Álvarez, J.C., Prado-Pano, B., Jiménez-Cisneros, B. Mobility of 4-nonylphenol and di(2-ethylhexyl) phthalate in three agricultural soils irrigated with untreated wastewater. Water Science and Technology. 2012; 66(2) 292-298.
44. Durán-Álvarez, J.C., Prado-Pano, B., Jiménez-Cisneros, B. Sorption and desorption of carbamazepine, naproxen and triclosan in a soil irrigated with raw wastewater: Estimation of the sorption parameters by considering the initial mass of the compounds in the soil. Chemosphere. 2012; 88(1) 84-90.
45. Siebe, C., Fischer, W.R. Effect of long-term irrigation with untreated sewage effluents on soil properties and heavy metal adsorption of Leptosols and Vertisols in Central Mexico. Z. Pflanzenernaehr. Bodenk. 1996; 159(4) 357-364.
46. CONAGUA. Estadísticas agrícolas de los Distritos de Riego. Año agrícola 2011-2012. 2013. http://www.conagua.gob.mx/CONAGUA07/Publicaciones/Publicaciones/SGIH-4-13.pdf (accessed 26 September 2014).
47. Lesser, L., Lesser, J.M, Arellano-Islas, S., González-Posadas, D. Balance hídrico y calidad del agua subterránea en el acuífero del Valle del Mezquital, México central. Revista Mexicana de Ciencias Geológicas. 2011; 28(3) 323-336. In Spanish.
48. Cifuentes, E. The epidemiology of enteric infections in agricultural communities exposed to wastewater irrigation: Perspectives for risk control. International Journal of Environment and Health. 1998; 8(3) 203-213.

49. Herre, A.; Siebe, Ch. Effects of irrigation water quality on organic matter, Cd, and Cu mobility in soils of central Mexico. Water Science and Technology. 2004; 50(2) 277-284.

50. Gallegos, E., Warren, A., Robles, E., Campoy, E., Calderon, A., Sainz, M. G., Bonilla, P., Escolero, O. The Effects of Wastewater Irrigation on Groundwater Quality in Mexico. Water Science and Technology. 1999; 40(2) 45–52.

51. Kimura, K., Amy, G., Drewes, J., Watanabe, Y. Adsorption of hydrophobic compounds onto NF/RO membranes: an artifact leading to overestimation of rejection. Journal of Membrane Science. 2003; 221(1-2) 89–101.

52. Zhang, Y., Causserand, C., Aimar, P., Cravedi, J.P. Removal of bisphenol A by a nanofiltration membrane in view of drinking water production. Water Research. 2006; 40(20) 3793-3799.

53. Chávez, A., Torner, F., Barrios, J.A., Maya, C. Lucario, E., Jiménez, B. Membrane process for spring water treatment in the Tula Valley. Comparative microbiology assessment. Water Science and Technology: Water Supply. In press.

54. CONAGUA (2014) Strategic projects: water supply, sewerage and sanitation. http://www.conagua.gob.mx/english07/publications/StrategicProjects.pdf (accessed 22 September 2014)

CHAPTER 5

Inhibition of Root Penetration in Subsurface Driplines by Impregnating the Drippers with Copper Oxide Particles

Seba Schifris[1] · Avi Schweitzer[1] · Eli Matan[1] · Gadi Borkow[2]

[1]Netafim Ltd.Tel AvivIsrael

[2].Cupron Inc.HerzeliaIsrael

ABSTRACT

Clogging of subsurface drip irrigation (SDI) systems by root penetration into the emitters results in inappropriate water supply, increased replacement rates of SDI systems and increased costs. We found that impregnation of SDI drippers with copper oxide particles inhibits root penetration very significantly. The inhibition of root penetration varied from 24 to 76 % of the control (60–80 % intrusion) depending on the copper oxide concentration of the drippers. The root penetration inhibition was demonstrated at two water flow rates, of 1 and 3.5 l/h, and with new and used drippers after 1750 irrigation hours. Inhibition of root penetration occurred also if sewage water was used. The inhibition of root penetration into drippers was demonstrated with lettuce and tomato plants and reached similar efficacy as compared to the widely used herbicide Stomp® root penetration inhibition treatment. The amount of copper that leached into the water was below detection limit (less than 0.006 ppm). No loss of copper oxide particles was detected in drippers through which 3350 l of water was passed (an amount of water that typically passes in SDI systems during 3–4 years of use), as determined by scanning electronic microscope and X-ray photoelectron spectrum analysis.

INTRODUCTION

Subsurface drip irrigation (SDI) is a very efficient irrigation system aimed at ensuring uniform and adequate supply of water to each plant according to its needs. Water is delivered directly to the roots of plants. This method prevents or minimizes evaporation and runoff and is especially useful in places where water is a limited resource.

A significant problem associated with SDI is the clogging of the emitters via root penetration. Roots tend to proliferate in favorable environments where water and nutrients are readily available, and the presence of a constant water and nutrient supply, emanating from the irrigation line emitters, favors root intrusion of the emitters. The root intrusion causes partial or total blockage of the dripper outlet. Depending on the aggressiveness of the root system and the type of the emitters, a number of adjoining emitters can be blocked, and in extreme cases, even a whole lateral, which in turn prevents water supply to downstream emitters.

One way to prevent root penetration is by treating the water or the drippers themselves with an herbicide, for example, with trifluralin (Sheval et al. 2008; Spera et al. 2006). The addition of the herbicide may necessitate an injection pump that releases the herbicide at a prescribed dosage and rate, or alternatively emitters may contain the herbicide. This provides a continuous slow release process of the herbicide as water passes through it. However, herbicide use carries risks that include environmental, ecological and human health effects (Ahemad and Khan 2009; Bernard et al. 2005; Eason et al. 2004). Another possibility of reducing root penetration has been the use of copper. Copper is known to have potent antibacterial, fungicidal and algaecidal properties (Borkow and Gabbay 2005; Murray-Gulde et al. 2002). Copper compounds are widely used today as algaecides and herbicides (Bishop and Rodgers 2012; Borkow and Gabbay 2005), and several soluble copper compounds have been shown to inhibit root elongation (Arduini et al. 1995; Meier et al. 2012). However, since these are chemicals that readily solubilize in water, they may have adverse environmental effects (de Andrade Waldemarin et al. 2012; Mastin and Rodgers 2000). Thus, non-soluble metallic copper pieces are being introduced in subsurface drip irrigation systems to protect the emitters from root intrusion (http://www.rainbird.com/landscape/products/dripline/XFS.htm).

In the current study, we introduced non-soluble copper oxide particles in subsurface drip irrigation emitter heads and investigated their capacity to reduce root penetration and obstruction of the irrigation systems.

MATERIALS AND METHODS

Introduction of copper oxide particles into emitter heads

We introduced the copper oxide particles into polyethylene by using the published platform technology of introducing copper oxide particles into different polymeric materials (Borkow and Gabbay 2004). As can be seen in scanning electronic microscope pictures and the X-ray spectra analyses in Figure 1, the copper oxide particles (the white dots seen in Figure 1c) are homogenously distributed throughout the high-density polyethylene (HDPE) matrix and constitute an integral part of the dripper's head. The concentration of copper oxide particles introduced into the matrix varied between 0 and 40 % weight/weight.

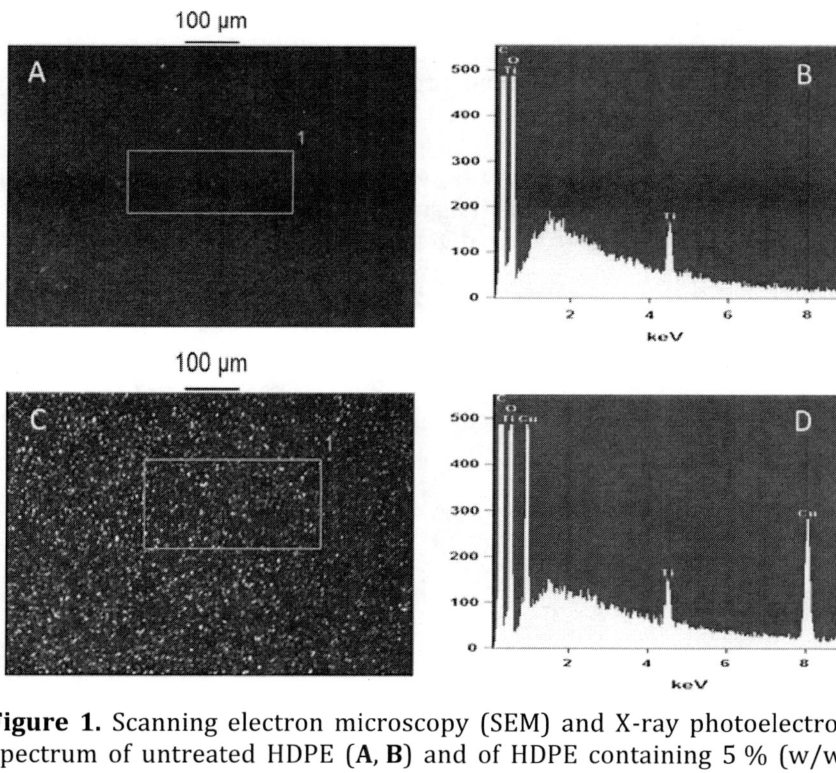

Figure 1. Scanning electron microscopy (SEM) and X-ray photoelectron spectrum of untreated HDPE (**A**, **B**) and of HDPE containing 5 % (w/w) copper oxide particles (**C**, **D**)

Determination of root intrusion into drippers

Experiments were generally performed as follows unless specified otherwise in the Results section: Individual, 4 m length, 0.5 m width and 0.35 m height, cases with an overall volume of 700 l were used as plots. Except in one case, the number of plots per experiment varied between 12 and 28 plots. The treatment of each plot was randomized. The cases were placed in a greenhouse in Kibbutz Hatzerim, Israel. The soil growth medium consisted of sand and compost at a 7/3 ratio. In each case, two subsurface irrigation lines (DripNet PC™ dripper lines or UniRam™ RC or UniRam™ AS, Netafim) were installed 15 cm below the surface. Each line had 40 drippers. The distance between each dripper was 20 cm, and each dripper flow rate, depending on the experiment, varied between 0.4 to 3.5 l/h. At least four plots per treatment were used, i.e., 160 drippers. A total of 40 plant seedlings of ~3 weeks of age (lettuce or tomato, as detailed in the "Results" section) were planted in each case in two rows, being the distance between each row 20 cm. Until plant establishment (~7 days from seeding), the plants were watered daily by using Coolnet Pro™ mist emitters in addition to the watering regime that varied among experiments as described in Figure 2. The watering regimes were designed to decrease gradually the amount of water applied in order to develop water stress and promote root intrusion into the drippers. During all experiments, the plants were treated via irrigation with a fertilizer composed of 6:4:4: nitrogen, phosphate and potassium, respectively, at 1.4 l/m^3daily application. At the end of the experiments, which varied between 2 and 4 months, the irrigation lines were removed, the drippers exposed and the root intrusion determined.

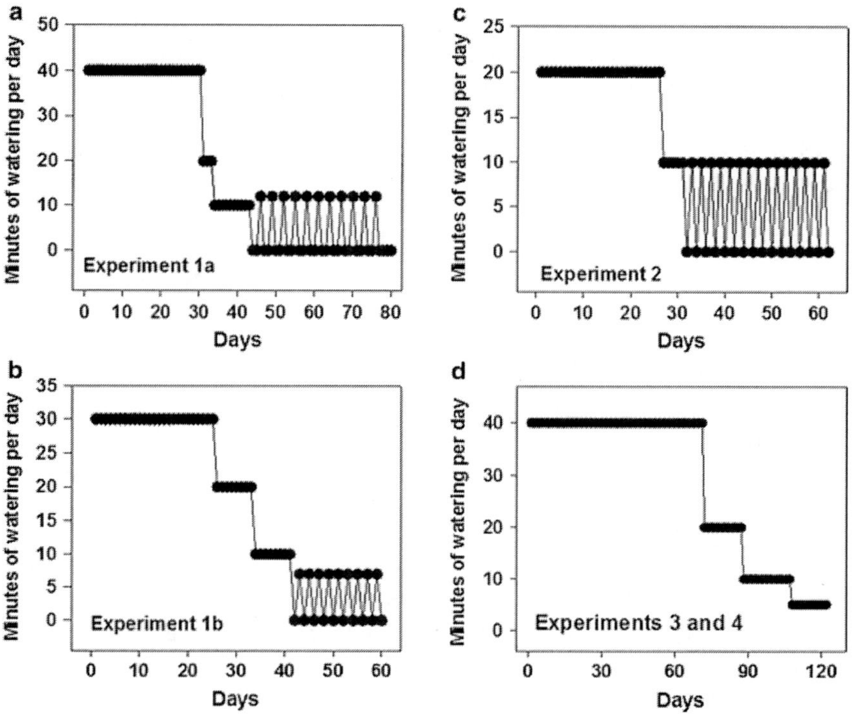

Figure 2. Watering regimens of the different experiments performed to encourage root intrusion into the drippers

Determination of copper elution from copper oxide-impregnated drippers

The elution of copper from the copper oxide-impregnated drippers was determined in two different experimental settings. Firstly, the amount of copper eluting from a dripper containing copper oxide was determined by passing 250 ml of tap water through the dripper at a flow rate of 1 l/h. The water passing through the dripper was collected for 15 min. The copper content of the collected 250 ml water, and that present in 250 ml of tap water that was not passed through the dripper, was determined by inductively coupled plasma mass spectrometry (ICP-MS, ARCOS, Spectro GMBH, Germany). Secondly, tap water was passed through a 15-m-long tube containing 50 copper oxide-impregnated drippers placed at a distance of 30 cm for a period of 140 days. A total of 3350 l of water, at a flow rate of 1 l/h, was circulated through the tubing. At the end of the experiment, the drippers were removed from the tubing and the presence of copper on the surface of eight drippers was examined by using scanning electronic microscopy (SEM; Jeol JMS 5410 LV scanning electron microscope, Japan) and the presence of copper coverage was determined by X-ray photoelectron spectrum analysis (Link IV, ISIS, Oxford Instruments, England). Eight copper oxide-containing drippers that were not exposed to water were used as reference.

RESULTS

Prevention of root intrusion

Experiment 1

We first tested the effect on root intrusion into drippers impregnated with two different copper oxide concentrations, 12.5 and 25 % w/w, as compared to the control drippers without copper ($n = 4$ cases of 40 drippers each, per group). Three-week-old lettuce seedlings (*Lactuca sativa*, cultivar "yellow leaves") were grown for 80 days according to the water irrigation regimen depicted in Figure 2a. As can be seen in Figure 3a, there was a clear trend of reduction in root intrusion into the drippers with the increase in the copper oxide concentration in the dripper's heads; from 54 % in the control to 43 % (24 % reduction, $p = 0.266$) and 31 % (43 % reduction, $p = 0.093$) in the drippers containing 12.5 and 25 % w/w copper oxide particles, respectively.

We tested again the efficacy of reducing the root penetration by the 25 % copper oxide-containing drippers as compared to the control drippers but enlarged the number of cases to 12 cases (480 drippers each) and used 3-week-old *Lactuca sativa*, *Noga* seedlings. Also in this experiment after 60 days of growth using the water regimen depicted in Figure 2b, there was a reduction in root intrusion into the drippers containing the copper oxide particles as compared to the control (35 % reduction, $p = 0.003$, Figure 3b). In addition, we tested the root penetration in four cases that contained 160 drippers without copper but that were treated with a widely used herbicide (Stomp® 400 SC, BASF, Ireland) via the irrigation system as instructed by the manufacturer. The drippers that were treated with Stomp® showed a similar overall reduction in root penetration as that seen in the copper oxide-containing drippers, but the variability was significantly higher (Figure 3b).

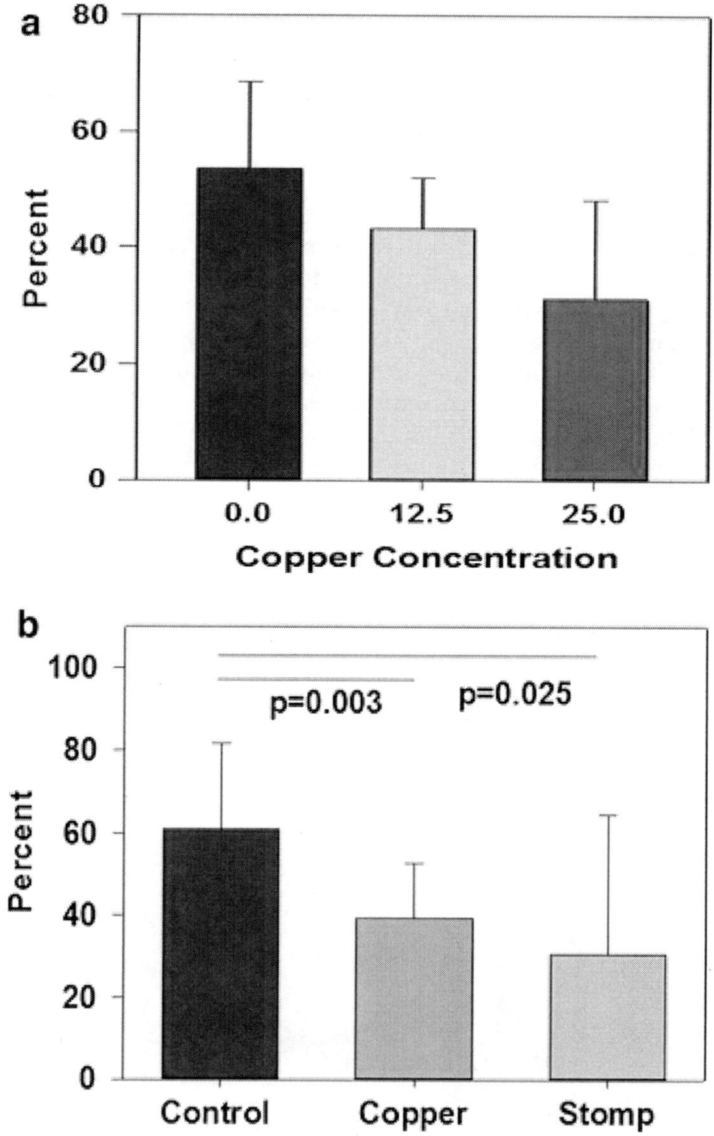

Figure 3. **A** Mean and standard deviation of percent *Lactuca sativa* (cultivar "yellow leaves" seedlings) root intrusion into drippers with different copper oxide concentration (n = 160); **B** mean and standard deviation of percent *Lactuca sativa* (*Noga* seedlings) root intrusion for each type of dripper tested or Stomp® treatment (n = 480). The p values of a t test between the groups tested are shown

Experiment 2

In this experiment, we examined whether there is a reduction in the efficacy of preventing root intrusion into the drippers containing 25 % copper oxide particles after 1000 l of water has passed through each dripper ($n = 4$ cases of 40 drippers each, per group). We used *Lactuca sativa*, cultivar "yellow leaves" 3-week-old seedlings. As depicted in Figure 4, the reduction in root penetration into the drippers containing copper oxide particles after 60 days of growth of growth using the water regimen depicted in Figure 2c was even more drastic than in the first experiment using the same seedlings, from ~75 % in the control to ~21 % in the copper-containing drippers (~72 % reduction, $p < 0.001$). Importantly, even in drippers through which 1000 l of water was passed before performing the experiment, there were clear significant reductions (~45 % reduction, $p < 0.001$) in the root intrusion levels into the drippers as compared to the control.

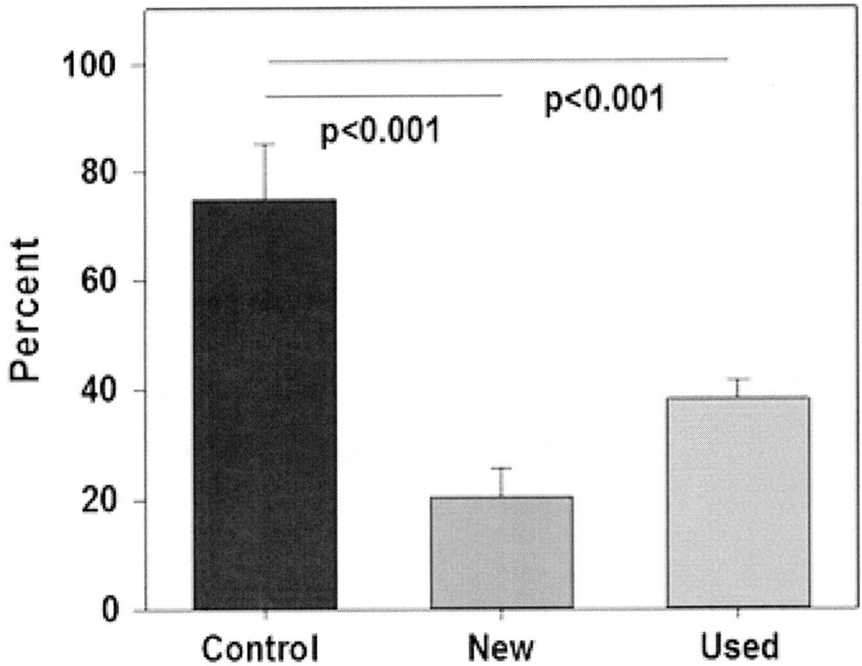

Figure 4. Mean and standard deviation of percent *Lactuca sativa* ("yellow leaves" seedlings) root intrusion into new drippers or drippers through which 1000 l of water was passed before performing the experiment ($n = 160$). The p values of a t test between the groups tested are shown

Experiment 3

In this experiment, we examined the effect of the water flow rate on the capacity to reduce root intrusion by the copper oxide impregnation. We used two types of UniRam™ AS drippers, with flow rates of 1 and 3.5 l/h. In each plot, 20 drippers were placed at a distance of 40 cm between each dripper. For each treatment, six plots were used with a total of 120 drippers per treatment. In each plot, two rows of 3-week-old tomato (*Lycopersicon esculentum*) seedlings were seeded with 40 cm distance between each seed in each row. The irrigation regime is described in Figure 2c. As can be seen in Figure 5, at a flow rate of 1 l/h, the presence of copper oxide particles in the drippers reduced the root intrusion into the drippers after 120 days by 50 % ($p = 0.026$), while at a flow rate of 3.5 l/h, the root intrusion reduction as compared to the control was reduced by 70 % ($p < 0.001$).

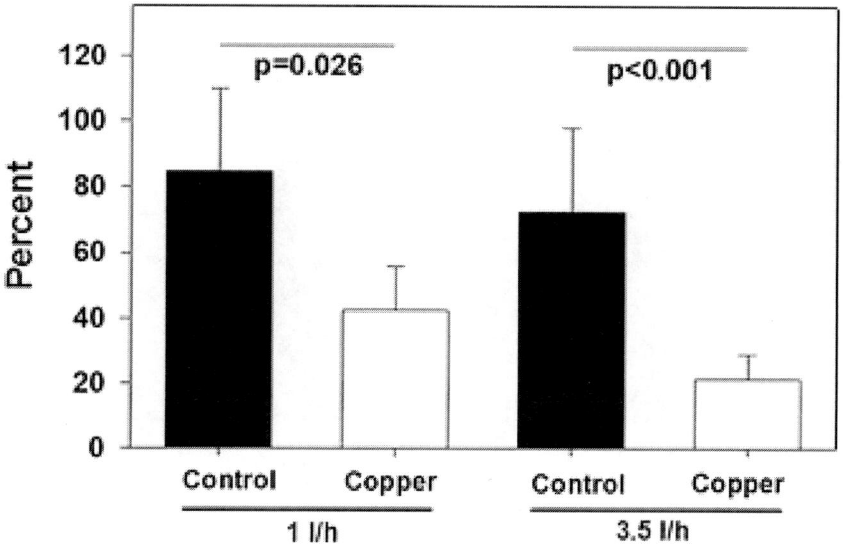

Figure 5. Mean and standard deviation of percent tomato (*Lycopersicon esculentum*) root intrusion at 1 or 3.5 l/h water flow rate ($n = 120$). The p values of a t test between the groups tested are shown

Experiment 4

In this last experiment, we examined whether the capacity of the drippers to inhibit root intrusion is affected when sewage water is used. We used UniRam™, either new or used drippers through which 1720 l of sewage water was first passed at a flow rates of 1 l/h, which customarily corresponds to 1.5–2 years of operation in the field. In each plot, 20 drippers were placed at a distance of 40 cm between each dripper. For each treatment, five plots were used with a total of 100 drippers per treatment. In each plot, two rows of 3-week-old tomato (*Lycopersicon esculentum*) seedlings were seeded with 40 cm distance between each seed in each row. The irrigation regime is detailed in Figure 2d. As can be seen in Figure 6, the presence of copper oxide particles in the new drippers reduced the root intrusion into the drippers by 76 %, from a mean of 73 to 17.5 % intrusion ($p < 0.001$), after 120 days of growth. Also the used drippers significantly reduced the root intrusion into the dripper, from a mean 78 to 43 % ($p < 0.001$), although as compared to the new drippers, there was reduced efficacy ($p = 0.013$).

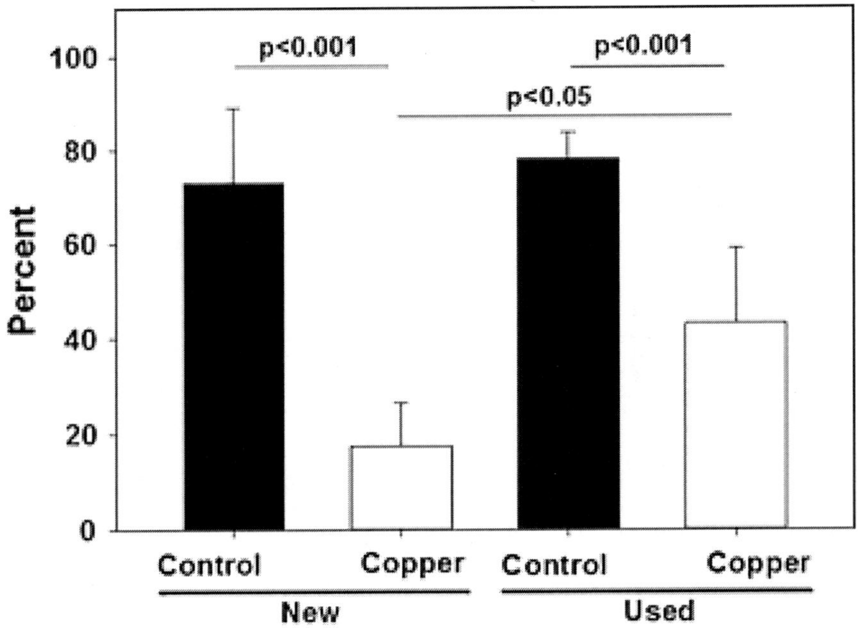

Figure 6. Mean and standard deviation of percent tomato (*Lycopersicon esculentum*) root intrusion for new drippers or drippers in which 1720 l of sewage water was first passed through them ($n = 120$). The p values of a t test between the groups tested are shown

Copper elution from drippers

The amount of copper in 250 ml of water before and after passing through a dripper containing copper oxide was less than 0.006 ppm (mg/l), which is the lower limit of detection of the assay. Copper oxide particles were present on the surface of the HDPE drippers containing 25 % copper oxide w/w, with no observable differences in the coverage in those not exposed or those exposed to 3350 l of water. The percentage of copper coverage on the surface of eight separate drippers though which 3350 l of water passed at a flow rate of 1 l/h during 140 days was 2 % lower in average than those found in eight separate drippers not exposed to water, as determined by SEM and X-ray photoelectron spectrum analysis (11.71 ± 0.53 vs 11.48 ± 1.59). 3350 l of water that passed through the drippers is equivalent to 3–4 years of irrigation using such flow rate irrigation system in a field.

DISCUSSION AND CONCLUSIONS

A typical life span of SDI could be 15–20 years and depends of several factors, such as the type of water used, the bioflora that develops within the irrigation system, the maintenance treatment and the level of root intrusion into the emitters (Lamm et al. 2011). In this study, we analyzed the capacity of reducing root intrusion into the emitters by incorporating microscopic copper particles during the manufacture of the emitters, as copper is known as a potent algaecide (Borkow and Gabbay 2005; Murray-Gulde et al. 2002).

We used copper oxide for two main reasons: (a) copper oxide is a very reactive copper compound with potent biocidal properties, which is being used successfully in many consumer, industrial and medical device products (Borkow and Gabbay 2004, 2006; Borkow et al. 2007, 2010a, b; Borkow and Monk 2012; Borkow 2012), and (b) copper oxide is a non-soluble copper compound. Obviously in such products, it is important that the active ingredient in the product remains in the product for the life of the product; in this case that it would not be lost into the running water.

As was demonstrated in this study, impregnating the copper oxide particles into the drippers indeed very significantly reduced the root intrusion into the drippers, as demonstrated in all five experiments conducted. As expected, the reduction in the root intrusion was copper dose dependent. The inhibition of root intrusion occurred both in tomato and lettuce plants with different root system, supporting the notion that the effect is universal and not plant specific, although additional trials should test this hypothesis. The efficacy reached is similar to the current standard, i.e., treatment with Stomp®. Replacing Stomp® is advantageous as Stomp®treatment is environmentally unfriendly and time consuming, and in some countries its use is forbidden. Irrigation water can vary very much. Importantly, even when using sewage water, rich in organic material that can interact with the copper and quench the algaecidal efficacy, a very significant reduction in root intrusion was observed. The inhibition of root intrusion occurred both in dripper flow rate of 1 and 3.5 l/h. Both of these flow rates are very common in drip irrigation systems.

In conclusion, in this study we demonstrated that incorporation of copper oxide particles into the dripper's heads of SDI lines significantly and consistently reduced root penetration into the drippers. The efficacy is maintained even after very prolonged use allowing for reducing the frequency needed for replacing SDI systems.

ACKNOWLEDGMENTS

This study was funded by Netafim Ltd.

REFERENCES

1. Ahemad M, Khan MS (2009) Toxicity assessment of herbicides quizalofop-p-ethyl and clodinafop towards Rhizobium pea symbiosis. Bull Environ Contam Toxicol 82:761–766
2. Arduini I, Godbold DL, Onnis A (1995) Influence of copper on root growth and morphology of *Pinus pinea* L. and *Pinus pinaster* Ait. seedlings. Tree Physiol 15:411–415
3. Bernard H, Chabalier PF, Chopart JL, Legube B, Vauclin M (2005) Assessment of herbicide leaching risk in two tropical soils of Reunion Island (France). J Environ Qual 34:534–543
4. Bishop WM, Rodgers JH Jr (2012) Responses of Lyngbya wollei to exposures of copper-based algaecides: the critical burden concept. Arch Environ Contam Toxicol 62:403–410
5. Borkow G (2012) Using copper to fight microorganisms. Curr Chem Biol 6:93–103
6. Borkow G, Gabbay J (2004) Putting copper into action: copper-impregnated products with potent biocidal activities. FASEB J 18:1728–1730
7. Borkow G, Gabbay J (2005) Copper as a biocidal tool. Curr Med Chem 12:2163–2175
8. Borkow G, Gabbay J (2006) Endowing textiles with permanent potent biocidal properties by impregnating them with copper oxide. JTATM 5:1–3
9. Borkow G, Monk AB (2012) Fighting nosocomial infections with biocidal non-intrusive hard and soft surfaces. World J Clin Infect Dis 12:77–90

10. Borkow G, Sidwell RW, Smee DF, Barnard DL, Morrey JD, Lara-Villegas HH, Shemer-Avni Y, Gabbay J (2007) Neutralizing viruses in suspensions by copper oxide based filters. Antimicrob Agents Chemother 51:2605–2607
11. Borkow G, Okon-Levy N, Gabbay J (2010a) Copper oxide impregnated wound dressings: biocidal and safety studies. Wounds 22:310–316
12. Borkow G, Zhou SS, Page T, Gabbay J (2010b) A novel anti-influenza copper oxide containing respiratory face mask. PLoS One 5:e11295
13. de Andrade Waldemarin KC, Alves RN, Beletti ME, Rantin FT, Kalinin AL (2012) Copper sulfate affects Nile tilapia (Oreochromis niloticus) cardiomyocytes structure and contractile function. Ecotoxicology 21:783–794
14. Eason A, Tim US, Wang X (2004) Integrated modeling environment for statewide assessment of groundwater vulnerability from pesticide use in agriculture. Pest Manag Sci 60:739–745
15. Lamm FR, Rogers DH, Alam M, O'Brien DM, Trooien TP (2011) Twenty-two years of SDI research in Kansas. In: Proceedings of the 23rd Annual Central Plains Irrigation Conference, Burlington, CO., February 22–23, 2011. pp 68–92
16. Mastin BJ, Rodgers JH Jr (2000) Toxicity and bioavailability of copper herbicides (Clearigate, Cutrine-Plus, and copper sulfate) to freshwater animals. Arch Environ Contam Toxicol 39:445–451
17. Meier S, Alvear M, Borie F, Aguilera P, Ginocchio R, Cornejo P (2012) Influence of copper on root exudate patterns in some metallophytes and agricultural plants. Ecotoxicol Environ Saf 75:8–15
18. Murray-Gulde CL, Heatley JE, Schwartzman AL, Rodgers JH Jr (2002) Algicidal effectiveness of Clearigate, Cutrine-Plus, and copper sulfate and margins of safety associated with their use. Arch Environ Contam Toxicol 43:19–27
19. Sheval EV, Kazhura YI, Poleshuk NA, Lazareva EM, Smirnova EA, Maximova NP, Polyakov VY (2008) Trifluralin-induced disorganization of microtubular cytoskeleton alters the development of roots in *Hordeum vulgare* L. Acta Biol Hung 59:465–478
20. Spera G, Rosati S, Rossi E, Scicchitano S (2006) Subsurface drip irrigation with micro-encapsulated trifluralin. Trifluralin residues in soils and cultivations. Commun Agric Appl Biol Sci 71:161–170

CHAPTER 6

Challenges and Constraints in Irrigation and Drainage Development: A World-Wide View

Marco Medici[1], Daniele De Wrachien[2]

[1]*Department of Industrial Engineering, University of Parma, Parma, Italy*

[2]*Department of Agricultural and Environmental Sciences, University of Milan, Milan, Italy*

ABSTRACT

In order to increase food production it is necessary to better understand all the factors underlying short term operations and long term strategies that lead to improve food security. In this regard any good effect deriving from agricultural research is strictly connected to a more accurate estimation of crops resources requirements, to major improvements in the use of irrigation and drainage systems, to a better understand of climatic variability. All the variables involved show strong dependency on both place and time. The failing of past, present and future systems can be essentially attributed to poor agricultural research planning, to incorrect agricultural systems design and resource allocation and to a not complete capacity to cope with short and long-term global climatic behavior.

Keywords: Challenges and Constraints, Irrigation and Drainage Development

1. INTRODUCTION

Irrigated agriculture is considered a key element in order to reach development objectives of achieving sustainable food security especially in developing countries. Especially as we are faced with the prospect of global population growth; the present world population of 6 billion will reach 8 billion by 2025 and 9.4 billion by 2050 [1] [2] : in this context, 8.2 billion will live in developing countries, of which 3 billion will be located in arid and semiarid environments.

In order to increase food production, it is necessary to better understand all the factors underlying short term operations and long term strategies that lead to improve food security. In this regard any good effect deriving from agricultural research is strictly connected to a more accurate estimation of crops resources requirements, to major improvements in the use of irrigation and drainage systems, to a better understand of climatic variability. All the variables involved show strong dependency on both place and time. The failing of past, present and future systems can be essentially attributed to poor agricultural research planning, to incorrect agricultural systems design and resource allocation and to a not complete capacity to cope with short and long-term global climatic behavior.

2. DEVELOPMENT IN IRRIGATION AND DRAINAGE

Of the world's total cultivated area (~1.5 billion ha) over the 73% is nowadays agriculturally exploited with no water management system and it is responsible for 40% of crop output, while irrigated land and drainage facilities occupy roughly 270 million ha (18%) and 130 million ha (~9%), being responsible for, respectively, 45% and 15% of crop output [3] [4] . However, despite its high intrinsic yield potential, irrigated agriculture it is now increasing at very slow rate, and this is due to many factors, including, among all, the significant slowdown in monetary investments, the loss of irrigated areas due to bad soil management, and urban encroachment.

In rainfed areas, that represent the largest majority of agricultural realities, yields of crops productivity vary dramatically to rainfall variability, which can be seen as the most significant natural variable. Some minimal improvements for rainfed agriculture can be at least achieved with improvements in water harvesting and watershed management. This is a current challenge for several humid and semiarid areas like northeastern South America, Sub-Saharan Africa (SSA), India and central Asian countries, each of them characterized by particular social, economic and environmental characteristics. For example, the general data of annual rainfall between 1961-1990 over Nigeria shows how the rainfall range in the humid south zone is triple with respect to the drier north areas, showing how the rainfall is in this case considerably affected by latitude [5] . This huge climatic variability is associated to high risk in crop production. Local farmers need risk reducing information at first, more than soil or crop management technologies.

In India about two third of the total rainfed area is arid and also it is characterized by the highest density of population compared to other arid regions at the same latitude. Also these regions are characterized by a high variability in rainfall: normally water harvesting is considered the best-suited solution, but after long drought periods groundwater becomes the only one source of water in most villages [6] ; periodically digging new wells and ponds is a common activity. In this context modern irrigation systems can be an intelligent solution for a country that already experienced the so called green revolution.

In some areas worldwide spread the rainfall can be sufficient for a period but short droughts and low productivity may happen as results of low water holding capacity of the soil. A substantial productivity loss attributable to soil degradation was estimated in SSA [2] . In this regard soil management technologies must be developed and adopted, being able to address the issue of enhancing/restoring soil structure, adopting conservation tillage, adopting efficient methods of water harvesting.

Generally important information and appropriate technologies already exist but often they are under-utilized due to policy and technology transfer constraints: new technologies that do not rely mainly on local sources of inputs are often associated to failures. At this regard attempts to introduce infrastructure dependent and irrigation based technologies in Africa did not lead to success [7] [8] .

A substantial effort need to be placed on resource management research about rainfed systems, timely and reliable climate forecast, and also intensive production systems. The share of irrigated and drained areas in food production will have to increase, and installing new facilities or modernizing existing systems, in order to increase crop yields, will be necessary.

3. THE ROLE PLAYED BY CLIMATE CHANGE

Current scientific research is focused on the enhanced greenhouse effect as the most likely cause of climate change in the short-term [9] . Local scenarios of future climatic conditions have been proposed to provide quantitative assessments about hydrologic results. These scenarios can be classified into three groups: hypothetical scenarios, climate scenarios based on general and regional circulation models (GCMs and RCMs), and scenarios based on reconstruction of past eras (paleo-climatic reconstruction) [10] . The scenarios of the second group have been widely utilized to reconstruct seasonal conditions of the change in temperature, precipitation and potential evapotraspiration at basin scale over the next century. They are complex three-dimensional computer-based models of the atmospheric circulation, which provide details of theoretical changes in regional climates for any part of the Earth. The standard approach has been to run the model with a nominal "pre-industrial" atmospheric CO_2 concentration (the control run) and then to rerun the model with doubled (or sometimes quadrupled) CO_2 (the perturbed run). This approach is known as "the equilibrium response prediction".

However, current results are not sufficiently reliable, being sometimes even in disagreement with empirical observations. Several state-of-the-art GCMs, greenhouse models in particular, have large uncertainties in defining climate scenarios; at this regard the dominant role of cumulus convection requires that warming in the tropical area upper troposphere has to be at least the double with respect to the surface; however empirical measurements deny it [11] [12] . What is known is that forecasts of anthropogenic global climate change have been unreliable so far, and in general, present climate alarmists respond to the evidence of the relatively warm recent decade. Yet, considering also that heat transferred between deep layers of the troposphere and the ocean surface occurs over a time scale from years to centuries, this demonstrates nothing in terms of temperature trends [13] [14] .

4. CONCLUDING REMARKS

Ÿ A substantial effort need to be placed on resource management research about rainfed systems in developing countries. The share of irrigated and drained areas in food production will have to increase, and installing new facilities or modernizing existing systems will be necessary. Appropriate technologies already exist but often they are under-utilized due to policy and technology transfer constraints. New technologies have to rely on local sources of inputs. It is also necessary to reduce vulnerability to environmental changes providing local farmers of higher incomes and more suitable technology.

Ÿ An integrated approach to irrigation and drainage systems development is necessary. There is a strong need for high quality and innovative research in soil science; efforts should focus on efficient use of fresh water resources and limit soil degradation.

Ÿ Nowadays the disturbing news about global warming need to be locally verified in order to perform reliable analysis. Empirical evidences demonstrate how the anthropogenic contribution to recent warming is surprisingly smaller than expected. On the other hand climate will always change as long as our planet will have a fluid envelope: attempts from scientists should focus simple approaches in order to reduce present persistent uncertainty and establish estimates of climate sensitivity that are fully focused and testable.

CITE THIS PAPER

Marco Medici, Daniele De Wrachien (2016) Challenges and Constraints in Irrigation and Drainage Development: A World-Wide View. *Voice of the Publisher*,**02**,9-12. doi: 10.4236/vp.2016.22002

REFERENCES

1. De Wrachien, D. and Feddes, R. (2003) Drainage and Land Reclamation in a Changing Environment: Overview and Challenges. Proceeding of the International Conference on Land Reclamation and Water resources Development, Mantua.
2. Lal, R. (2000) Soil Management in the Developing Countries. Soil Science, 165, 57-72.
3. Schultz, B. (2001) Irrigation Drainage and Flood Protection in a Rapidly Changing World. Irrigation and Drainage, 50, 4.
4. Smedema, L.K. (2000) Global Drainage Needs and Challenges. The Role of Drainage in the Today's World. Proceedings of the 8th International Workshop on Drainage, New Delhi.
5. Jagtap, S.S. and Chan, A.K. (2000) Agrometeorological Aspects of Agriculture in the Sub-Humid and Humid Zones of Africa and Asia. Agricultural and Forest Meteorology, 103, 59-72.
6. ingh, R.B. and Kumar, A. (2014) Vulnerability of Agriculture to Climate Change in Arid Regions: A Case Study of Western Rajasthan, India. Vulnerability of Land Systems in Asia, 77-90.
7. Goldman, R.H. and Block, S. (1993) Proceedings from the Symposium on Agricultural Transformation in Africa, ARAP II. Technical Report No. 137, Abt Associates, Cambridge.
8. Garb, Y. and Friedlander, L. (2014) From Transfer to Translation: Using Systemic Understandings of Technology to Understand Drip Irrigation Uptake. Agricultural Systems, 128, 13-24.
9. Environmental Protection Agency (EPA) (2013) Overview of Greenhouse Gases. Carbon Dioxide Emissions. http://www.epa.gov/climatechange/ghgemissions/gases/co2.html
10. De Wrachien, D. and Goli, M. (2015) Global Warming Effects on Irrigation Development and Crop Production: A World-Wide View. Agricultural Sciences, 6, 7.
11. Lindzen, R.S. (2007) Taking Greenhouse Warming Seriously. Energy and Environment, 18, 937-950.

12. Douglass, D.H., Christy, J.R., Pearson, B.D. and Singer, S.F. (2008) A Comparison of Tropical Temperature Trends with Model Predictions. International Journal of Climatology, 28, 1693-1701.
13. Tsonis, A.A., Swanson, K. and Kravtsov, S. (2007) A New Dynamical Mechanism for Major Climate Shifts. Geophysical Research Letters, 34, L13705.
14. Lindzen, R. (2011) A Case against Precipitous Climate Action. Energy and Environment, 22, 747-751.

CHAPTER 7

A Study on Planned and Applied Irrigation Modules in Irrigation Networks: A Case Study at Büyük Menderes Basin, Turkey

Cengiz Koç

Department of City and Regional Planning, Faculty of Architecture, Muğla Sıtkı Koçman University, Muğla, Turkey

ABSTRACT

In this study, irrigation modules calculated in planning and actualized operational stage of the irrigation networks are examined. Irrigation module used irrigation networks is a constant discharge parameter, meeting the needs of irrigation water requirement smonthly of crops in one hectare of irrigation area and it is a constant discharge flowing continuously for a month. Extent of the overlapping between the irrigation planning module and the operation module actualized during the operational stage of the irrigation network depends on changes in the cropping patterns, differences in the effects of field irrigation methods used by farmers on the capacity of the constructed system, the increases or decreases in the water demands depend- ing on the irrigation period, as well as the extent of sustainability according to the selected operation method. A2 irrigation area of Aydın plain irrigation network, locating in the Büyük Menderes basin, Turkey is selected as study area, with an area of 2500 ha. Irrigation planning module calculated for this network is q = 1.16 l/s/ha and it has been designed as per the supply demand operation method. For the study; actualized irrigation module in the operation stage has been compared with the planning irrigation module by using Excel software and taking parameters such as actual crop pattern and percentage distributions, actualized irrigated areas, irrigation networks and water distribution, water intake of irrigation networks which have been calculated without operation losses, as well as with 5%, 10%, and 15% operation losses. The July operation module calculated for the examined irrigation network generally conforms to the planning module, as it has received the values close to or below the value of irrigation planning modules.

Keywords: Irrigation Network, Irrigation Module, Operational Method, Irrigation Water, Turkey

1. INTRODUCTION

A well planned distribution network which has been turned into a project for irrigation purposes, should deliver the necessary amount of water needed within the desired timing, without causing problems to the operational and management requirements. Creating suitable projects of irrigation systems in accordance reveals some important results. These would be classified as health of the local people, wasted natural resources, water pollution, operational security and economic factors. Economic factors include the economic life of the irrigation system, the economic benefits of irrigation as well as the irrigation costs. Irrigation systems made with weak projects, where the water is not used homogeneously, causes water and the chemicals applied with water to go to waste. As a result, some areas are over-watered, while some areas are not sufficiently irrigated. Even though irrigation managers may apply the right amount of water, especially due to water stress in incompletely irrigated areas, there will be yield and quality reductions, and in extremely irrigated areas (due to the washing of water and chemicals), it can lead to reduction in efficiency and quality.

The most important investment share in irrigation projects which require large investments for agricultural fields is the construction of water transmission and distribution channels. In our country, the ongoing construction of irrigation networks by public institutions are the necessary facilities to meet water needs in a balanced manner for the development of crops in agricultural areas. The function of main, secondary (backup) and tertiary water transmission and distribution systems is to transport irrigation water to the plant from the water intake point. To examine the water source in detail and to acquire the necessary design parameters generally requires identifying the distribution areas of plants which are to be grown with plant density and to prepare the starting location of the networks. This case also encompasses the required structure of the type, capacity, the size of main channels, regional conditions, the water level, flow control, as well the shape and size of the dominant area [1] .

Differences in the crop patterns within the area served by the water resources can bring about irregularities in the total project area. Peak resource utilization for different parts of the project area may occur at different times. In this manner, the specified resource plans, mainly from resource requirements, can vary quite a lot and constantly uses monthly data source [2] . Since the plant density and their variations can cause differences in resource requirements, this initiates some flexibilities to be maintained in the irrigation program magnitude. Therefore, during the planning stages of a project, the normal estimate is that in order to operate effectively it takes many years. The resulting increase in resource requirements has to be covered by improvements in the irrigation operations of fields and improvement in water savings.

Different strategies are used in water distribution and control of irrigation networks throughout the world. Outdated irrigation network design types are observed in both developed and developing countries. Many years ago many irrigation projects in the United States, Western Europe and Australia were inefficient in terms of energy and water use, and required urgent modernization [3] . In most developing countries, an inflexible water distribution system is used where the flow rate and the frequency is fixed to the water supply. This application is still applied in various irrigation systems. In this scenario, water users do not have much flexibility in terms of continuity of water supply. They use it when water is available, regardless of whether plants need water or not, and thus they will be wasting water [4] . There are several types of water distribution methods available. However, fundamentally they can be divided into two basic factors, including demand and supply methods. Although formally the demand method is utilized in Turkey, the supply method is actually used as well in many projects. Irrigation networks are designed according to the crop patterns set in the planning report, but in practice they may show annual changes in crop patterns and create significant problems during the operational processes [5] . When physical, socio-economic and technical knowledge base is taken as a baseline; different distribution types, such as rotational, continuous flow and demand distribution types are implemented for years. The rotation program in irrigation channels are used in large irrigation schemes of developing countries such as China, Pakistan and India [6] .

This study, which was conducted in western of Turkey, is at the Aydin plains irrigation A2 project area which is located in the Büyük Menderes Basin. The investigation has focused on the overlapping of planning irrigation module with the operational irrigation modules calculated with crop patterns, without operational losses and with 5% to 10% operational losses considered. The planning irrigation module (q_P= 1.16 l/s/ha) is calculated with preplanned crop patterns, crop areas, the water consumption of plants and irrigation losses taken as a base material.

2. MATERIAL AND METHOD

2.1. Material

In this study, A2 İrrigation Area, which was constructed by Public Water Works and which is located in the Lower Great Menderes basin irrigation area in Aydin was taken. The study area is located at 37°51' north latitude and at 32°52' east longitude, in a region bounded near about 3 km south of the center of Aydın province and between Büyük Menderes River, which includes the area bounded by İncirliova Osmanbükü village in the section extending to the land (Figure 1). A2 gross irrigated area is 2500 ha, and the net area is 2158 ha. An important part of the parcels that make up the irrigation area has a surface area smaller than 10 decare. The effective crop pattern in the irrigated area is mainly cotton, while corn, grain, tomatoes, forage and II. product corn plants are cultivated. The cultivation of maize products in the planning stages consist of 48% cotton, 16% corn, 13% vegetables, 10% forage crops, fruit 6%, 4% II corn products and 3% maize seedlings. A2 Irrigation area in Aydin Plain is located in the flood area of the Büyük Menderes River. The flow rate of the Büyük Menderes River is very low and the materials which are transported in the river are clay-silt size. Therefore, geological structure of the region and the study area is made of alluvium material [7] .

The region is influenced by the Mediterranean climate and as a result winters are mild and very wet, and summers are hot and dry. Over the years the highest tempera ture recorded was 43.8°C, the lowest temperature recorded was 3.5°C, while the average temperature was 17.7°C. The average height above sea level of the irrigation area is 32.50 m, and the estimated annual rainfall is 656.1 mm.

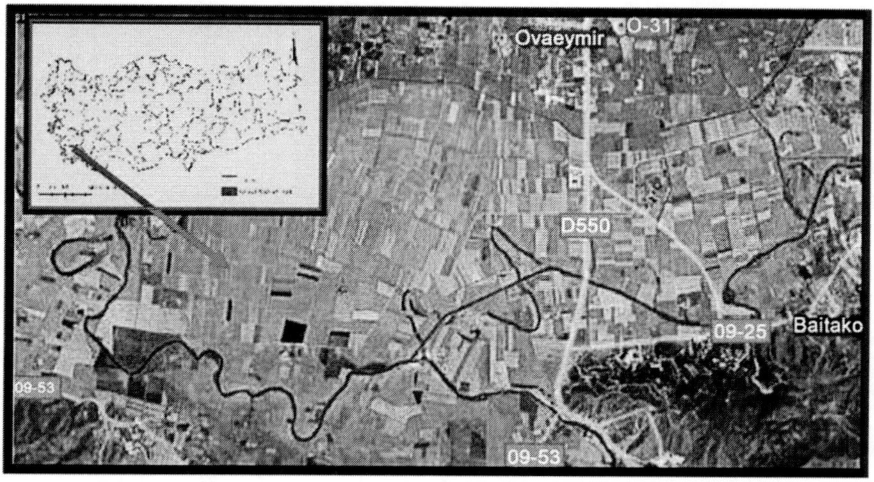

Figure 1. The location of A2 irrigation area in Turkey [8] .

The Water supply comes from Büyük Menderes River, and the storage facilities used are Adıgüzel and Kemer dams. Water consumption values of planned design and actual plant application in the irrigation field was calculated by Blaney-Criddle method [9] .

The total length of the A2 main irrigation channel in the study area is 11,351 m, 9007 meters comprises of classic trapezoidal channels, canals, and 2344 meters comprises of small channels. The main channel flow is 2780 m^3/s. 28 pieces of secondary (backup) channel was built with a total length of 56,395 m connected to the main channel A2. Tertiary total number of channels is 34 and the total length is 39,001 meters. All of the secondary and tertiary canals were built in the flume system. Table 1 gives the main irrigation channel sections and hydraulic properties for A2 main irrigation channel [10] . The irrigation efficiency of irrigation projects in A2 is 54%. Water transmission efficiency was determined as 90%, and the field irrigation efficiency was determined to be 60% [8] . Average washing water requirement needed for the research field was calculated as 0.24 [11] .

2.2. Method

2.2.1. Irrigation Planning Module

Irrigation module is the most important element of an irrigation network sizing. This is because the design capacity of the channel is calculated by the formula $Q = A_{net} \times q$ (l/s).The irrigation module which depicts the need for water is determined according to parameters such as irrigation of the field;

cropping patterns, planting area rate, evapotranspiration, climatic conditions, soil properties, irrigation losses (transmission + field irrigation losses); and it is defined as the amount of water in liters per second given in a continuous manner for 24 hours and 30 days to the terrain during the month where maximum demand occurs.

Table 1. Hydraulic properties and cross sections pertaining to A2 main irrigation channel.

Cross section No	Slope (j)	Flow Rate (Q) (m^3/s)	Km							Base Width (m)	Water Height (m)
1	0.0001	2.778	0	+	025	-	1	+	874	d	1.28
2	0.0001	2.692	1	+	874	-	2	+	607	2.20	1.26
3	0.0001	2.608	2	+	607	-	4	+	000	2.20	1.24
4	0.0001	1.938	4	+	000	-	5	+	536	1.60	1.00
5	0.0001	1.569	5	+	536	-	7	+	580	1.40	0.94
6	0.0001	1.161	7	+	580	-	8	+	000	1.40	0.85
7	0.0001	1.161	8	+	000	-	9	+	007	1.20	0.85
8	0.0001	0.820	9	+	000	-	11	+	351	Type 1000 Cahnnel (Exit Type)	

If the irrigation water which flows for a month is due to the need of the farming irrigation, then this module is called the farm irrigation scheduling module. If this water which flows for a month is due to diversion of the irrigation water need, then this module is called the irrigation planning module. The irrigation planning module is calculated as 1.16 l/s/ha for the study area [12] .

Planned farming irrigation module (l/s/ha):

$$\frac{\mathrm{Ut} \times 10^4}{(30 \text{ veya } 31) \times 24 \times 60 \times 60} = \mathrm{qç}$$

Planned irrigation module (l/s/ha):

$$\frac{Uk \times 10^4}{(30 \text{ veya } 31) \times 24 \times 60 \times 60} = qp$$

Ut × 10^4: Monthly farm irrigation water requirement per hectare (mm)

$Uk \times 10^4$: Monthly diversion irrigation water requirement per hectare (mm)

2.2.2. Operational Irrigation Module

Operational irrigation module is the function of the irrigation water to be utilized per the real crop pattern, the water with and without losses, monthly processes and the actual irrigated area in the irrigation region.

$$\frac{W_k}{(30 \text{ veya } 31) \times 24 \times 60 \times 60} \times 1000 = S$$

Operational irrigation module (l/s/ha)

$$\frac{S}{A_f} = qi$$

W_k, amount of water intake to the channel (m^3/month); S, amount of water intake to the channel (l/s); A_f, actual irrigated area in the irrigation region; qi, operational irrigation module for that month (l/s/ha)

3. STUDY DATA AND DISCUSSION

In this research, the conformity of the irrigation module which is calculated according to actual plant patterns, the irrigation module calculated per intake water without losses, the irrigation modules as per 5% to 10% losses during operation, versus the planning irrigation module (q_P = 1.16 l/s/ha) considered crop pattern in the planning stages of the irrigation network, planting area, evapotranspiration and irrigation losses (transmission + field) is calculated. It can also be said that the irrigation planning module and the actual irrigation module has been compared for the project area. After the operation has started in the project area, the calculated irrigation module per the years 1999-2014 for the month of July (q_T) 1.01 to 1.15 l/s/ha, for the month of August (q_A) is 0.61 to 1.09 range is realized (Figure 2). During the study, the years of 2007-2008 have not been taken into consideration in the calculation of the irrigation module as they have been very dry years with heavy drought. The highest operational irrigation has taken place in July 2002 within all the years studied. However, this value is below the irrigation planning module. In August 2002 it has been calculated as 1.09 l/s/ha. Planning for the highest plant irrigation module is calculated based on water consumption in July, as it is the time where the highest water consumption takes place. However, in order to make an accurate assessment, the irrigation modules for August is also calculated.

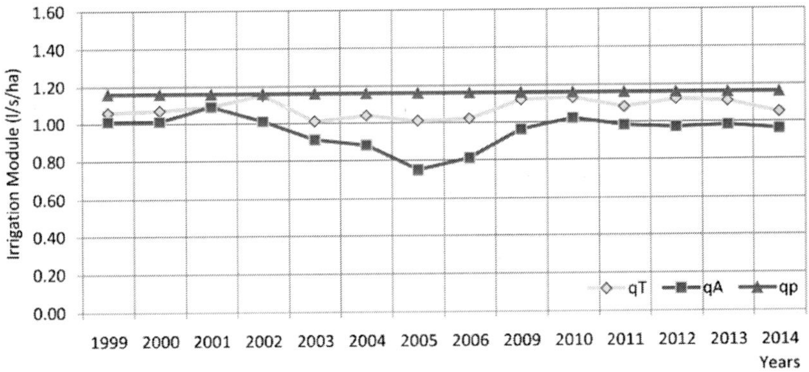

Figure 2. Irrigation module calculated as per crop patterns.

It is observed that there is no problem in terms of irrigation module values, which have been calculated according to the planned and actual cropping patterns. However, having a little higher planned module versus the realized cropping patterns has caused the channel capacity to be increased, but in terms of construction economics, this has led to only very few negative aspects. However, since planning for irrigation module is based on the July irrigation module, the 1.15 and 1.16 l/s/ha values have an overlap with a ratio of 0.99, thus causing construction economics to have negligible negative effect.

Without taking into account the channel operational losses (the main channel drain outlet, secondary and tertiary canal breaks), and under the assumption that all of the water taken into the channel is utilized in irrigation; the calculated irrigation modules for the year examined is found for the month of July as (q_T) 1.07 to 1.25 l/s/ha, and for as August (q_A) 0.98 to 1.11 l/s/ha (Figure 3). Since the calculations are conducted with July planning module taken into account, the July 2001 irrigation channel module with value of 1.25 l/s/ha is slightly above the planning module. The irrigation modules for the years of 1999-2000-2001-2002-2003-2009 for the month of July have given values above the value of the planning module. However, it can be stated that when some amount of water taken into the channel drain outlets, and also to secondary and tertiary canal breaks, then the high July modules will not be creating a problem. The highest module that took place in August for the years under the highest module is realized as 1.16 l/s/ha, which is below the planned irrigation module. Since the water taken into the channel can't be utilized without some losses, then it won't be a problem for the irrigation modules calculated under these conditions.

The operational irrigation module calculated with the assumption of 5% operational losses in waters taken into the channel, has shown a variation of 1.02 - 1.19 l/s/ha with (q_T) in the month of July and (q_A) has shown a variation of 0.90 - 1.12 l/s/ha in the month of August (Figure 4).

For the examined period, just in July 2011, the operational irrigation module with the value of 1.19 l/s/ha has been realized as 1025 times more than the planning module. This value can be omitted since it is not an amount which can create a trouble for the operations. When the calculated data is taken into account, there is no problem observed between the calculated irrigation module at planning stage and the irrigation module calculated with 5% operational losses. Since the irrigation network is designed as per the operational demand method, the problems that may crop up can be eradicated by reflecting the current demand to the irrigation planning during the operation of the system. Furthermore, the changes in the plant patterns during the agricultural farming at the irrigation area may prevent demand pileup.

The operational irrigation module which has been calculated with the assumption of 10% operational losses in waters taken into the channel has shown a variation of 0.97 - 1.10 l/s/ha with (q_T) in the month of July. Furthermore, (q_A) has shown a variation of 0.91 - 1.06 l/s/ha in the month of August (Figure 5).

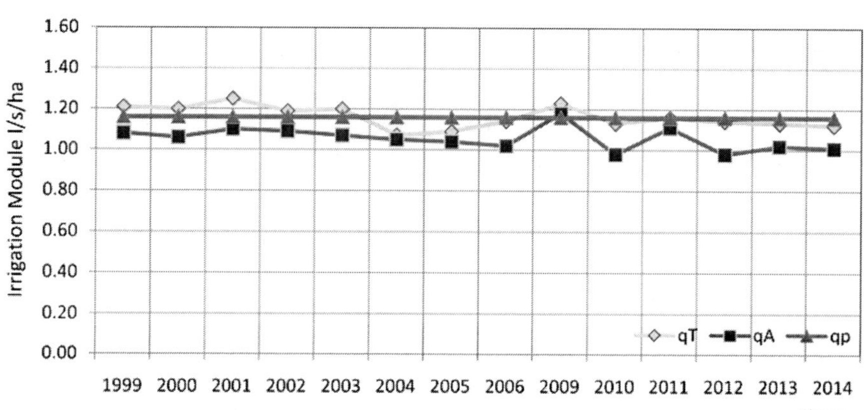

Figure 3. Module calculated as per ıntake of water without losses.

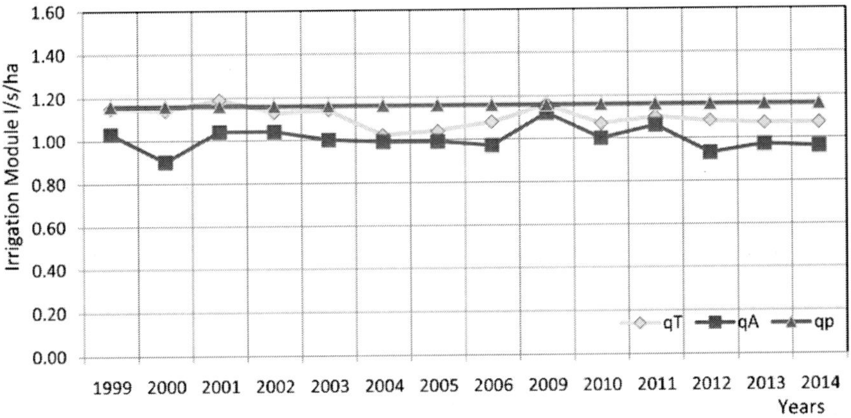

Figure 4. Calculated module with 5% losses during water ıntake.

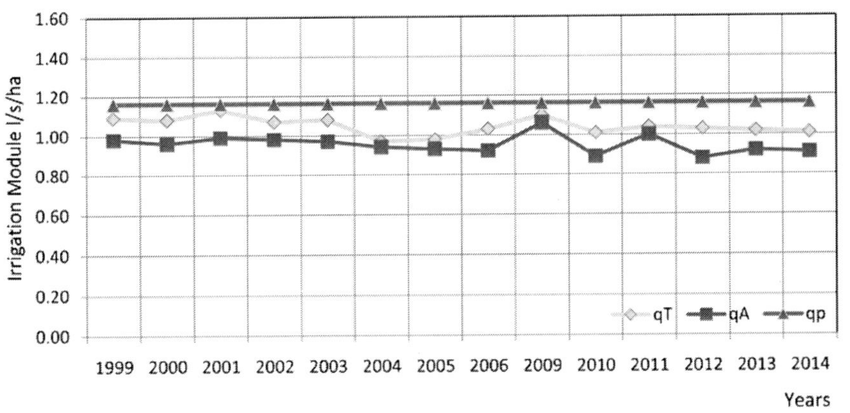

Figure 5. Calculated module with 10% losses during water ıntake.

The calculated operational module values have been realized below the planning module values. As per this assumption, there is no problem observed with irrigation operations. However, a design which is incorporated with this assumption can reduce the construction costs to a certain extent. When these assumptions are taken into account, it becomes necessary to have a more planned water distribution, which reduces the channel losses. The planning by taking into account water intake channel capacity can remove problems related to module and design flow.

4. CONCLUSIONS AND RECOMMENDATIONS

There are no problems found in between the calculated irrigation module at the planning stage and the operational irrigation module calculated with the actual cropping patterns, the intake water taken into the channel without losses as well as the water intake with 5% and 10% operational losses. In this context, there is also no problem found in terms of the design flow of the entire channel system in the irrigation network. This case leads to the conclusion that the work done during the planning stages of the irrigation network were very attentive and worth appreciating. This reinforces the statement once again that good planning always leads to a good project, and a good project creates the good conditions for its operation. The overlapping of planning values with the operational values shows the success of planning and it allows for the realization of the expected benefits from the system within the expected time frame.

Whether the calculated operation irrigation module is lower or higher than the planned irrigation module carries an importance related to irrigation system operations, system design (capacity, dimensions) and for construction costs. If it is lower than the planned irrigation module, then this can cause a lower amount of water to be dispersed to unit area per unit of time. This situation will mean that the crops growing in the project area will not have sufficient amount of irrigation water. If it is higher than the planning module, then this causes more water to be released to the unit area per unit time. This situation can be caused by the change of crop cultivation pattern over time in the project area or due to reduction of agricultural lands caused by various reasons. This condition can limit the irrigation of plants grown within the project area to be lesser then the designated intervals. Therefore, it is not desirable for the operational irrigation module to have either higher or lower values then the planned irrigation module. In these types of situations, in accordance with the overlap ratio, the problems can be eliminated by taking precautions within the flexibility of operational irrigation programs and demand management.

Since the irrigation planning module is related to plant patterns in the project area as well as to the crop patterns, its density and crop fields; it is essential to carefully identify the best crop pattern for the project area and it must be determined close to real conditions. Local cultivation habits and plants which don't conform to market conditions should not be taken into consideration.

Since the planned irrigation system directly determines the system design and in this context, it affects the construction costs which require careful field and office work. Specifically, more parameter based on the calculation of water consumption values for plant-cultivating plants considered in the project area should be encouraged to use the empirical method. Farming habits of the person conducting agriculture and socio-economic structures in the region should be taken into consideration for the determination of the efficiencies of field irrigation method and efficiency of irrigation water application. These are the direct project criteria, and thus this must be executed kind of thinking must be abandoned and better suggestions as per the region and conditions present.

Implementation of land consolidation and land leveling work in the field of research, and to design the tertiary intervals at a maximum range of 200 m in this context provides a positive contribution to the overlap of the planning and operation of the module. Therefore, the implementation of land consolidation and land leveling before the construction of irrigation networks in the irrigation project area will increase the success of the project [13] .

The calculation of operations feedback module for irrigation scheduling criteria would provide feedback which will assist in the change of planning or project criteria or it will contribute to making it better. As in every business system, feedback and performance data must be calculated, it should be evaluated, and then it should especially be compared with the planning criteria.

The irrigation planning module which is calculated by taking into account certain assumptions which are true, attentive, and related to climate change and business conditions, actually is a module utilized for the determination of the direct channel capacity (design flow) and for better evaluation of the available financial resources for the construction of irrigation networks. Thus, it will have an effect on irrigation network construction economy and it will help to provide the expected benefits for the desired duration.

The changes (reduction-increase) in the actual irrigated areas in the study area affect the operational irrigation module, and this situation can cause discomfort or as an alternative provide relief to the irrigation network management.

The accurate and precise measurement of irrigation waters of the channel will allow the calculation of a healthier operational module value. Therefore, it will contribute to achieving the correct and reliable data for the frequency of water intake by remote sensing methods.

The appropriate irrigation module has to be calculated as depending on whichever farm irrigation methods will be implemented in the irrigation of the area (border, furrow, sprinkler, drip). If there is going to be a change in the method, then a new module must be recalculated. Since the water transmission and field application efficiency of irrigation can change, as a result the irrigation module will also vary.

CITE THIS PAPER

Koç, C. (2016) A Study on Planned and Applied Irrigation Modules in Irrigation Networks: A Case Study at Büyük Menderes Basin, Turkey. Computational Water, Energy, and Environmental Engineering, 5, 112-122. http://dx.doi.org/10.4236/cweee.2016.54011

REFERENCES

1. Koc, C. (1990) Design and Operation of Irrigation Systems. General Directorate of State Hydraulic Works (DSI). Ministry of Forestry and Hydraulic Works. DSI Technical Bulletin, 72, 28-34.
2. Burt, C.M. (1999) Essential Water Delivery Policies for Modern On-Farm Irrigation Management. In: Irrigation Scheduling: From Theory to Practice. Proceedings ICID/FAO Workshop, September 1999, Rome, Water Report No. 8 FAO, Rome, Italy.
3. Plusquellec, H. (2002) How Design Management and Policy Affect the Performance of Irrigation Projects. Emerging Modernization Procedures and Design Standards. FAO. Bangkok.
4. Reddy, M. (2004) Technical Management of Irrigation Systems. In: Operation of Irrigation Conveyance Systems. International Capacity Building Program on Participatory Irrigation Management. INPIM Train Module. 5-15 December 2004, Menemen.
5. Tekinel, O. (2001) Participatory Approach in Planning and Management of Irrigation Systems. Advanced Short Course on Appropriate Modernization and Management of Irrigation Systems, 2-15 September, 2001, Kahramanmaras University, Turkey.
6. Maffiotra, S.P. (1982) Warabandi System and Its Infrastructure. Central Board of Irrigation and Power. New Delhi.
7. DSI (State Hydraulic Works) (2002-2006) Report of Planning Irrigation Water Distribution in Aydin Plain Network. General Directorate of State Hydraulic Works (DSI). Ministry of Forestry and Hydraulic Works. 21. Rural Directorate of State Hydraulic Works. Aydin. 125 p.
8. Google Earth Pro 6.2.1.6014 (beta) (2011) Eagan Region, ON Turkey 37°51' North Latitude and at 32° 52' East Longitude Borders and Labels; Places Layers. NOAA, Digital Globe 2013.
9. Koc, C. (1998) Buyuk Menderes Havzasi Sulama Sebekelerinde Organizasyon-Yonetim Sorunlarive Yeni Yonetim Modelleri Uzerine Arastirmalar. Doktora Tezi. Ph.D. Thesis, (Unpublished), Ege Universitesi Ziraat Fakultesi. Tarimsal Yapilarve Sulama Anabilim Dali, Izmir, 183 s.
10. DSI (State Hydraulic Works) (1992) Aydin Plain Network Facility Introduction Sheet. General Directorate of State Hydraulic Works (DSI). Ministry of Forestry and Hydraulic Works. 21. Rural Directorate of State Hydraulic Works. Aydin. 105 p.
11. Koc, C. (2008) The Environmental Effects of Salinity Load in Great Menderes Basin Irrigation Schemes. Environmental Monitoring and Assessment, 146, 479-489.
12. DSI (State Hydraulic Works) (1984) Planning Report of Aydin Plain Irrigation. General Directorate of State Hydraulic Works (DSI). Ministry of Forestry and Hydraulic Works. 21. Rural Directorate of State Hydraulic Works. Aydin. 450 p.
13. Koc, C. (2006) An Investigation on Water Delivery Methods Actualized in Practice and Suggested in Project of Irrigation Schemes. International Symposium on Water and Land Management for Sustainable Irrigated Agriculture in Memory of Prof. Dr.

Osman Tekinel. Sütcü Imam University, RIHN, CIHEAM, TüBITAK, Cukurova University, Adana, Abstract Page 289 Turkey.

CHAPTER 8

Diurnal Effects on The Efficiency of Drip Irrigation

Warwick R. Adams[2] · Ketema T. Zeleke[1,2]

[1] Graham Centre for Agricultural Innovation (An Alliance Between NSW Department of Primary Industries and Charles Sturt University), Wagga Wagga, Australia

[2] School of Agriculture and Wine Sciences, Charles Sturt University, Wagga Wagga, Australia

ABSTRACT

Drip irrigation is widely used in viticulture and other field crops. However, there has been little evidence to date regarding the net effects of diurnal factors on drip irrigation and whether or not irrigating early morning compared to in the afternoon changes the water use efficiency. A field experiment was undertaken to investigate the diurnal factors and to quantify any effects on irrigation efficiency. It showed that, for soil depths >15 cm, drip irrigation efficiency is not affected by the time of day of watering—there were no diurnal effects and the results did not support a long-held belief that watering in the heat of the day should be avoided. At shallow depths (from 0 to 15 cm), afternoon irrigation actually resulted in a higher moisture content in this layer, with some of that difference persisting over the following 72 h. This finding also has the potential to be applied to irrigation scheduling for salad vegetables and similar shallow-rooted crops where, by irrigating in the afternoon rather than early morning, water usage could be reduced while at the same time potentially increasing the yield.

INTRODUCTION

The majority of winegrowing regions worldwide are located in semi-arid and arid regions with low rainfall, meaning that irrigation is required. Climate change is expected to result in higher temperatures and higher evapotranspiration rates in those regions, putting additional pressure on limited water supplies for domestic use, industry and agriculture (Monaghan et al. 2013). Dwindling water supplies, increasing drought frequency and uncertainties associated with a changing climate mean that the irrigated viticulture sector needs to improve water efficiency.

Trickle ('drip') irrigation, a widely utilised irrigation method in viticulture, involves application of water at a slow rate from regularly spaced point sources (or emitters) above the soil surface. It is a relatively efficient means of irrigation, especially when combined with techniques such as regulated deficit irrigation, partial rootzone drying, subsurface drip irrigation and application of mulches. However, one aspect that requires further investigation is the potential for water savings by irrigating at only particular times of the day, to take advantage of diurnal factors. A long-held belief, dating back to the fourteenth century, has been that 'you should not water in the heat of the sun' (Power 1928) on the basis that this wastes water, but the evidence for any such claim needs to be tested.

The efficiency of irrigation depends on the losses which take place during and following the irrigation—losses above-ground (wind drift and evaporation losses or WDEL), at the soil surface, from plant root uptake (transpiration) and from percolation within the soil. Under windy conditions, a considerable portion of the spray droplets from sprinkler systems can be carried away from the spray area and lost. Yazar (1984) found that WDEL ranged from 1.5 to 16.8% of the total sprinkled volume, with wind velocity and vapour pressure deficit being the most significant factors. Importantly, they found that air temperature is not a factor itself and can be omitted and this finding was supported by Playán et al. (2005). In order to minimise WDEL, night sprinkler irrigation has long been practised in many areas of the world, as night wind speeds are generally lower and humidity is higher. For drip irrigation on the other hand, because of the small vertical distance between the emitter and the soil surface, the low droplet velocity and relatively large droplet sizes, losses from WDEL are minimal, meaning that overall losses will be from other causes.

The soil physics of the coupled transport of water, water vapour and heat in a soil are well understood for both isothermal and non-isothermal conditions, as are the mechanisms of evaporation. Soil water movement in unsaturated soils was initially studied under isothermal conditions based on Richards' 1931 equation (Deb et al. 2011). Early non-isothermal studies demonstrated that moisture movement in response to thermal gradients through an unsaturated soil occurs mainly in the vapour phase. Philip and De Vries (1957) subsequently developed a moisture migration model (PDV) within homogeneous soil temperature profiles to account for the effects of temperature gradients on moisture migration. Nearly 60 years later, the PDV model still remains the basis for most soil–atmosphere continuum modelling, although today's version of the that model has been modified through numerous studies such as those by Milly (1982, 1984, 1986), Sophocleous (1979), Braud et al. (1995), Shurbaji et al. (1995) and Nassar and Horton (1997).

Because of the complexity of the coupled water and heat transport in the unsaturated zone and the difficulties associated with field measurements, especially near the soil surface, numerical models are often used to analyse these processes (Deb et al. 2011), along with modelling tools such as Hydrus (PC-Progress 2015). However, there has been little focus on the dynamics which occur during the infiltration of irrigation water where the conditions are non-isothermal, vary by the time of day due to diurnal factors and also undergo quite rapid changes in the soil heat and moisture fluxes as the irrigation wetting front proceeds downwards through the soil. The result is that there is little evidence to date regarding the net effects of diurnal factors on drip irrigation, especially whether irrigating overnight or early morning, compared to in the afternoon, will result in higher water use efficiency, either during the irrigation period itself or over subsequent days.

A field experiment was undertaken to investigate the diurnal factors and to quantify any effects both during and after water application by drip irrigation. The findings were also applied to the case of drip irrigation of shallow-rooted crops such as salad vegetables.

MATERIALS AND METHODS

Location

The experimental site was located in a vineyard at Belgrave Park Pty Ltd (Bega, NSW, Australia), at latitude 36.41°S, longitude 149.86°E and an elevation of 150 m above sea level. The climate of the site, from a viticultural perspective, is classified as warm and moderately maritime, with a mean January temperature (MJT) of 20.6 °C (Bureau of Meteorology 2013b) and a summer-dominant rainfall pattern with an annual rainfall of 960 mm (Figure 1). The annual pan evaporation for the area is 1360 mm (Bureau of Meteorology 2013a), with the aridity classified as low.

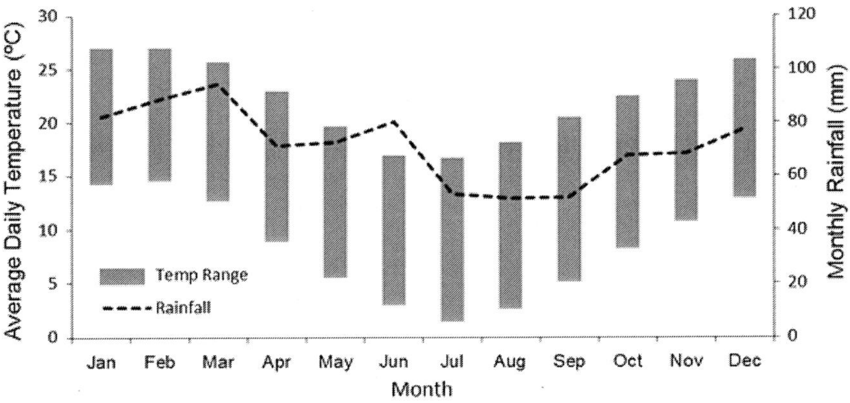

Figure 1. Monthly temperatures and rainfall totals, Bega, NSW, Australia

The topology of the area has rolling to undulating low hills on granodiorite (Tulau 1997). The soil comprised moderately deep, moderately well-drained leached yellow earth with:

A Horizon:

Sandy clay loam, 30 cm deep, with a pH of 6.5.

B Horizon:

Decomposed granite, orange to light brown, structured light clay, with the clay content increasing with depth, partly dispersive and with a variable, acidic pH.

Experimental design: Part A

The objective of Part A of this experiment was to measure the effect of time of day on the volume of drip irrigation water that is needed to infiltrate and wet the soil (to various soil depths) and the main features of the design are shown in Figure 2.

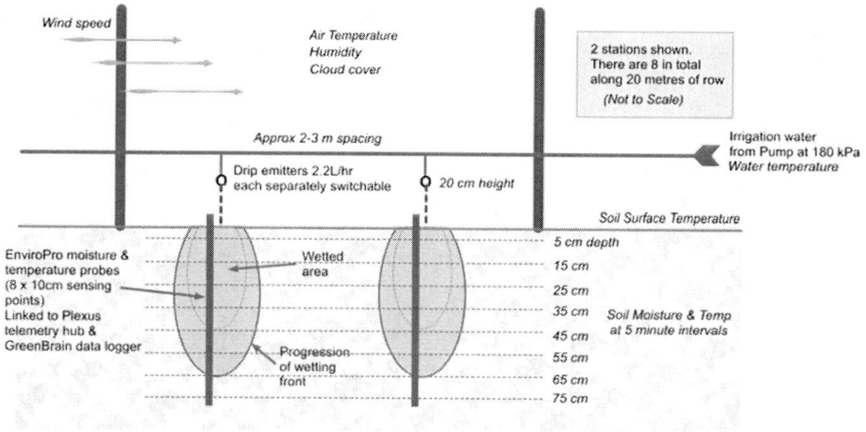

Figure 2. Experimental design—main features of the drip irrigation set-up and soil sensor probes

The experimental site comprised bare soil, in order to eliminate the effect of plant transpiration, but retained all other vineyard soil characteristics such as grapevine roots, other plant roots and worm channels. The experiment used a series (8) of above-ground 2.2 L/h low-pressure Toro Turbo Plus II drip emitters (Toro Australia Pty Ltd. 2003) spaced along a short section of vineyard row at between 2- and 3-m intervals, in order to prevent any interaction between the wetting zones. The water supply pressure was regulated to 180 kPa, which was well within the pressure compensation range of the emitters.

Each emitter had an 80 cm long Enviropro subsurface soil moisture and temperature probe (Measurement Engineering Australia 2013) buried vertically alongside, providing for measurements at eight soil depths (5, 15, 25, 35, 45, 55, 65 and 75 cm). The probes were connected to a Plexus mesh-networking radio system, comprising 4 field stations and a hub (Measurement Engineering Australia 2015) which in turn transmitted the readings to a web-based Green Brain database. This provided both a visual interface to view real-time temperature and moisture data from each station, as well as the ability to export raw data to a CSV file. The Enviropro probes measured volumetric soil moisture content from 0 to 50% with an accuracy of ±2% and temperature from −10 to 60 °C with an accuracy of ±1 °C. The probes were preloaded with a calibration based on a standardised calibration media (fine sand) (Apcos Pty Ltd 2015). Even though the values returned in other soil types would differ in accordance with the average particle size distribution in the soil, this was not a consideration for this experiment where all measurements were differential (A.M. compared to P.M.). The various temperatures (ambient, soil surface, irrigation water) and relative humidity were also recorded.

The system was set up with a data-logging interval of 5 min. All emitters and their associated probes (stations) were positioned within a total horizontal distance of 20 m in order to minimise any differences in soil characteristics or profile. There was also a small horizontal separation of 10 cm between the emitter and probe centrelines at each station in order to prevent any irrigation water running directly down the outside surface of the probes.

Since the hypothesis was that the volume of water required would depend on the time of the day that the irrigation took place, the experiment comprised two separate irrigation start times (factor levels) with each factor level applied to a random subset of 4 of the 8 stations. The respective irrigation start times were as follows:

(a)Early morning, starting at 06:00 (Eastern Standard Time). This is the period after the soil fluxes have stabilised overnight, but before the surface and shallow soil temperatures begin rising (Figure 3).

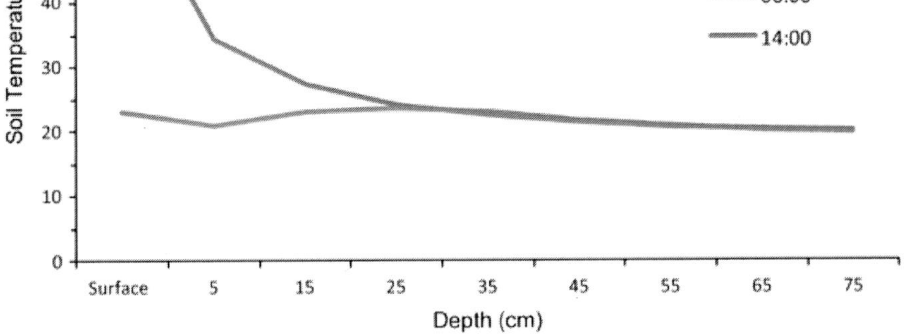

Figure 3. Typical soil temperature profile, without irrigation, at 06:00 and 14:00 (representative day during December 2015)

(b)Early afternoon, starting at 14:00. This is approximately the period of day where the soil surface temperature and shallow soil temperature gradients are highest, sometimes exceeding 6 °C/cm between the surface and 5 cm depth at this site.

The measurement was specifically of the volume of water needed to be applied for the wetting front during irrigation to reach 35 cm depth, at A.M. compared to P.M., calculated as:

- A.M. Volume Required = Emitter Discharge Rate × A.M. Time to reach 35 cm depth
- P.M. Volume Required = Emitter Discharge Rate × P.M. Time to reach 35 cm depth and the null hypothesis was
- A.M. Volume Required = P.M. Volume Required

The 35-cm depth limit was chosen because there is very little diurnal temperature fluctuation below that depth, for example just 1.1 °C at 35 cm and even less at greater depths (Figure 4). The irrigation was run for a fixed period (90 min), long enough to ensure that field capacity (FC) had been reached to at least 60 cm depth, but limiting the volume of water applied in order to prevent excessive percolation to greater depths.

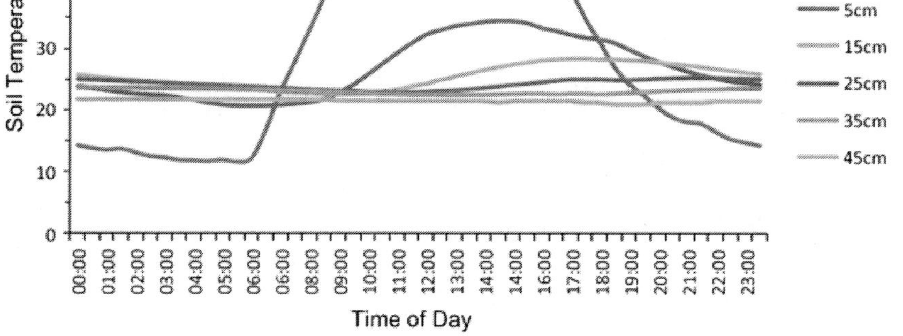

Figure 4. Diurnal fluctuations of site soil temperature at various depths (31 December 2015)

If there was a difference in the volume of water needed, then that could be attributed to differences in the losses (of any type). In order to eliminate most of the uncontrolled variables (for example surface evaporation), the experimental design was bounded by both space and time—in essence it comprised a cylindrical soil column 20 cm diameter and 35 cm deep, with the primary hypothesis being tested over an irrigation period of just 90 min.

Each replication comprised all 8 stations, with some being watered at 06:00 (A.M.) and the remainder at 14:00 (P.M.), according to a predetermined randomisation (Table 1). At the start of each run (A.M. or P.M.), the appropriate emitters were turned on or off. Replications only took place on days where the soil moisture content at 5 cm depth was in the range of 13–19% v/v, that is in the 'dry' moisture range and equivalent to approximately −200 to −400 kPa water tension for a loam soil (van Genuchten 1980; Measurement Engineering Australia 2016).

Table 1. Randomised block design for the experiment (comprising eight irrigation stations, with two factors)

Run	Station number								Experiment part
	1	2	3	4	5	6	7	8	
1	P.M.	P.M.	A.M.	P.M.	P.M.	A.M.	A.M.	A.M.	A
2	A.M.	A.M.	P.M.	A.M.	A.M.	P.M.	P.M.	P.M.	A
3	P.M.	A.M.	P.M.	P.M.	A.M.	A.M.	P.M.	A.M.	A
4	A.M.	P.M.	A.M.	A.M.	P.M.	P.M.	A.M.	P.M.	A
5	A.M.	P.M.	P.M.	A.M.	A.M.	P.M.	A.M.	P.M.	B
6	P.M.	A.M.	A.M.	P.M.	P.M.	A.M.	P.M.	A.M.	B
7	A.M.	A.M.	P.M.	A.M.	A.M.	P.M.	P.M.	P.M.	B

Part of the experiment was to also measure moisture content (at each station and depth) over the ensuing 72 h as redistribution and drying took place. To ensure consistency, the volume of irrigation water applied at each station during each run was kept constant at 3.3 L/station (equating to the 90-min timed application).

Vertical infiltration time as a measure of losses

The experiment measured any difference in the total water losses between A.M. and P.M. by measuring the amount of water needed to be applied to wet the soil to a depth of 35 cm. That is, if the losses were different, then the water volume would be different. Since the experiment was an investigation of diurnal effects only, there was no requirement to identify the sources of losses nor to apportion them. The application rate at each emitter was constant (2.2 L/h), so the volume of water applied was directly related to the elapsed time since start of irrigation.

Subsequent redistribution and drying

Part of the experiment was to compare the soil moisture content reached at various depths at the end of each irrigation run, A.M. and P.M. Then to investigate the decline in moisture content at each of those depths over the subsequent 72 h, as soil moisture distribution and drying took place. In order to achieve this, it was important that each station received the same amount of water during each irrigation in order to establish a consistent baseline. This was the purpose of having a timed 90-min irrigation, which applied 3.3 L of water at each station. At the end of each irrigation period, the soil moisture contents were at their peak, and over the ensuing 72 h period the moisture content at each station and soil depth continued to be measured at 30-min intervals.

Variations in the antecedent moisture content (before irrigation) would affect the peak moisture values reached at the end of the irrigation so, in order to eliminate this effect, (a) all measurements were of the *additional* soil moisture content (that is the current moisture content at any point in time, less the moisture content at the start of irrigation) and (b) each irrigation was only performed when the soil water content at 5 cm depth was in the range of 13–19% v/v.

Other variables

Other variables in the experiment included evaporation, sunlight intensity and cloud cover, ambient temperature, relative humidity, wind speed, rainfall, evaporation, plant transpiration and the temperature of the irrigation water. Of these, plant transpiration and the effect of rainfall were excluded through the experimental design.

(a)*Soil characteristics* Despite the experimental row being restricted to a horizontal distance of <25 m, the soil characteristics would still vary somewhat between stations (for example because of rocks, worm channels, plant roots and the like) and the effect of these variations was mitigated by the use of a paired design so that the A.M. and P.M. applications could be compared for each separate station.

(b)*Antecedent water content* In a soil, plant available water (PAW) cycles between a lower point of permanent wilting point (WP) and upper point of FC. The soils undergo three stages of drying (Hillel 1980) with Stage 3 (being very dry soils) resulting in a much lower rate of evaporation from the surface. For this experiment, the initial near-surface conditions prior to each irrigation were kept within a soil moisture range of 13–19% v/v (with a mean of 14% v/v at 5 cm depth), that is in the Stage 2 range. At depths of 15 cm and greater, the antecedent moisture content was in the range of 76–86% of FC. (The measured FC's at the experimental site and the mean moisture contents at the start of irrigation runs are shown in Table 2). The irrigation raised the moisture level to FC or greater at each of these depths.

Table 2. Field capacities and initial soil moisture contents at various depths at the experimental field site

Horizon	Depth (cm)	Soil type	Field capacity	Soil moisture at start of irrigation (mean of all runs)	
			Soil moisture (vol%)	Moisture (vol%)	Proportion of FC
A horizon	5	Sandy clay loam, pH 6.5	30	15	0.46
	15		50	39	0.76
	25		51	43	0.83
B horizon	35	Structured light clay, variable acidic pH	49	42	0.86
	45		49	–	–

According to Jury and Letey (1979), a large number of separate experimental studies of water transport in response to thermal gradients have shown that the movement is independent of water content over a wide range. This meant that the antecedent water content was not important to this experiment.

(c)*Evaporation* In this experiment, because of the short 90-min duration of irrigation, evaporation during irrigation itself could be ignored (especially as the soil surface temperature cooled rapidly to 35 °C or less at the drip area within 40 min of the start of irrigation). For example, the experimental site had an annual pan evaporation of 1360 mm. For a summer afternoon, compared to the same period in the morning, this equated to an order of 0.25–0.5-mm difference in evaporation during the irrigation period—or just 1–2% of the applied 3.3 L of water (being approximately 26 mm).

Experimental design: Part B

Part A of this experiment focussed on downward infiltration and the effect of diurnal factors on this. However, the effectiveness of drip irrigation in a field situation is three dimensional—including both downward (vertical) and horizontal (lateral) wetting. Thus, the size and shape of the wetted profile under each emitter are an important factor and this was investigated as Part B of the experiment, by measuring the horizontal expansion of the wetting front at various depths and times.

In Part B, at each of the 8 experimental stations the emitter was repositioned so that there was a horizontal separation of 25 cm from the drip line to the centreline of the Enviropro probe. In this way, it was possible to measure how long it took for the wetting front to reach a radius of 25 cm at various soil depths (5, 15, 25, 35 and 45 cm). There have been various studies into wetting patterns and their calculation. In particular, the equations of Philip (1984) have been found to give good predictions by Revol et al. (1997) and Thorburn et al. (2003), and a useful computer application was developed by Cook et al. (2003).

For the purposes of this experiment, the horizontal infiltration distance was calculated by approximating the wetted soil profile as a prolate ellipsoid, in which case the lateral radius increases as the cube root of the elapsed irrigation time. In this manner, the dripper profile at any point in time (3 h was used in the experiment) can be derived from the time taken to reach a radius of 25 cm (at various depths) and the results correlate well with the solution of Revol et al. (1997). Since any approximations apply equally to A.M. and P.M., the profiles can be directly compared and contrasted.

Data analysis

A randomised block design (Quinn and Keough 2002; McKillup 2012) provides a means of isolating treatment effects from spatial variations. In this experiment, each station within the experimental block comprised a drip emitter/probe pair, and the treatment factors—A.M. or P.M.—were applied randomly, with each station receiving one treatment. Hence, a complete experimental replication comprised all eight stations, with half (randomised) receiving A.M. irrigation and the other half P.M. irrigation (Table 1). There were four replications conducted (each with a different randomisation), and an analysis of variance (ANOVA) was run on the 32 result sets.

A paired sample *t* test (McKillup 2012) is intended for experiments where the same variable or variables are measured on each experimental unit under two different conditions. The greatest spatial variation between the stations in this experiment was due to soil characteristics, vine roots, worm channels, rocks and antecedent moisture content. By using a paired *t* test, comparing the difference of the results between consecutive replications for each unit, elimination of the soil-related differences could be achieved. The null hypothesis in this case is that the difference between the values of the two replications at each station is zero.

Inclusion of such a paired *t* test was made possible by having each alternate replication pattern as the inverse of the immediately preceding one; hence, for example, if the first replication irrigated station 3 at 06:00, then the second replication irrigated it at 14:00, and this design is shown as part of Table 1.

All data analyses were performed using Minitab© statistical analysis software (Minitab Inc. 2016).

RESULTS AND DISCUSSION

Part A: Irrigation efficiency

Diurnal soil temperatures

For a typical day during the experiment, the fluctuations in soil temperature over a 24-h period are shown in Figure 4, with corresponding values in Table 3. Note that the experimental runs were only made on sunny days where there was little or no cloud cover (the 5th, 14th, 24th and 30th December 2015).

Soil temperatures during irrigation

The mean temperature of the irrigation water being applied was in the range of 18 °C (at 06:00) and 21 °C (at 14:00) and, as it moved downward from the soil surface, it transferred heat to or from the soil at each depth, subsequently transporting it to lower depths.

At 06:00, the surface soil temperature was beginning to rise and the effect of applying 18 °C irrigation water was simply to slow the rate of heating, as shown in Figure 5.

Table 3. Magnitude of daily soil temperature fluctuations at various depths (31st December 2015)

Depth (cm)	Daily temperature fluctuation (°C)	Lag of peak temperature (h, compared to surface)
Surface	52.4	0
5	13.9	1.5
15	5.8	4.5
25	2.5	8.0
35	1.1	11.5
45	0.7	14.0
55	0.6	16.5
65	0.5	21.0
75	0.3	27.0

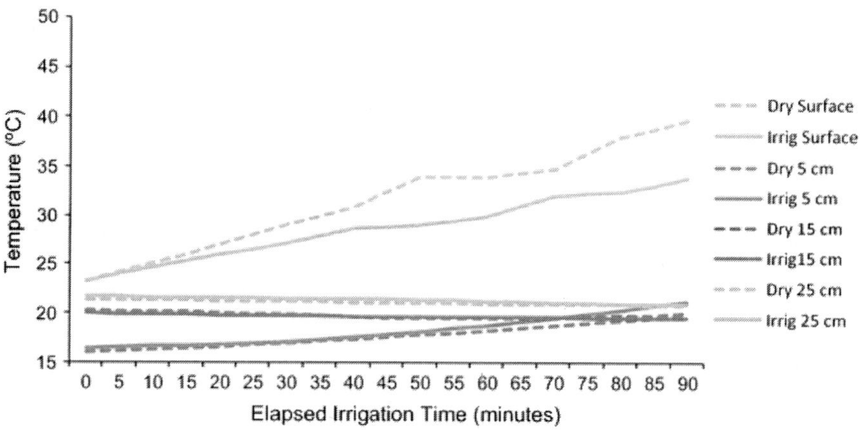

Figure 5. Soil temperature changes during the 06:00 irrigation period, measured at different depths in both the irrigated and un-irrigated (dry) soil (mean of all runs)

At 14:00, the soil surface temperature had typically reached 49 °C or more (Figure 6) and, during the P.M. irrigation, part of this heat was transferred to the 21 °C irrigation water, which then transported the heat downward to lower levels in the soil as it infiltrated downward. The result was that the soil surface cooled quite rapidly to 35 °C within the first 40 min of irrigation and thereafter continued cooling to reach 30 °C by the end of the irrigation period. The infiltrating water at 5 cm depth in turn cooled the soil to a temperature similar to the surface temperature. However, at 15 cm depth, since the initial soil temperature was 25 °C—cooler than the infiltrating water by the time it reached that depth—the result was that the soil at 15 cm increased its temperature from 25 to 28.2 °C by the end of irrigation. These changing temperature fluxes at all levels had a significant effect on both the infiltration speed and on the final moisture content at the various depths.

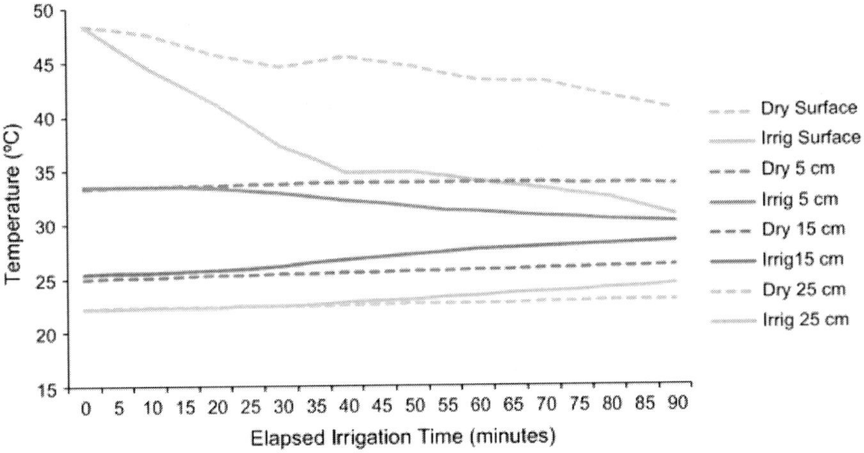

Figure 6. Soil temperature changes during the 14:00 irrigation period, measured at different depths in both the irrigated and un-irrigated (dry) soil (mean of all runs)

Infiltration

Figure 7 shows the mean vertical infiltration times of the eight stations over four replications. Over the range of soil depths investigated in this experiment, the downward infiltration progressed quite quickly, reaching 35 cm after a mean time of 42 min. The experimental hypothesis was that the time (and hence the water volume) taken to reach 35 cm depth should be the same for both A.M. and P.M.

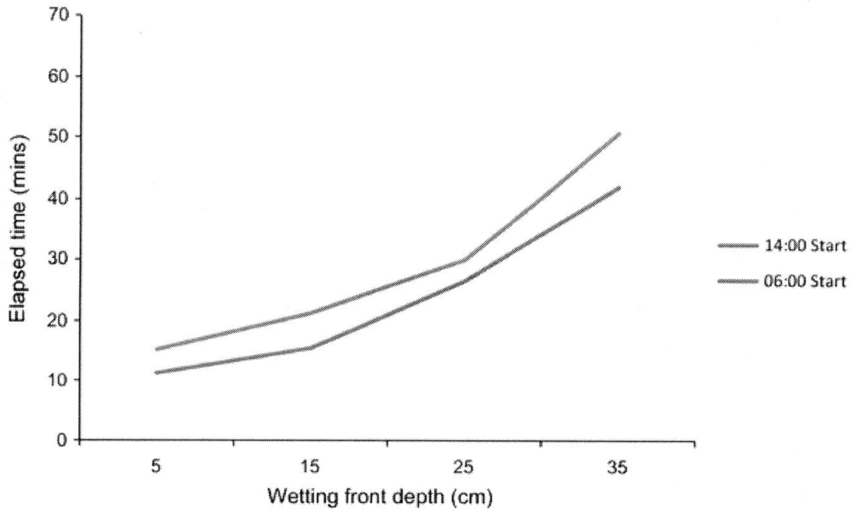

Figure 7. Vertical infiltration rate of irrigation wetting front to various depths during irrigation, comparing irrigation at 06:00 to that at 14:00 (mean of all runs)

Figure 7 shows that the P.M. irrigation resulted in infiltration that was slightly more rapid at all depths than that of A.M. A one-way ANOVA was run to compare the effect of time of day (A.M. and P.M.) on the time to infiltrate to 35 cm and hence the volume of drip irrigation water needed to be applied to reach that depth. The ANOVA showed that, at the $p < 0.05$ level, there was no evidence to suggest any difference between the A.M. and P.M. means [$F(1, 37) = 1.0$, $p = 0.323$], and that the amount of water needed to be applied to reach a depth of 35 cm at 06:00 compared to 14:00 was the same.

With regard to near-surface depths though, the situation was different. For 5 cm depth, there was actually less water needed to reach that depth at 14:00 than at 06:00, with a one-way ANOVA showing a difference at the $p < 0.05$ level between A.M. and P.M. [$F(1, 38) = 4.61$, $p = 0.038$]. A paired sample t test (by station) showed this even more clearly with A.M. (M = 13.75, SD = 4.65) and P.M. (M = 10.94, SD = 2.02) conditions; $t(16) = 2.52$, $p = 0.023$.

The soil moisture content increased at each depth as the wetting front progressed downwards. The most noticeable difference between A.M. and P.M. moisture content (Figure 8) was that there was a more rapid build-up at the 5 and 15 cm depths during P.M. irrigation, and that the final moisture contents at those depths were also higher, for example by 9% at 5 cm depth. More importantly, the incremental moisture content immediately after irrigation (that is the final moisture% – initial moisture%) was higher by 18% at 5 cm and 9% at 15 cm by P.M. irrigation compared to A.M.

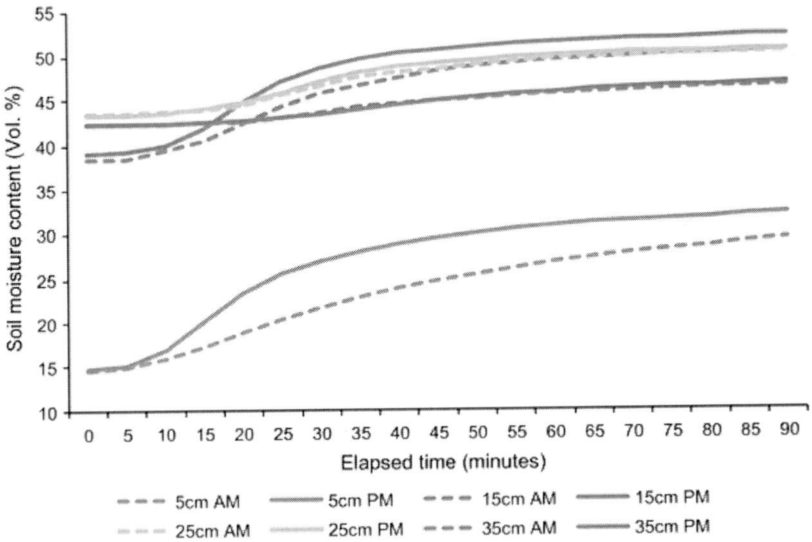

Figure 8. Changes in soil moisture content at various depths during irrigation, comparing A.M. (06:00) and P.M. (14:00) irrigations (mean of all runs)

Moisture redistribution after irrigation

Following each irrigation, the soil moisture contents were monitored as moisture redistribution and soil drying occurred over the following 72 h (measured from the respective irrigation start times). As shown in Figure 9, immediately after irrigation the peak moisture contents at 5 cm and 15 cm depths were higher for P.M. irrigation compared to A.M., which seemed both surprising and counterintuitive. A paired sample t test showed this difference was significant at the 95% confidence level for both those depths, with the results at 5 cm depth being [A.M. ($M = 29.48$, $SD = 6.10$) and P.M. ($M = 10.94$, $SD = 2.02$) conditions; $t(16) = -4.8$, $p = 0.000$], and for 15 cm depth being [A.M. ($M = 50.373$, $SD = 2.849$) and P.M. ($M = 52.099$, $SD = 2.928$) conditions; $t(16) = -7.29$, $p = 0.000$].

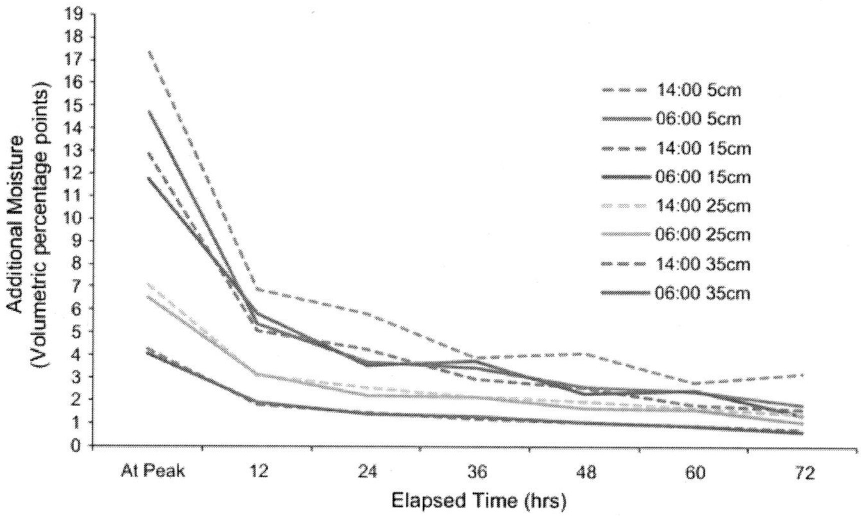

Figure 9. Retained moisture content at various depths, over the 72-h period following irrigation, for 06:00 and 14:00 irrigation (compared with antecedent moisture content, mean of all runs)

Much of this initial difference in peak moisture between A.M. and P.M. was retained during all periods throughout the ensuing 72 h at those depths. Even though the rate of soil moisture loss is generally higher at higher soil water content, Figure 9 shows that this factor is not significant, as the difference between the moisture levels at 5 cm depth after 24 h is similar to that immediately after irrigation. At greater depths than 15 cm, there was no significant difference in either peak moisture content (A.M. compared to P.M.) or in the retained moisture after 72 h.

Part B: Horizontal wetting

The wetted profile of drip irrigation comprises both vertical and horizontal wetting. With regard to this experiment, even if there was no difference in vertical infiltration by A.M. irrigation compared to P.M., there could be a difference in the respective horizontal spreads. This would result in a difference in the wetted profile and/or its volume A.M. and P.M.

The extent of horizontal wetting was measured as a Part B of the experiment (Figure 10). This shows that, at depths 15 cm and greater, there was no apparent difference in the wetted radius after 3 h of irrigation (between P.M. and A.M.) and that the wetted profile was similar at those depths. For example, at 25 cm depth (the point of maximum wetted diameter) an ANOVA showed there was no evidence to suggest a difference between A.M. and P.M. [$F(1, 22) = 0.11, p = 0.743$]. With A.M. application though, there was a significant 'flaring' at the soil surface and down to 5 cm depth, when compared to P.M.

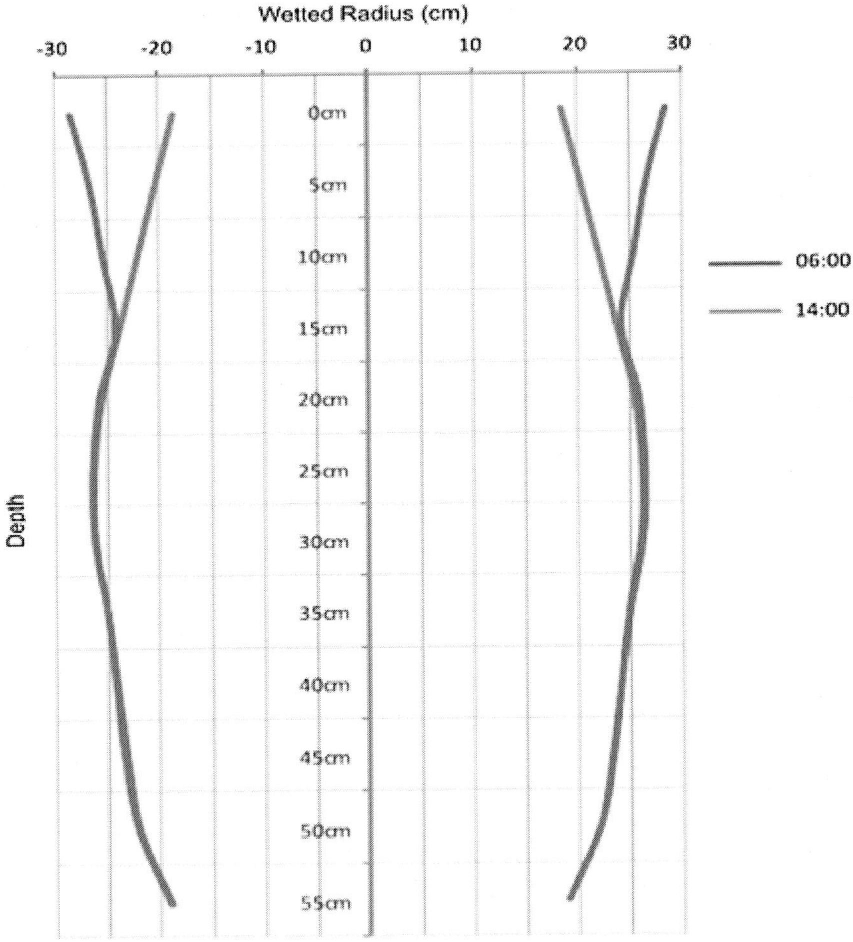

Figure 10. Horizontal spreading at various depths after 3 h of irrigation, for irrigation start times of 06:00 and 14:00 (mean of all runs)

Overall diurnal effect on efficiency

Many growers in the USA traditionally irrigate during predawn or early morning hours (Warren and Bilderback 2001) and, anecdotally, the practice appears to be similarly widespread in Australia. With regard to existing research studies regarding irrigation scheduling, it is rare to find any reference to the time of day of scheduling.

The experimental results of the current investigation have shown that there is no evidence that the efficiency of drip irrigation is affected by the time of day at which it is applied. This means that for deep-rooted crops such as grapevines, irrigation scheduling can be completely flexible—scheduled at any time and not be limited to night time or cooler periods of the day. This is important for minimising the total elapsed time required for irrigating a vineyard, especially for capacity-constrained irrigation systems or heat wave conditions where a water application needs to be completed quickly. The results showed that, at more shallow depths, there were significant beneficial effects from irrigating in the afternoon (as compared to morning)—the peak moisture content achieved by the irrigation was higher and the amount of moisture retained over the ensuing period (for example the next 3 days) was also greater. This is analysed and discussed in more detail in the following sections.

Diurnal effects on infiltration dynamics

Temperature, liquid and vapour fluxes

The soil surface temperature varied widely in a daily cycle, reaching peaks of 50–65 °C or more in the early afternoon and falling to 10 °C or less overnight. At depths of more than 45 cm or so, the temperature was stable from day to day (Figure 4). This meant that there were very strong heat fluxes over the first few centimetres of soil depth at particular times of day; the flux changed throughout the day; and there was a temperature 'wavefront' that slowly proceeded downwards through the soil through the hotter parts of day and into the evening.

The transfer of heat in soil is due to radiation transfer (important just at or near the soil surface); convective heat transfer (caused by the water percolation through the soil, especially where there are large differences between the soil temperature and water temperature, such as the P.M. irrigation in this experiment); and conduction (which occurs when liquid water is converted to water vapour through evaporation or condensation processes in the soil). Deb et al. (2011) found that surface energy components and contributions due to heat conduction, convection and vapour transfer were more pronounced with irrigation, and differences between thermal profiles of wet and dry soil conditions before, during and after irrigation were due to the latent heat exchanges. As can been seen from the above, the application of cool irrigation water into a hot soil layer will result in quite different infiltration dynamics depending on the soil temperatures, the temperature gradients between depths and the difference between the temperature of the irrigation water and the soil temperature, and the combination of these factors is reflected in the experimental results between A.M. and P.M.

Soil vapour movement

The transport of moisture in the soil by evaporation and then by re-condensation can be a significant contributor to the net moisture movement in the soil and, because of the large latent energy needed to vaporise water, the water vapour transports a significant amount of energy as it evaporates and re-condenses (Cahill and Parlange 1998). Since vapour pressure is related to the soil temperature (Jury et al. 1991), water vapour will be transported from regions of higher temperature to those of lower temperature—meaning that at some times of the day the temperature gradient will result in this vapour migration being downwards rather than upwards, and not necessarily out through the soil surface to the atmosphere above. Because the soil temperature and heat flux at shallow depths follow a daily cycle with a peak each afternoon, there is a corresponding diurnal variation in water and vapour movement—both in magnitude and direction—and evaporation. The heat flux varies with both depth and the time of day, so the corresponding temperature gradient continually changes in intensity and direction. For example, at night when the shallow soil temperatures become lower than those at greater depths, water vapour in the soil will be transported towards the surface, and in the afternoon the reverse applies and it will tend to move further downward into the soil (Figure 11). Zeng et al. (2008) found that the thermal water vapour and liquid flux were dominant in uppermost soil layer at night, whereas the isothermal liquid water dominated during the day and in the deeper soil layer and this is evident in Figure 11. This situation was also investigated by Saito et al. (2005) who found that at noon, even though a large downward thermal water vapour flux occurred due to a large downward temperature gradient, the isothermal liquid water and water vapour fluxes are upward because of an upward pressure head gradient.

The figure is from Zeng et al. (2008). With permission

As reported by Parlange et al. (1998), there has never been a satisfactory reconciliation of the Philip and De Vries (1957) theory of water vapour movement in soils with the short-term field observations subsequently made by others (Rose 1968; Jackson et al. 1973; Cass et al. 1984) who found that the vapour flow is perhaps an order of magnitude larger than the theory predicts. Wierenga et al. (1969) reported that 40–60% of the heat flux in the top 2 cm of a bare field soil, and up to 20–25% of the total heat flux at 25 cm depth, was due to water vapour flow and Grifoll et al.

(2005) showed that for a near-surface dry soil layer, diffusion and dispersion of water vapour are the significant water transport mechanisms.

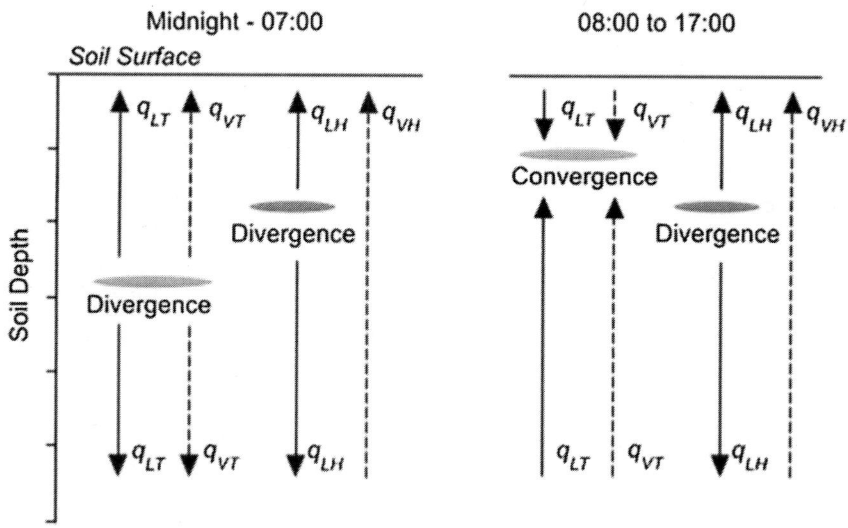

Figure 11. Schematic illustration of the variation (in direction and depth) of liquid and vapour fluxes between **A** night to early morning and **B** through the daylight hours. q_{LT} thermal liquid, q_{VT} thermal vapour, q_{LH} isothermal liquid, q_{VH} isothermal vapour.

In this experiment, during irrigation the temperature of the initially cool irrigation water itself changed as it travelled downward through the soil, gaining or losing heat depending on the temperature of the soil at each depth, and transferring that heat to (or from) the soil at lower depths, causing the heat flux in the soil to be modified rapidly and substantially. As the soil probes recorded temperature and moisture values at each depth every 5 min during the irrigation, the corresponding temperature and moisture gradients between each depth can be calculated. These gradients are not the same as thermal and isothermal fluxes, but it is reasonable to relate the temperature gradient to the thermal liquid and vapour fluxes, and the moisture gradient to the isothermal liquid and vapour fluxes which are shown in Figure 11.

Infiltration

From the previous sections, it can be concluded that the rate of infiltration from irrigation (for any particular soil characteristics) will be determined primarily by the soil temperature and moisture gradients, the differences between those A.M. and P.M. and, importantly, the manner in which they change during the irrigation period itself. The experimental results are discussed in this context below.

(a)Temperature changes during A.M. irrigation. At 3 cm depth, there was initially a downward temperature gradient which increased as irrigation progressed because of the rising soil surface temperature, as shown in Figure 12. At 10 cm depth, the temperature gradient was initially slightly upward and this changed over the irrigation duration to become slightly downward because of the increasing soil temperature above that depth. The overall effect was that moisture contents at all depths increased steadily during the period of irrigation (Figure 14).

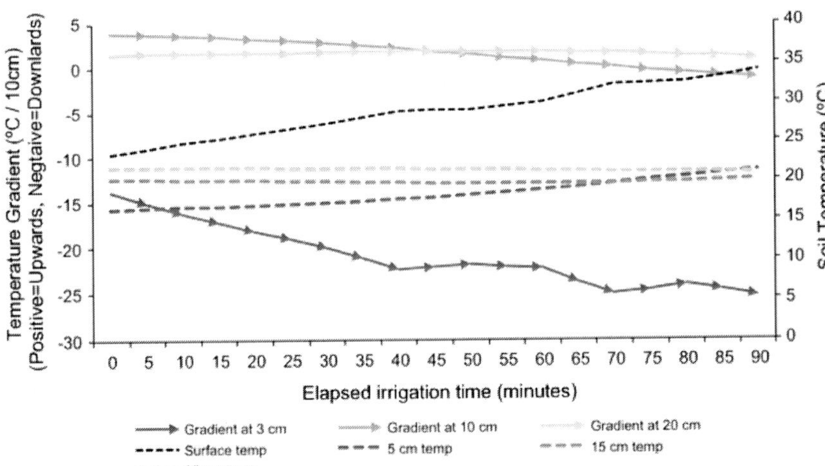

Figure 12. Changes in temperature and temperature gradients at various depths during A.M. irrigation (06:00, mean of all runs)

(b)Temperature changes during P.M. irrigation. The temperature gradients and their effects were more complex in this case (Figure 13) and combined to result in the peak moisture contents at depths up to 15 cm being higher with P.M. irrigation than with A.M.

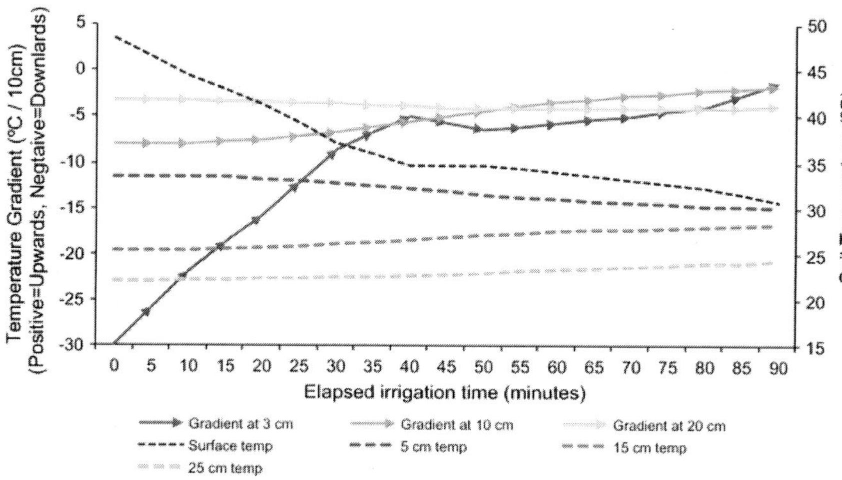

Figure 13. Changes in temperature and temperature gradients at various depths during P.M. irrigation (14:00, mean of all runs)

It can be seen that, at the start of irrigation, because of the initially high soil surface temperature, there was a very strong downward temperature gradient at 3 cm depth. This gradient decreased rapidly during irrigation, because of the cooling effect of the colder irrigation water at the soil surface and as it percolated downward, halving in value within 20 min of irrigation and thereafter continuing to decline to nearly zero after 40 min. At 10 cm depth, there was initially a slight downward gradient, but this reduced steadily to zero over the irrigation period, again through the cooling effect of the infiltrating water. When P.M. was compared to A.M., it can also be seen that the gradients at 10 cm depth were in opposite directions.

The overall effect was that, at the start of P.M. irrigation, there was initially a rapid downward movement of water (liquid and vapour) due to the strong downward heat flux, but this then slowed as the temperature gradients reduced at shallow depths. It is hypothesised that this caused a more rapid build-up of moisture content at the 5 and 15 cm depths compared to the A.M. situation at the same elapsed time (as shown in Figure 8) and was a contributing factor to the peak moisture contents at those depths being higher with P.M. irrigation.

Even though the experimental design could not directly measure non-isothermal liquid and vapour fluxes, the moisture gradients and the changes in those were measured. During A.M. irrigation, the moisture gradient at 3 cm depth (Figure 14) increased rapidly downward in the initial 10 min of irrigation and thereafter declined slowly over the remaining irrigation. At 10 cm depth, there was an upward moisture gradient which did not change significantly, but at greater depths moisture gradients were close to zero. The situation during P.M. irrigation was very similar to that of A.M. (Figure 15).

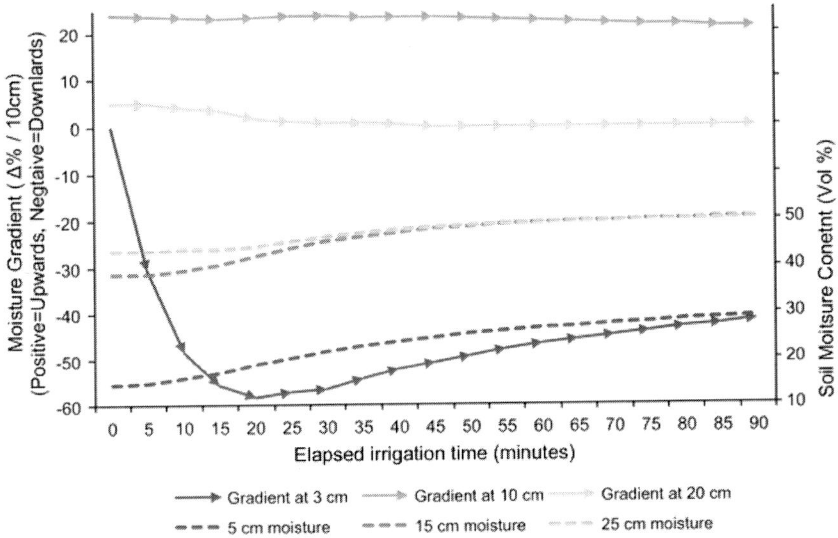

Figure 14. Changes in moisture content and moisture gradients at various depths during A.M. irrigation (06:00, mean of all runs)

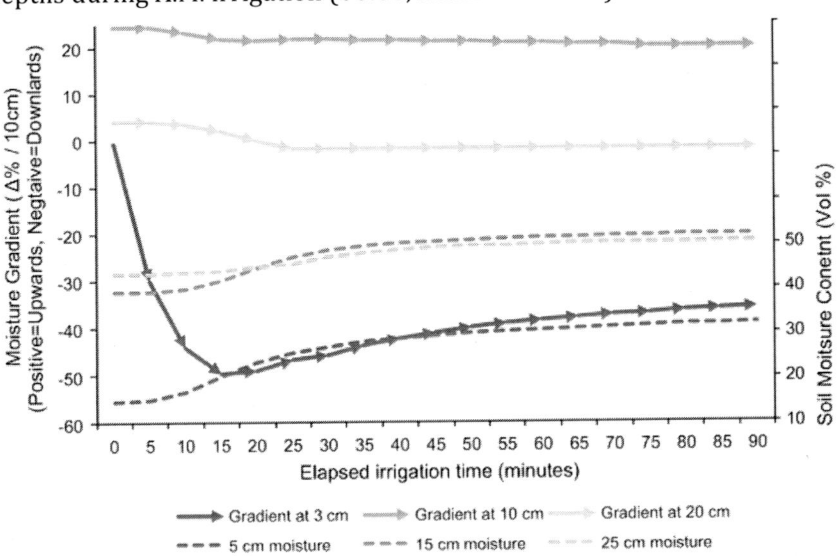

Figure 15. Changes in moisture content and moisture gradients at various depths during P.M. irrigation (14:00, mean of all runs)

Since the overall moisture gradients were similar for both A.M. and P.M. and were dominated by the strong and varying temperature gradients, it was concluded that the moisture fluxes were not a significant contributor to differences between the A.M. and P.M. results. This finding also reinforced the view that antecedent soil moisture was not a significant factor in the experimental design.

This field experiment showed that there are a complex series of factors which change dynamically over the period of irrigation, particularly the effect of cool irrigation water into a relatively hot soil and the role of the irrigation water in the heat transfer, and a follow-up 3-D computer simulation would be worthwhile in order to better understand these dynamics.

Horizontal spread

The results showed that the horizontal spreading of irrigation water, and hence the shape of the wetted profile, was the same for both A.M. and P.M. at depths of 15 cm and greater. However, there was an effect at depths up to 15 cm, with A.M. watering having a significantly greater horizontal spread than P.M. (Figure 10). This can be explained by that fact that the temperature fluxes in the P.M. were directed strongly downward rather than horizontally, whereas in the A.M. any downward soil temperature fluxes were initially small and similar in magnitude to those in a horizontal direction, and these resulted in a higher proportion of lateral spreading. The greater A.M. near-surface horizontal spread would also result in some reduction in the soil moisture content at those depths, since the same amount of applied water was being distributed through a larger soil volume compared to P.M.

Effects on redistribution and drying

For grapevines and other deep-rooted plants, even though P.M. irrigation increases the soil moisture content at shallow depths compared to A.M., this is of limited benefit to such crops. However, the situation is different for shallow-rooted crops such as salad vegetables (cucumbers, lettuce, tomatoes and the like) which have, in the past, largely been watered by sprinkler or furrow irrigation. For example, spray irrigation is still the principal irrigation type for salad crops in the UK (Doyle and Erickson 2008).

There are a growing number of cases of food contamination worldwide by pathogens such as *Escherichia coli* and *Salmonella* with outbreaks linked to lettuce, spinach and tomatoes (Heaton and Jones 2008). Groundwater, surface water and human wastewater are commonly used for irrigation, and poor-quality water is one way that fruit and vegetables can become contaminated (Steele and Odumeru 2004). There is an increasing trend to use drip irrigation, either surface or subsurface, as a means of both increasing water use efficiency and minimising the transfer of pathogens to the leaves or fruit (by reducing or eliminating splashing, for example).

The current experiment identified some opportunities to improve drip irrigation scheduling for shallow-rooted crops such as salad vegetables, with associated water savings. For crops such as tomatoes, it has been shown by Warren and Bilderback (2002) that applying water at the time of the day when photosynthesis is highest also increases yield—irrigation at midday or in the P.M. produced a 46–56% greater top dry weight than earlier morning irrigation. The current experiment showed that, at depths down to 15 cm, there is an increase in peak moisture content with P.M. irrigation (as compared to A.M.) and that the effect also carries forward over the subsequent days as the soil dries (Figure 9). This has the potential to be applied to irrigation scheduling for salad vegetables and similar crops, which typically only require irrigation to these shallow depths, to both increase yield and reduce the amount of irrigation water required. For example, by directly applying the experimental results of the rate of drying after irrigation at 5 cm and 15 cm depths (the data shown in Figure 9) to create an irrigation schedule (Figure 16), the interval between irrigations can be increased from 6 to 8 days by simply switching to P.M. irrigation rather than A.M. (providing that there is appropriate emitter spacing). This change would still keep the soil moisture content between 20% and FC at all times, but could potentially reduce the water use by up to 25%.

Of course, since the experiment was conducted on a bare site without plant transpiration, the irrigation frequencies shown in Figure 16 need to be modified for cropping situations with their associated transpiration. By using the same data (Figure 9), but also applying an arbitrary cropping factor (which has the effect of reducing the soil moisture levels more rapidly from day to day), a hypothetical irrigation schedule can be calculated (Figure 17) which shows that the interval between irrigations could be 3 days (for A.M. irrigation) and 4 days (for P.M. irrigation), respectively. This demonstrates that there is still a 25% potential saving in the total water usage while keeping the soil moisture in the range of 20% to FC (that is irrespective of the amount of plant transpiration, the relative improvement in scheduling by moving to P.M. irrigation rather than A.M. remains constant).

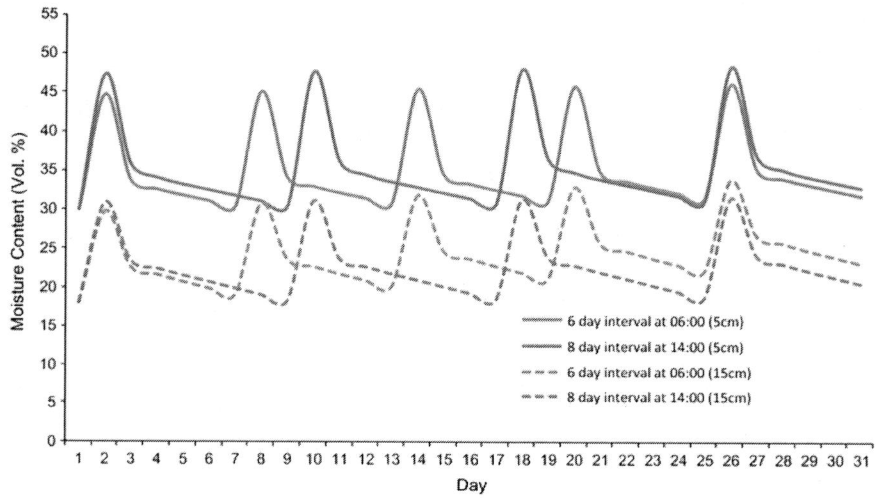

Figure 16. Differences in the interval between successive irrigations, due to watering at 14:00 compared to 06:00, bare soil conditions and without plant transpiration (typically 8 days compared to 6 days)

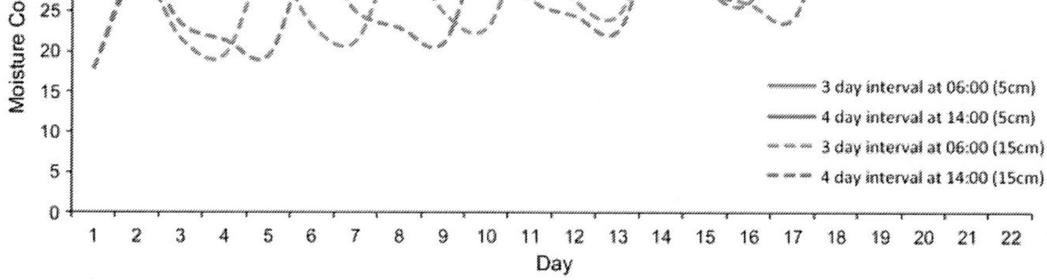

Figure 17. Hypothetical irrigation interval, including an allowance for crop transpiration, due to watering at 14:00 compared to 06:00 (typically 4 days compared to 3 days)

CONCLUSION

The experiment showed that, for the soil depths relevant to field crops such as grapevines, drip irrigation efficiency is not affected by the time of day of watering, that is, that there are no diurnal effects. For drip irrigation, the belief that irrigation should not be applied in the heat of the day is not supported by the results of this study. This means that, for deep-rooted crops, drip irrigation scheduling can be completely flexible, can be conducted equally efficiently at any time of the day or night and not be limited to cooler periods of the day. This is important for minimising the total elapsed time needed to irrigate a large vineyard, especially where there are capacity constraints on the irrigation systems. At shallow soil depths, down to 15 cm, irrigating at 14:00 resulted in a higher peak moisture content being reached (a 9–18% increase in moisture added) and much of this difference persisted over the following 72 h of moisture distribution and soil drying. The shape of the wetted soil profile was also investigated and showed that, even though the time of day of irrigation had no effect on the radial wetted distance at depths of 15 cm and greater, at near-surface depths A.M. irrigation did result in a greater wetted radius.

These findings regarding the benefits of P.M. irrigation at shallow depths suggest that they have the potential to be applied to irrigation scheduling for salad vegetables and similar crops, which typically take up most of their plant water from depths <25 cm. By irrigating in the afternoon, rather than early morning, the frequency of irrigation could be reduced without impacting soil moisture content in the rootzone, resulting in significant savings in irrigation water. At the same time, the crop yield could potentially be increased through better matching of moisture availability with daily peak photosynthesis periods.

There are opportunities for further research, in particular by using 3-D computer simulations to model both the diurnal effects (considering heat, water and vapour flows) and the irrigation dynamics (resulting from the application of cool water into a hot soil) and also to further investigate the potential of P.M. irrigation for shallow-rooted crops.

REFERENCES

1. Apcos Pty Ltd (2015) User manual—enviropro sub-surface soil probe, 5.1 edn. ETS Envirotek Solutions, Lyrup, South Australia. http://www.envirotek.com.au/images/ep200/User_Manual_EnviroPro_Series_E_SDI12_1.pdf

2. Braud I, Dantas C, Vauclin M, Thony L (1995) A simple soil plant atmosphere transfer model (SiSPAT): development and field verification. J Hydrol 166(3):213–250
3. Bureau of Meteorology (2013a) Average annual, monthly and seasonal evaporation. Retrieved 1 Nov 2015, from http://www.bom.gov.au/jsp/ncc/climate_ averages /evaporation/index.jsp?#maps
4. Bureau of Meteorology (2013b) Monthly climate statistics (Bega). Retrieved 7 Feb 2016, from http://www.bom.gov.au/climate/averages/tables/cw_069002.shtml
5. Cahill AT, Parlange MB (1998) On water vapor transport in field soils. Water Resour Res 34(4):731–739
6. Cass A, Campbell GS, Jones TL (1984) Enhancement of thermal water vapor diffusion in soil. Soil Sci Soc Am J 48(1):25–32
7. Cook F, Thorburn P, Fitch P, Bristow K (2003) WetUp: a software tool to display approximate wetting patterns from drippers. Irrig Sci 22(3–4):129–134
8. Deb SK, Shukla MK, Sharma P, Mexal JG (2011) Coupled liquid water, water vapor, and heat transport simulations in an unsaturated zone of a sandy loam field. Soil Sci 176(8):387–398
9. Doyle MP, Erickson MC (2008) Summer meeting 2007—the problem with fresh produce: an overview. J Appl Microbiol 105(2):317–330
10. Grifoll J, Gasto JM, Cohen Y (2005) Non-isothermal soil water transport and evaporation. Adv Water Resour 28(11):1254–1266
11. Heaton JC, Jones K (2008) Microbial contamination of fruit and vegetables and the behavior of enteropathogens in the phyllosphere: a review. J Appl Microbiol 104(3):613–626
12. Hillel D (1980) Environmental soil physics—fundamentals. Academic Press, New York
13. Jackson RD, Kimball BA, Reginato RJ, Nakayama FS (1973) Diurnal soil–water evaporation: time-depth-flux patterns. Soil Sci Soc Am J 37(4):505
14. Jury WA, Letey J (1979) Water vapor movement in soil: reconciliation of theory and experiment 1. Soil Sci Soc Am J 43:823–827. https://www.soils.org/publications /sssaj/abstracts/43/5/823
15. Jury WA, Gardner WR, Gardner WH (1991) Soil physics. Wiley, New York
16. McKillup S (2012) Statistics explained: an introductory guide for life scientists, 2nd edn. Cambridge University Press, Cambridge
17. Measurement Engineering Australia (2013) Enviropro. Retrieved 8 Feb 2016, from http://mea.com.au/upload/Brochures/SoilMoistureMonitoring/EnviroPro_W eb.pdf
18. Measurement Engineering Australia (2015). Plexus. Retrieved 14 Feb, 2016, from http://mea.com.au/upload/Brochures/SoilMoistureMonitoring/Plexus_Web_2 .pdf
19. Measurement Engineering Australia (2016) Comparing soil moisture tension and soil moisture content. Retrieved 8 Feb 2016, from http://mea.com.au/soil-plants-climate/soil-moisture-monitoring/learning-centre/comparing-soil-moisture-tension-and-soil-moisture-content
20. Milly PCD (1982) Moisture and heat transport in hysteretic, inhomogeneous porous media: a matric head-based formulation and a numerical model. Water Resour Res 18(3):489–498
21. Milly PCD (1984) A simulation analysis of thermal effects on evaporation from soil. Water Resour Res 20(8):1087–1098
22. Milly PCD (1986) An event based simulation model of moisture and energy fluxes at a bare soil surface. Water Resour Res 22(12):1680–1692
23. Minitab Inc. (2016) Minitab 17. Retrieved 22 Feb 2016, from https:// www.minitab.com/en-us/products/minitab/
24. Monaghan JM, Daccache A, Vickers LH, Hess TM, Weatherhead EK, Grove IG, Knox JW (2013) More "crop per drop": constraints and opportunities for precision irrigation in European agriculture. J Sci Food Agric 93(5):977–980

25. Nassar IN, Horton R (1997) Heat, water, and solution transfer in unsaturated porous media: I—theory development and transport coefficient evaluation. Transp Porous Media 27(1):17–38
26. Parlange MB, Cahill AT, Nielsen DR, Hopmans JW, Wendroth O (1998) Review of heat and water movement in field soils. Soil Tillage Res 47(1–2):5–10. http://www.sciencedirect.com/science/article/pii/S016719879800066X
27. PC-Progress (2015) Hydrus 2D/3d. Retrieved 21 Mar, 2015, from http://www.pc-progress.com/en/?hydrus-3d
28. Philip JR (1984) Travel times from buried and surface infiltration point sources. Water Resour Res 20(7):990–994.
29. Philip JR, De Vries DA (1957) Moisture movement in porous materials under temperature gradients. Eos Trans Am Geophys Union 38(2):222–232
30. Playán E, Salvador R, Faci JM, Zapata N, Martínez-Cob A, Sánchez I (2005) Day and night wind drift and evaporation losses in sprinkler solid-sets and moving laterals. Agric Water Manag 76:139–159
31. Power E (trans) (1928) The goodman of Paris (Le Ménagier de Paris), a treatise on moral and domestic economy by a citizen of Paris. Harcourt, Brace, & Co., New York, pp 195–204. http://search.proquest.com.ezproxy.csu.edu.au/docview /757325001/fulltextPDF/6D589545B92042BEPQ/1?accountid=10344
32. Quinn G, Keough M (2002) Experimental design and data analysis for biologists. Cambridge University Press, Cambridge.
33. Revol P, Clothier BE, Mailhol J, Vachaud G, Vauclin M (1997) Infiltration from a surface point source and drip irrigation: 2. An approximate time-dependent solution for wet-front position. Water Resour Res 33(8):1869–1874
34. Rose CW (1968) Water transport in soil with a daily temperature wave. I. Theory and experiment. Soil Res 6(1):31–44
35. Saito H, Šimůnek J, Mohanty BP (2005) Numerical analysis of one-dimensional coupled water, vapor and heat transport in the vadose zone using HYDRUS. In: Proceedings of the workshop on HYDRUS applications, p 22
36. Shurbaji A-RM, Phillips FM, Campbella AR, Knowlton RG (1995) Application of a numerical model for simulating water flow, isotope transport, and heat transfer in the unsaturated zone. J Hydrol 171(1):143–163
37. Sophocleous M (1979) Analysis of water and heat flow in unsaturated-saturated porous media. Water Resour Res 15(5):1195–1206
38. Steele M, Odumeru J (2004) Irrigation water as source of foodborne pathogens on fruit and vegetables. J Food Prot 67(12):2839–2849
39. Thorburn PJ, Cook FJ, Bristow KL (2003) Soil dependent wetting from trickle emitters: implications for systems design and management. Irrig Sci 22(3–4):121–127
40. Toro Australia Pty Ltd. (2003) Toro turbo plus II drip emitters. Retrieved 8 Feb 2016, from http://www.toro.com.au/product/toro-turbo-plus-ii-drip-emitters?c3=drip-3&c4=online-emitters
41. Tulau MJ (1997) Soil landscapes of the Bega-Goalen point. Department of Land and Water Conservation, Sydney. Retrieved from http://www. Environment.nsw.gov.au/soils/bega.htm
42. van Genuchten MT (1980) A closed-form equation for predicting the hydraulic conductivity of unsaturated soils. Soil Sci Soc Am J 44(5):892
43. Warren SL, Bilderback TE (2001) Irrigation timing: Effect on plant growth, photosynthesis, water-use efficiency and substrate temperature. In: International symposium on growing media and hydroponics, vol 644, pp 29–37
44. Warren SL, Bilderback TE (2002) Timing of low pressure irrigation affects plant growth and water utilization efficiency. J Environ Hortic 20(3):184–188
45. Wierenga PJ, Nielsen DR, Hagan RM (1969) Thermal properties of a soil based upon field and laboratory measurements. Soil Sci Soc Am J 33(3):354–360
46. Yazar A (1984) Evaporation and drift losses from sprinkler irrigation systems under various operating conditions. Agric Water Manag 8:439–449

47. Zeng YJ, Wan L, Su ZB, Saito H, Huang KL, Wang XS (2008) Diurnal soil water dynamics in the shallow vadose zone (field site of China University of Geosciences, China). Environ Geol 58(1):11–23

CHAPTER 9

Potential of Waste Water Use for Jatropha Cultivation in Arid Environments

Arisoa M. Rajaona [1,*], Nele Sutterer [2] and Folkard Asch [1]

[1]*Department of Plant Production and Agroecology in the Tropics and Subtropics Section: Crop Waterstress Management, University of Hohenheim, Garbenstrasse 13, 70599 Stuttgart, Germany*

[2]*Institute for Applied Material Flow Management, Trier University of Applied Sciences, Campusallee 9926, 55768 Neubrücke, Germany*

ABSTRACT

Water is crucial for socio-economic development and healthy ecosystems. With the actual population growth and in view of future water scarcity, development calls for improved sectorial allocation of groundwater and surface water for domestic, agricultural and industrial use. Instead of intensifying the pressure on water resources, leading to conflicts among users and excessive pressure on the environment, sewage effluents, after pre-treatment, provide an alternative nutrient-rich water source for agriculture in the vicinity of cities. Water scarcity often occurs in arid and semiarid regions affected by droughts and large climate variability and where the choice of crop to be grown is limited by the environmental factors. Jatropha has been introduced as a potential renewable energy resource since it is claimed to be drought resistant and can be grown on marginal sites. Sewage effluents provide a source for water and nutrients for cultivating jatropha, a combined plant production/effluent treatment system. Nevertheless, use of sewage effluents for irrigation in arid climates carries the risk of salinization. Thus, potential irrigation with sewage effluents needs to consider both the water requirement of the crop and those needed for controlling salinity build-up in the top soil. Using data from a case study in Southern Morocco, irrigation requirements were calculated using CROPWAT 8.0. We present here crop evapotranspiration during the growing period, required irrigation, the resulting nutrient input and the related risk of salinization from the irrigation of jatropha with sewage effluent.

Keywords: *Jatropha curcas* L.; sewage water; salinity; Morocco; irrigation requirement; N-P-K balance

1. INTRODUCTION

Agricultural systems are increasingly under pressure from competition for water from urban areas and industrial activities. Both increasing population and climate change additionally constrain global and regional water resources [1]. Most affected by this are arid and semi-arid regions. Taking the Maghreb region as an example, largely depending on precipitation as the sole water input [2], the supply of both water for drinking and for irrigation is equally important [3]. Morocco, classified as a water scarce country [4], with about 32 million inhabitants (July 2010) [3], needs to manage its water supply in a way that satisfies the needs of human consumption, agriculture, industry and tourism. The world bank projected a widening gap between water availability and water withdrawal for this country [4]. Inter-annual fluctuations in water availability aggravate climate change impacts and, thus, will lead to a decrease of 10 to 15% of available water by 2050 [4].

Agriculture contributed 15% to Morocco's GDP in 2009 [3], but requires 80% of the available fresh water. In Morocco, water availability differs significantly between regions. The Northern part receives about 1000 mm precipitation per year, whereas the Southern part only receives about 200 mm a^{-1} (e.g. Tan-Tan in the Presaharan zone). In Southern Morocco, animal production on grasslands is the most important agricultural activity. Crop farming only amounts to 2% of the agricultural land and is mainly situated in close proximity to the cities [5]. Since the country's freshwater resources are limited and partly over-exploited (exploitation rate of groundwater = 114% water [6]), using non-conventional water sources, such as treated sewage effluents (TSE), represents a potential alternative for irrigation, since TSE contain considerable amounts of nutrients (N-P-K) and are generally available close to cities. Globally, the amount and availability of wastewater is the only source of water that will increase in the coming years. In Morocco, 700 mil m^3 in 2011, (predicted to increase to 900 mil m^3 in 2020) of water from settlements were discharged without treatment into the environment and just about 70 mil m^3 are being re-used each year [7]; thus, there is a large potential for increased used of waste water in agriculture in Morocco. However, the use of TSE in agricultural production can lead to the build-up of soil salinity, the leaching of nutrients into the groundwater [8] and may present a health risks due to pathogens if not properly treated [9]. Nevertheless, water and nutrient supply can be ensured if the crop to be irrigated is carefully selected with regard to potential salinization and pollution of the groundwater. Thus, potential crops need to be salt tolerant, adapted to arid areas and, due to the health risk, ideally should be used for non-food products.

Morocco's energy supply is based almost entirely on the import of fossil fuel with fluctuating prices and potentially negative effects for the environment [10]. In this context, the production of biofuels for energy, particularly in the rural and remote areas, becomes increasingly interesting. Among the biofuel crops, *Jatropha curcas*, a multipurpose tropical oil crop, claimed to be suitable for growth under adverse conditions, such as drought, low nutrient supply and salinity [11], may be a suitable option for biofuel production from plantations irrigated with wastewater. According to a worldwide study on jatropha, no production areas have been identified yet in Morocco [12]. Whereas for countries like Egypt, encouraging results for oil tree/shrub species have been reported, the assessment of jatropha cultivation in Morocco and Tunisia did not yield congruent results [12,13], indicating the need for more research before embarking on large-scale plantations. Therefore, the study reported here used the Southern region of Morocco as a case study to investigate the possibility of using waste water for jatropha cultivation. It constitutes, thus, an early attempt to estimate potential biomass production, the input requirements in terms of water and fertilizer and the environmental effects on jatropha growth. For this, the model CROPWAT 8.0 [14] was used to estimate the water requirements of jatropha. The necessary physiological and phenological data on jatropha were derived from literature and from our own research data. The irrigation water properties were provided from a waste water treatment plant in Benslimane, Morocco. The case study used the town of Tan-Tan situated in a semi-arid environment where jatropha can be grown, which produces about 45 liters of waste water per capita per day, resulting in approximately 1 mil m^3 per year [15]. Based on climate information for Tan-Tan [16], CROPWAT 8.0 was parameterized to estimate the water use and the potential soil salinization on an irrigated jatropha plantation.

2. RESULTS AND DISCUSSION

2.1. Water Requirements

Cumulative crop evapotranspiration (ET_c) was estimated to total 780 mm per growing season (March–September) (Figure 1). The calculations were based on the climate data for Tan-Tan (Table 1) and the assumed crop coefficient values (K_c) (Table 2). Since this study aims at the potential use of available wastewater for large-scale jatropha cultivation in Tan-Tan, assessment of ET_c, effective rainfall [14] and irrigation requirements (IR) are presented on a monthly basis (Figure 1), instead of daily time steps that are common for drip irrigation systems. The average daily ET_c varied between 1.7 and 5.4 mm day^{-1} during the growing season. Effective rain during the growing period totaled 15 mm (Figure 1). IR increased with the advancing season and was highest at mid-season. Total minimum IR over the growing period reached 765 mm (Figure 1) without considering the leaching fraction. In detail, the IR was in the range of 51 mm (=1.7 mm day^{-1}) in September to 160 mm (=5.3 mm day^{-1}) in June.

The use of secondary TSE is increasingly discussed in Morocco [15]. Irrigating energy crops, such as *Jatropha curcas* with TSE in close proximity to wastewater treatment plants, offers additional possibilities to further purify the water, reduces risks of pollution through uncontrolled dumping of waste water and provides additional energy resources through biofuel production. To evaluate the sustainability of such a system, the water requirements and the nutrient balance of a TSE-irrigated jatropha plantation in Southern Morocco were exemplarily studied. The results provide a basis for identifying some environmental and agronomic impacts of such a system.

The area of a jatropha plantation that could be irrigated with a given amount of TSE depends on the gross irrigation requirements (IR_g), which in turn depend on the crop evapotranspiration (ET_c) and the leaching fraction (LF). Assuming a LF of 0 or 0.5, the IR_g would amount to 1787 and 2761 mm respectively. Thus, the amount of effluent produced by the treatment plant in Tan-Tan would allow irrigation of an area of 41 to 64 ha for jatropha cultivation. Assuming an average seed yield of 1.52 t ha^{-1} [17] with a 28% oil content [18], 17 to 27 tons of oil could be produced.

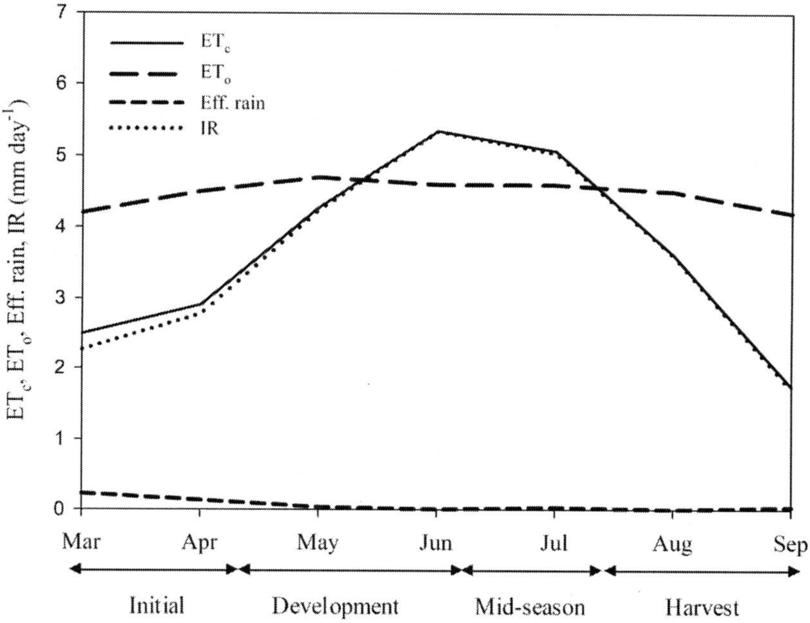

Figure 1. Daily crop evapotranspiration (ET_c), reference evapotranspiration (ET_o), effective rainfall (Eff. Rain) and basic irrigation requirements (IR) of jatropha in the reference climate during one growing period (March–September).

The water requirements calculated here are based on several parameters, such as K_c or salinity tolerance levels, whose values were derived from previous studies either conducted in different environments or in greenhouses and, thus, still need to be validated under semi-arid field conditions. However, the results for the crop evapotranspiration (the plant's water demand) are in line with several other reports [19,20]. In contrast, the ET_c calculated here, differs considerably from a study on jatropha plants in Egypt in spite of similar climatic and soil conditions (sandy texture, low organic carbon content) [13]. In the Egyptian study, different irrigation treatments were related to seed yields. The highest yield was obtained by irrigating the trees with 100% ET_o, ($ET_c = ET_o$), implying a K_c of 1. Surprisingly, the gross irrigation requirements during the 7.5 months cropping season were only 0.178 m^3 per tree [13]. By extrapolating the individual IR_g to an assumed planting density of 2500 trees ha^{-1} seasonal, IR_g amounts to just 45 mm. This is, compared to the annual transpiration water use of 144 mm of the four-year-old, non-irrigated South African plants [21], 70%

lower than the pure transpiration water loss measured in South Africa during nine months and equals only 11% of the ET_c modeled by Jongschaap *et al.* (2009) [22].

Table 1. Monthly means of minimum temperature (T_{min}), maximum temperature (T_{max}), air humidity (RH_{air}), wind speed (Wind), sunshine duration (Sun) and rainfall (Rain). Solar radiation (Rad) and reference evapotranspiration (ET_o) (calculated with CROPWAT 8.0) in Tan-Tan, Morocco.

Month	T_{min}	T_{max}	RH_{air}	Wind	Sun	Rad	ET_o	Rain
	(°C)	(°C)	(%)	(m s^{-1})	(h day^{-1})	(MJ m^{-2} day^{-1})	(mm day^{-1})	(mm day^{-1})
January	9.6	21.1	70	3.0	6.2	12.1	2.6	0.4
February	10.5	21.8	56	3.0	6.8	14.8	3.5	0.6
March	13.0	23.7	56	3.0	6.8	17.3	4.2	0.2
April	13.7	23.8	60	2.9	7.1	19.7	4.5	0.1
May	15.6	24.4	59	2.7	6.9	20.3	4.7	0.0
June	16.6	25.7	59	2.6	5.6	18.5	4.6	0.0
July	17.6	27.0	59	2.4	5.2	17.7	4.6	0.0
August	18.0	29.1	67	2.4	5.9	18.1	4.5	0.0
September	17.4	28.3	66	2.3	6.2	17.0	4.2	0.1
October	15.5	27.3	62	2.3	6.8	15.4	3.8	0.4
November	13.3	25.6	66	2.6	6	12.3	3.1	0.8
December	9.8	21.5	71	2.9	5.9	11.2	2.5	1.2
Average	14.2	24.9	63	2.7	6.3	16.2	3.9	0.31

Table 2. Input parameters for calculating crop evapotranspiration of a potential jatropha plantation in the reference climate of Tan-Tan Morocco with CROPWAT 8.0.

Input parameters	Development stage				
	Initial	Development	Mid-season	Late season	Total
Duration (day)	43	60	30	75	208
Crop coefficient value (K_c)	0.6	n.i.	1.2	0.4	n.i.
Rooting depth (m)	0.3	n.i.	1.2	n.i.	n.i.
Crop height (m)	n.i.	n.i.	3	n.i.	n.i.
Critical depletion fraction	0.4	n.i.	0.4	0.4	n.i.
Yield response factor	0.5	0.5	1	1	3

n.i.: no input.

Gush and Moodley (2007) [23] reported basal crop coefficients (K_{cb}) of 0.04–0.26 for four-year-old trees and 0.15–0.76 for 12-year-old trees. They divided the observed daily transpiration values by calculated ET_o-values to calculate K_{cb}. Under South Africa's climatic conditions with high temperatures, high VPD and annual precipitation of about 550 mm, it is likely that the plants at least occasionally suffered from water deficit. Despite jatropha's hardiness, its photosynthetic activity and transpiration rates have been shown to decrease during water scarcity [24,25]. Hence, the obtained values from rain-fed jatropha plants reflect its water use under non-standard conditions and are most likely lower than K_{cb} obtained from jatropha plants with optimum water supply. For irrigated conditions as in the present study, K_{cb} derived under standard conditions according to Allen *et al.* (1998) are more appropriate, since drought stress effects are excluded [26].

A recent study on biomass production and allocation of jatropha seedlings under drought stress and optimal conditions resulted in K_{cb}-values for immature jatropha plants ranging from 0.51 to 0.60 [24]. These coefficients exceeded those calculated by Gush and Moodley (2007) [23] for four-year-old jatropha, but were more or less similar to that of 12-year-old plants. In both studies, K_e (soil evaporation coefficient) was not mentioned, and therefore, the data were used as a base for the K_c estimations in this study, since data entry in CROPWAT 8.0 requires the total K_c and does not distinguish between K_e and K_{cb}. To accurately calculate water use and irrigation requirements of jatropha, the evaporation component represented by K_e needs to be determined over a wider range of environments and reflected in the K_c input for CROPWAT 8.0. Along this line, Jongschaap *et al.* (2009) [22], for example, used the data of Gush and Moodley (2007) [23] to model the ET_c of a South African jatropha plantation. They simulated a total crop evapotranspiration of 405 mm for one growing season (8.5 months). This is 48% less than the ET_c calculated in this study for jatropha in Morocco for a seven-month growing period (Table 3), indicating that the environmental influence, particularly vapor pressure deficit, on the water requirements of such a plantation is not fully captured by K_c alone. ET_c resulting from this study represents the potential water requirements of a jatropha plantation solely for Southern Morocco under field conditions. In order to allow transfer of these results to other regions, ET_c calculation should follow a dual coefficient approach [26] to better reflect additional environmental effects on water use, particularly when irrigation systems other than drip irrigation may be used. Thus, K_e, K_{cb} and K_c, the respective yield response factors, and rooting depth, need to be established across a larger variety of systems and environments.

Table 3. N, P, K, Na^+ and K^+ inputs (kg ha^{-1}) from effluent irrigation at different leaching fractions (*LF*) corresponding to both crop evapotranspiration (ET_c, mm) and gross irrigation requirement (IR_g, mm) in mm per growing season (March–September).

LF	0	0.05	0.10	0.15	0.20	0.25	0.30	0.35	0.40	0.45	0.50
ET_c	779	820	866	917	974	1039	1113	1199	1299	1417	1559
IRg	1787	1838	1895	1959	2030	2111	2204	2311	2436	2584	2716
N	76	80	84	89	94	101	108	116	126	137	151
P	7	7	8	8	9	9	10	11	12	13	14
K	152	160	169	179	190	203	217	234	253	276	304
Na^+	803	845	892	944	1003	1070	1147	1235	1338	1459	1605
K^+	152	160	169	179	190	203	217	234	253	276	304

TSE contains relatively large amounts of nutrients (Table 3) that are potentially available for plant production, while at the same time, the nutrients will be removed from the effluent, thus increasing the quality of the remaining water.

2.2. Soil Salinity

Jatropha is known to survive extended drought spells [27]. However, it is not particularly efficient in water use, and seedlings are only moderately tolerant to salinity [28]. Furthermore, seed yield and growth of jatropha have been shown to be strongly reduced by salt stress [11,29]. According to Ayers and Westcot (1985) [30], a *LF* between 0.2 and 0.4 would allow maintaining average long-term soil salinity in the root zone between 2.4 and 3.7 dS m^{-1}, which is in the range for moderately sensitive to moderately tolerant crops [30]. However, long term irrigation with TSE may lead to salt accumulation in the root-zone [31,32].

In order to calculate the salt build-up in the soil, different leaching fractions (*LF*) were used to modify water use results from CROPWAT 8.0. This allowed calculating the development of the soil electrical conductivity (EC) and the resulting additional water requirements to control soil salinity build-up. Varying the leaching fraction from 0.05 to 0.5 at the soil surface and at 30 cm depth resulted in soil EC of 1.4 and 2.8 dS m^{-1}, respectively (Figure 2). At 60 cm depth, soil salinity was still below the average salinity of the whole profile. At 90 cm below ground, EC varied from 2 to 9 dS m^{-1} for the highest (0.5) and the lowest (0.05) leaching fraction, respectively, and was higher than the average EC of the whole profile. At 120 cm, soil salinity increased exponentially with decreasing leaching factor. Depending on the leaching fraction, IR_gincreased from 1787 mm (without leaching) per growing season to 2716 mm (LF = 0.5) for 100% ET_c (Table 3).

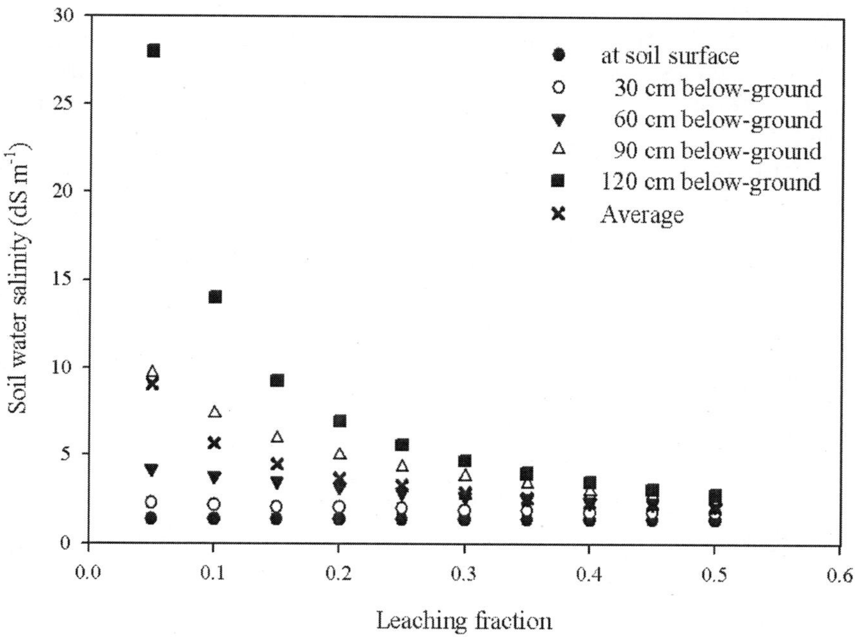

Figure 2. Simulated average soil water salinity (expressed as electrical conductivity) at the soil surface, 30 cm, 60 cm, 90 cm and 120 cm below ground and average root-zone salinity resulting from different leaching fractions.

In irrigated agriculture, the concentrations of total dissolved solids and the proportion of sodium (Na) in irrigation water have long been recognized as key factors for classifying water as suitable, problematic or unsuitable [33]. Irrigation with TSE often leads to elevated soil Na concentrations [8,34] that may have negative effects on crop growth and yield [30,35], since they adversely affect water availability and may lead to toxic levels of Na in plant tissues.

Leaf K/Na ratio is considered a reliable indicator of yield loss [36], taking into account that Na uptake may reduce internal K availability. Silva *et al.* (2009) [37] reported that, due to severe ionic imbalances, toxicity symptoms (interveinal leaf necrosis) in young jatropha occurred if the K/Na ratio was below 0.5. A study on adult jatropha [29] reported, K/Na ratios of 1.74 and 0.72 under 2 and 5 dS m^{-1} of soil salinity, respectively, induce a reduction in canopy transpiration. Therefore, the salinity level of 1.3 dS m^{-1} of the TSE in Morocco (available data were reported for the wastewater treatment plant in Bensilmane) (within the range of 0.12 to 2.65 for 0 to 300 mmol L^{-1} NaCl in the nutrient solution [29]), would still be acceptable for irrigation of adult jatropha, considering the entire soil profile (Figure 2). Reducing the *LF* to 0.4, EC at 120 cm already reached 5 dS m^{-1}, whereas a *LF* of 0.3, 5 dS m^{-1} is already reached within the defined root zone at given irrigation intensities. A *LF* smaller than 0.5, therefore, has to be considered as critical for the growth of jatropha under irrigation with TSE in the Tan-Tan area.

2.3. Nutrient Requirement

As irrigation requirement increased with including a leaching fraction, so did the nutrient supply through TSE. The nutrient input of N, P and K was calculated using the water quality data of Kouraa *et al.* (2002) [38]. Average concentrations in the effluent were multiplied by the basic irrigation requirements plus *LF*:

$$Input = C * AW$$

where C is the matter concentration (mg L^{-1}) and AW is the applied water (L).

Total mineral N (N_{min}), mineral P (PO_4-P) and (K) increased from 80 to 151, from 7 to 14 and from 160 to 304 kg ha^{-1}, respectively, with an increasing *LF* from 0.05 to 0.5 (Table 3). With minimum leaching factor (0.00), nutrient load corresponded to 76kg N ha^{-1}, 7 kg P ha^{-1} and 152 kg K ha^{-1} during the growing season. In detail, nutrient delivery from the effluent was highest in June, with an N, P and K input of 16, 1 and 31 kg ha^{-1}, respectively. Until September, it decreased constantly to a monthly input of 5 kg N ha^{-1}, 0.5 kg P ha^{-1} and 10 kg K ha^{-1} (Figure 3).

Assuming a potential seed yield of 1.52 t ha^{-1} [17], with a total aboveground biomass of 7.7 t ha^{-1} reported by Gosh *et al.* (2007) [17], of which approximately 75% are seasonally produced new biomass [39] jatropha would absorb in total 178 kg N ha^{-1}, 10 kg P ha^{-1} and 147 kg K ha^{-1}, respectively (Table 4).

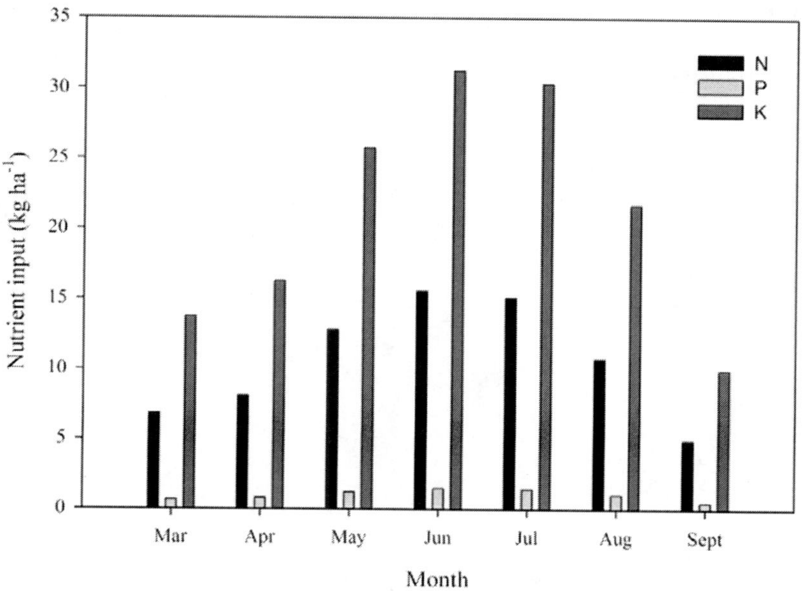

Figure 3. N-P-K input from effluent irrigation (IR = 779 mm, LF = 0) during one growing period of jatropha in Tan-Tan, Morocco.

At present, there are no precise fertilizer recommendations for jatropha available. The plant is known to respond well to nutrient supply [40]. According to the biomass partitioning suggested by Jongschaap *et al.* (2007) [41] the share of the fruits is 1.7 t DM, which is in the range of moderate yields compared to the wide range of seed yield reported (0.2 to 7 t ha^{-1}) [40,41].

The IR of a mature jatropha (more than three years old [24]) plantation under Southern Moroccan climate was calculated to be 779 mm per growing season (LF = 0). This would theoretically deliver a total amount of 76 kg of N, 7 kg of P and 149 kg of K per hectare and growing season. An irrigation amount of 1113 mm per growing season (LF = 0.3) would deliver 108, 10 and 217 kg of N, P and K, respectively, per hectare. If an irrigation amount of 1559 mm per growing season (LF = 0.5) was applied, the corresponding N-P-K amounts would increase to 151 kg N, 14 kg P and 304 kg K, respectively.

Table 4. Estimated biomass partitioning of jatropha (% of aboveground biomass) and the corresponding N, P and K stocks in the seasonal newly developed parts (estimated at 75% of the total dry weight of standing biomass [39]) of the jatropha plant (percentage and element weigh, kg) [17,39,42,43] based on above ground biomass (DM) of 7.7 t ha^{-1} including seed yield of 1.52 t.

Total biomass	Partitioning	N	P	K	N	P	K
	(%DM)		%			kg ha^{-1}	
Wood	58	3.34	0.09	2.87	112	3	96
Leaves	15	4.70	0.15	3.77	41	1	33
Fruit							
Coat	7	0.19	00.4	2.35	1	0	9
Seed	20	2.15	0.50	0.73	25	6	8
Total					178	10	147

It was calculated that a jatropha above ground DM production of 7.7 t ha^{-1} would fix a total amount of 178 kg N ha^{-1}, 10 kg P ha^{-1} and 147 kg K ha^{-1} in the biomass during the growing season (Table 4). Compared to the potential input of N, P and K from 779 mm of TSE (=ET_c) ($LF = 0$), there would be a lack of N (−102 kg) and P (−3 kg) a surplus of K (+5 kg). Assuming LF values between 0.3 and 0.5 with an irrigation of 1113 mm to 1559 mm, respectively, plant demand for K largely would exceed the requirements and P supply would largely be satisfied, whereas supply of N would still not be sufficient. This irrigation amount would exceed the minimum water requirement (500–600 mm) to obtain moderate seed yield [22], but soil salinity will be kept at a tolerable level. The question is if the plants are able to take up the entire nutrient load of the effluent. According to Janssen *et al.* (2005) [44] the nutrient value of TSE for crops can be estimated from target yield, soil fertility and use efficiency of the added nutrients. For N, losses by leaching and volatilization can be considerable, whereas accumulation is negligible. Elevated concentrations of mineral N, especially NO_3-N, have been found in the soil solution after TSE irrigation [45,46,47]. Leal *et al.* (2010) [48] reported N concentrations of up to 388 mg L^{-1}in the soil solution of TSE irrigated sugarcane throughout the growing season. However, the corresponding N concentrations in the effluent were 32 mg L^{-1}. In contrast, the TSE used in this study has N_{min} concentrations of only 10 mg L^{-1}. Regarding the relatively low nutrient contents of the TSE here and assuming irrigation according to the ET_c, no great leaching risks are expected.

Phosphorus, in contrast, is not easily leached, but may accumulate strongly under irrigation with high pH TSE, leading to an increase of soil pH [8,49]. For K, both leaching and accumulation is found, but since it is easy to mobilize again, it can be considered as plant available [44]. In order to accurately assess the nutrient availability, the amount of nutrients lost through leaching and the overall suitability of TSE to supplement nutrient management in bioenergy plantations, more information is needed relating irrigation with TSE to soil properties and plant nutrient requirements.

3. MATERIALS AND METHODS

3.1. Location of the Case Study Area

Tan-Tan province (−11.15°W, 28.45°N, 200 m.a.s.l.), located in the region of Guelmim-Es Semara, was chosen to represent the reference climate for Southern Morocco. This region is characterized by semi-arid climate with a cold winter and a hot and dry summer (average minimum and maximum monthly temperature of 14 and 25 °C, respectively) (Table 1) and an annual total rainfall of 112 mm typically occurring from October to February (average of 60 years data in monthly value [16]), where the number of rainy days is at most 30.

The soils in the Southern regions of Morocco are predominantly sandy. Major soil orders are Entisols, Aridisols and dunes [50]. In the FAO Classification [51], this corresponds to Regosols, Lithosols, Rankers, Yermosols and Xerosols. In this region, erosion by wind is common, because of vegetation scarcity and secondary salinization takes place. In the valley, saline soils are frequent.

3.2. Model Parameterization

In order to be able to accurately estimate water use and water extraction from the soil under cultivation with jatropha, it was necessary to estimate the parameters describing evapotranspiration and water uptake specific for jatropha. CROPWAT 8.0 needs rooting depths and crop coefficients (K_c), according to phenological stages, as inputs in order to estimate evapotranspiration. These parameters were not available for jatropha within the model and, therefore, were derived either from earlier experiments [29] or from the literature. Although experiments

[24] were performed under high CO_2concentration in a greenhouse, the results are in agreement with the results reported by Gush and Modley [23]; therefore, the K_c value for the initial and development stage were taken from Achten *et al.* (2010) [24] for field grown adult jatropha. Additionally, the coefficient value (K_c) during mid- and late-season, and the duration of growth stages were taken from Abou Kheira and Atta (2008) [13]. Planting date was set to the 1st of March. The dual crop coefficient procedure suggested by Allen *et al.* (1998) [52] was applied to convert basal crop coefficients into a single crop coefficient. Additionally, rooting depth was set to 0.3 and 1.2 m [52] for initial and mid-season, respectively, and critical depletion fraction to 0.4 [24] (40% plant available water [53]). The yield response factor (to water supply [14]) was set to 0.5 (initial stage) and 1 (all other stages). Sandy soils are common in the Southern regions of Morocco; the default values from the "UNSODA sand" soil characterized in the CROPWAT 8.0 database were selected. This soil type is characterized by a total available soil moisture of 180 mm m^{-1}, maximum infiltration rate of 120 mm day^{-1}, a maximum rooting depth of 120 cm, an initial soil moisture depletion of 20% and an initial available soil moisture of 144 mm m^{-1}.

3.3. Calculations

3.3.1. Irrigation Requirements (*IR*)

Knowledge of the water requirements of jatropha under Southern Morocco's typical climatic conditions is crucial to identify the potential effects of the treated sewage effluent (TSE) irrigation on the soil-plant system. Crop characteristics are described in Table 2, and *IR* was calculated using CROPWAT 8.0 model [14].

3.3.2. Leaching Fractions (*LF*)

Depending on water quality and the crop sensitivity to salinity, leaching fraction (*LF*) varies. This fraction is relevant to estimate *IR* and, therefore, salt accumulation and nutrient load in the soil, especially in the root-zone. In this study, a wide range of *LF* (0.05, 0.1, 0.15, 0.2, 0.25, 0.3, 0.35, 0.4, 0.45 and 0.5) was assumed, independently of the given electrical conductivity (EC) value of the TSE used as irrigation water (Table 2) and of the salinity tolerance of adult jatropha that had been rarely studied.

3.3.3. Gross Irrigation Requirement (IR_g)

The gross irrigation requirements (IR_g) can be calculated according to Phocaides (2000) [54]:

$$IRg = ((ET_c - P_e)K_r + IR) / E_a$$

where

- IR_g is the gross irrigation requirement (mm),
- ET_c is the average crop evapotranspiration (mm),
- P_e is the effective rainfall (mm),
- K_r is the reduction factor for crop cover (= 0.85 [13]),
- IR is the basic irrigation requirement (mm),
- and E_a is the irrigation efficiency (=0.8 corresponding to micro-irrigation system [54].

The water requirement necessary to generate a certain *LF* for high frequency sprinkler or drip irrigation can be calculated according to Ayers and Westcot, 1985 [30]:

$$IR = \frac{ET_c}{1 - LF}$$

where

- IR Applied water (mm),
- ET_c Crop evapotranspiration (mm),
- LF Leaching fraction.

3.4. Nutrient Requirements and Water Quality

Fertilizer application effects on seed yield of field-grown jatropha at adult stage had been roughly estimated [41]. To assess the nutrient requirement for adult jatropha, we compared the nutrient content (N, P and K) of TSE in Morocco and that of the jatropha plant, assuming an above ground biomass production of 7.7 t ha^{-1}, including 20% of the seed yield (corresponding to 1.52 t ha^{-1}) [17,41]. Above ground biomass partitioning has been adapted from Rajaona *et al.* (2011) [39], allocating 15% of dry mass (DM) to leaves and the rest to coat, stem and branches. We could not integrate the absorption of nutrients by below ground biomass (roots) of jatropha, as these data have never been published. Within the wide range of nutrient composition of jatropha components [41], a complete data set from Saturino *et al.* (2005) [42] and Pacheco *et al.* (2007) [43] has been selected.

We used the water quality data reported by Kouraa *et al.* (2002) [38] from a combined stabilization pond system in Benslimane, Morocco, to represent a modern type of waste water treatment plant (Table 5). In this combined system, pre-treatment is followed by anaerobic, aerated and facultative treatments; then, water is retained in storage reservoirs. This procedure allows the production of a high effluent water quality and the scheduling of irrigation independently from the daily outflow rates of the waste water treatment plants [38].

Table 5. Average values for physical and chemical characteristics of treated sewage effluent (Range in TSE) from different studies in the world [8,34,55,56,57]. Average values of treated sewage effluent (TSE) from a wastewater treatment plant in Benslimane, Morocco [38].

Constituent	Unit	Range in TSE	TSE
pH		7.8–8.1	8.4
Electrical Conductivity	(dS m^{-1})	1.0–3.1	1.4
Total Suspended Solids	(mg L^{-1})	–	28
Biochemical Oxygen Demand	(mg L^{-1})	10–80	16
Chemical Oxygen Demand	(mg L^{-1})	30–160	53
Mineral Nitrogen	(mg L^{-1})	10–50	9.7
PO_4-P	(mg L^{-1})	4.2–9.7	2.8
K^+	(mg L^{-1})	10–40	19.5
Cl^-	(mg L^{-1})	–	224
Ca^{2+}	(mg L^{-1})	20–120	7.5 (meq L^{-1})
Mg^{2+}	(mg L^{-1})	10–50	5.2 (meq L^{-1})
Na^+	(mg L^{-1})	50–250	103
Sodium adsorption ratio			2.1

3.5. Soil Salinity

Soil salinity has been calculated according to the FAO Irrigation and Drainage Paper 29 using the average electrical conductivity of the irrigation water (EC_{sw} average) and the leaching fraction in the root-zone [30]. Five points in the root zone were used to characterize salt accumulation: at soil surface, 30, 60, 90 and 120 cm below ground.

4. CONCLUSIONS

In the global context, it is evident that effluent reuse in agriculture will increase in water scarce countries in the near future, particularly in the vicinity of cities. This is necessary to release precious freshwater resources for the drinking water supply of the growing urban areas and to ensure irrigation. In this study, the water requirements of a jatropha plantation in Southern Morocco irrigated with TSE were determined. Depending on the leaching fraction needed to control salinity build-up in the soil, a surface of 41–64 ha (6 to 9 m^2 $inhabitant^{-1}$) could be irrigated. Since jatropha has been reported to be salt sensitive, the use of waste water while controlling soil salinity has to be considered, even if the nutrient and water supply can be satisfied. The N content of TSE at the considered irrigation requirement is not sufficient to produce a moderate seed yield, while P and K demand can be easily satisfied. The feasibility of such an irrigation system depends to a large extent on environmental factors, such as climate, soils and the overall water availability, and thus, transferability of the model results presented here needs to be studied further.

REFERENCES AND NOTE

1. Sala, O.E.; Stuart Chapin, F., III; Armesto, J.J.; Berlow, E.; Bloomfield, J.; Dirzo, R.; Huber-Sanwald, E.; Huenneke, L.F.; Jackson, R.B.; Kinzig, A.; et al. Global biodiversity scenarios for the year 2100. Science 2000, 287, 1770–1774.
2. Abdelfadel, A.B.; Driouech, F. Climate change and its impacts on water resources in the Maghreb region. Available online: http://www.arabwater council.org /administrator/Modules/Events/IWRA%20Morocco%20Paper (accessed on 21 November 2012).
3. Coqk, S. Link Global Markets: Trade Information Packet—Kingdom of Morocco; World Trade Center: San Diego, Country, 2011; p. 77.
4. The World Bank. Project Performance Assessment Report: Morocco—Water Resources Management Project; Independent Evaluation Group (World Bank): Washington, DC, USA, 2009; p. 67.
5. Berkat, O.; Tazi, M. Country pasture/forage resources profiles Morocco. Available online: http://www.fao.org/ag/AGP/AGPC/doc/Counprof/Morocco/morocco.htm (accessed on 21 November 2012).
6. Frenken, K. L'irrigation en Afrique en chiffres: Enquête AQUASTAT—2005; Organisation des Nations Unies pour l'Alimentation et l'Agriculture: Rome, Italy, 2005; p. 652.
7. United Nations Economic Commission for Africa. Water Resources Development in North Africa: Summary of Subregional Report; UNECA: Tanger, Morocco, 2005.
8. Da Fonseca, A.F.; Herpin, U.; De Paula, A.M.; Victória, R.L.; Melfi, A.J. Agricultural use of treated sewage effluents: Agronomic and environmental implications and perspectives for Brazil. Sci. Agric. 2007, 64, 194–209.

9. Shon, H.K.; Vigneswaran, S.; Snyder, S.A. Effluent Organic Matter (EfOM) in Wastewater: Constituents, Effects, and Treatment. Crit. Rev. Env. Sci. Tec. 2006, 36, 327–374.
10. Ministère de l'énergie des Mines de l'Eau et de l'environnement Secteur de l'énergie, chiffre clés. Available online: http://www.mem.gov.ma/ ChiffresCles /Energie/ChiffreEnergie.htm (accessed on 21 November 2012).
11. Silva, E.N.; Ribeiro, R.V.; Ferreira-Silva, S.L.; Viégas, R.A.; Silveira, J.A.G. Comparative effects of salinity and water stress on photosynthesis, water relations and growth of Jatropha curcas plants. J. Arid Environ. 2010, 74, 1130–1137.
12. GEXSI. Global Market Study on Jatropha; Jatropha alliance: London, UK, 2008; p. 187.
13. Abou Kheira, A.A.; Atta, N.M.M. Response of Jatropha curcas L. to water deficit: Yield, water use efficiency and oilseed characteristics. Biomass Bioenergy 2009, 33, 1343–1350.
14. Swennenhuis, J. CROPWAT 8.0; Food and Agriculture Oganization of the United Nations: Rome, Italy, 2009.
15. KFW—Entwiklungsbank. Morocco: Drinking Water Supply Guelmim/Tan Tan and Youssoufia/Chemaia; KFW—Entwiklungsbank: Frankfurt am Main, Germany, 2006.
16. Grieser, J. CLIMWAT 2.0; Food and Agriculture Oganization of the United Nations: Rome, Italy, 2006.
17. Ghosh, A.; Patolia, J.S.; Chaudharry, D.R.; Chikara, J.; Rao, S.N.; Kumar, D.; Boricha, G.N.; Zala, A. In Response of Jatropha curcasunder Different Spacing to Jatropha De-Oiled Cake; FACT Foundation: Wageningen, The Netherlands, 2007.
18. Kaushik, N.; Kumar, K.; Kumar, S.; Kaushik, N.; Roy, S. Genetic variability and divergence studies in seed traits and oil content of Jatropha (Jatropha curcas L.) accessions. Biomass Bioenergy 2007, 31, 497–502.
19. Daey Ouwens, K.; Francis, G.; Franken, Y.J.; Rijssenbeek, W.; Riedacker, A.; Foidl, N.; Jongschaap, R.E.E.; Bindraban, P.S. Position Paper on Jatropha curcas State of the Art, Small and Large Scale Project Development; Foundation FACT: Wageningen, The Netherlands, 2007.
20. Maes, W.H.; Trabucco, A.; Achten, W.M.J.; Muys, B. Climatic growing conditions of Jatropha curcas L. Biomass Bioenergy 2009, 33, 1481–1485.
21. Holl, M.A.; Gush, M.B.; Hallowes, J.; Versfeld, D.B. Jatropha curcasin South Africa: An Assessment of Its Water Use and Bio-Physical Potential; WRC: Pretoria, South Africa, 2007.
22. Jongschaap, R.E.E.; Blesgraaf, R.A.R.; Bogaard, T.A.; van Loo, E.N.; Savenije, H.H.G. The water footprint of bioenergy from Jatropha. curcas L. Proc. Natl. Acad. Sci. USA 2009, 106, E92–E92.
23. Gush, M.B.; Moodley, M. Water Use Assessment of Jatropha curcas; Report No. 1497/1/07 (Chapter 4). In An Assessment of Its Water Use and Biophysical Potential; Holl, M.A., Gush, M.B., Hallowes, J., Versfeld, D.B., Eds.; Water Research Commission: Pretoria, South Africa, 2007.
24. Achten, W.M.J.; Maes, W.H.; Reubens, B.; Mathijs, E.; Singh, V.P.; Verchot, L.; Muys, B. Biomass production and allocation in Jatropha curcas L. seedlings under different levels of drought stress. Biomass Bioenergy 2010, 34, 667–676.
25. Rajaona, A.M.; Brueck, H.; Asch, F. Leaf gas exchange characteristics of Jatropha as affected by Nitrogen supply, leaf age and atmospheric vapor pressure deficit. J. Agron. Crop Sci. 2012. in press.
26. Allen, R.G.; Pereira, L.S.; Raes, D.; Smith, M. Crop Evapotranspiration: Guidelines for Computing Crop Water Requirements Irrigation and Drainage; Paper No. 56; Food and Agriculture Organization of the United Nations: Rome, Italy, 1998.
27. Dos Santos, C.M.; Verissimo, V.; Wanderley Filho, H.C.d.L.; Ferreira, V.M.; Cavalcante, P.G.d.S.; Rolim, E.V.; Endres, L. Seasonal variations of photosynthesis, gas exchange, quantum efficiency of photosystem II and biochemical responses of Jatropha curcas L. grown in semi-humid and semi-arid areas subject to water stress. Ind. Crop Prod. 2013, 41, 203–213.

28. Li, G.T. The photosynthesis and water use efficiency of eight garden tree species. For. Res. 2002, 15, 291–296.
29. Rajaona, A.; Seckinger, C.; Brueck, H.; Asch, F. Effect of salinity on canopy water vapor conductance of young and 3-years old Jatropha curcas L. J. Arid Environ. 2012, 87, 35–41.
30. Ayers, R.S.; Westcot, D.W. Water Quality for Agriculture Irrigation and Drainage; Paper No. 29 Rev. 1; Food and Agriculture Organization of the United Nations: Rome, Italy, 1985.
31. Lado, M.; Ben-Hur, M. Treated domestic sewage irrigation effects on soil hydraulic properties in arid and semiarid zones: A review. Soil Till. Res. 2009, 106, 152–163.
32. Balks, M.R.; Bond, W.J.; Smith, C.J. Effects of sodium accumulation on soil physical properties under an effluent-irrigated plantation. Soil Res. 1998, 36, 821–830.
33. Tanji, K.K.; Kielen, N.C. Agricultural Drainage Water Management in Arid and Semi-Arid Areas; Irrigation and Drainage Paper No. 61; FAO: Rome, Italy, 2002.
34. Prescod, M.B. Wastewater Treatment and Use in Agriculture; Irrigation and Drainage Paper No. 47; Food and Agriculture Organization of the United Nations: Rome, Italy, 1992.
35. Kijne, J.W.; Rathapar, S.A.; Woperis, M.C.S.; Sahrawat, K.L. How to Manage Salinity in Irrigated Lands: A Selective Review with Particular Reference to Irrigation in Developing Countries; International Water Management Institute: Colombo, Sri Lanka, 1998; p. 44.
36. Asch, F.; Dinkuhn, M.; Dörffling, K.; Miezan, K. Leaf K/Na ratio predicts salinity induced yield loss in irrigated rice. Euphytica 2000, 113, 109–118.
37. Da Silva, E.N.; Silveira, J.A.G.; Fernandes, C.R.R.; Dutra, A.T.B.; de Aragão, R.M. Ion uptake and growth of physic nut under different salinity levels. Rev. Cienc. Agron. 2009, 40, 240–246.
38. Kouraa, A.; Fethi, F.; Fahde, A.; Lahlou, A.; Ouazzani, N. Reuse of urban wastewater treated by a combined stabilisation pond system in Benslimane (Morocco). Urban Water 2002, 4, 373–378.
39. Rajaona, A.; Brueck, H.; Asch, F. Effect of pruning history on growth and dry mass partitioning of Jatropha on a plantation site in Madagascar. Biomass Bioenergy 2011, 35, 4892–4900.
40. Achten, W.M.J.; Verchot, L.; Franken, Y.J.; Mathijs, E.; Singh, V.P.; Aerts, R.; Muys, B. Jatropha bio-diesel production and use. Biomass Bioenergy 2008, 32, 1063–1084.
41. Jongschaap, R.E.E.; Corré, W.J.; Bindraban, P.S.; Brandenburg, W.A. Global Jatropha curcas Evaluation, Breeding and Propagation Programme; Wageningen UR: Wageningen, The Netherlands, 2007; p. 66.
42. Saturnino, H.M.; Pacheco, D.D.; Kakida, J.; Tominaga, N.; Gonc, A.N.P. Cultura do pinhao-manso (Jatropha. curcas L.). In EPAMIG-Informe Agropecuário; EPAMIG: Belo Horizonte, Brazil, 2005; Volume 26, pp. 44–78.
43. Pacheco, D.D.; Saturino, H.M.; Santos, D.A.; da Souza, R.P.D.; de Almeida Junior, A.B.; Robeiro, D.P.; Antunes, P.D. Avaliçao nutritional de pinhão-manso em função de cálcio e magnésio usado como corretivos de acidez de solo. Powerpoint presentation for EPAMIG-Informe Agropecuário. 2007.
44. Janssen, B.H.; Boesveld, H.; Rodriguez, M.J. Some theoretical considerations on evaluating wastewater as a source of N, P and K for crops. Irrig. Drain. 2005, 54, S35–S47.
45. Johns, G.G.; McConchie, D.M. Irrigation of bananas with secondary treated sewage effluent. I. Field evaluation of effect on plant nutrients and additional elements in leaf, pulp and soil. Aust. J. Agric. Res. 1994, 45, 1601–1617.
46. Speir, T.W.; Van Schaik, A.P.; Kettles, H.A.; Vincent, K.W.; Campbell, D.J. Soil and stream-water impacts of sewage effluent irrigation onto steeply sloping land. J. Environ. Qual. 1999, 28, 1105–1114.
47. Barton, L.; Schipper, L.A.; Barkle, G.F.; McLeod, M.; Speir, T.W.; Taylor, M.D.; McGill, A.C.; Van Schaik, A.P.; Fitzgerald, N.B.; Pandey, S.P. Land application of domestic

effluent onto four soil types: Plant uptake and nutrient leaching. J. Environ. Qual. 2005, 34, 635–643.

48. Leal, R.M.P.; Firme, L.P.; Herpin, U.; da Fonseca, A.F.; Montes, C.l.R.; dos Santos Dias, C.T.; Melfi, A.J. Carbon and nitrogen cycling in a tropical Brazilian soil cropped with sugarcane and irrigated with wastewater. Agric. Water Manag. 2010, 97, 271–276.
49. Stewart, H.T.L.; Hopmans, P.; Flinn, D.W.; Hillman, T.J. Nutrient accumulation in trees and soil following irrigation with municipal effluent in Australia. Environ. Pollut. 1990, 63, 155–177.
50. USDA NRCS Global soil region map. Available online: http://soils.usda.gov/use/worldsoils/mapindex/order.html (accessed on 21 November 2012).
51. IUSS Working group WRB. World Reference Base for Soil Resources 2006, First Update 2007; World Soil Resources Report No. 103; Food and Agriculture Organization of the United Nations: Rome, Italy, 2007.
52. Krishnamurthy, L.; Zaman-Allah, M.; Marimuthu, S.; Wani, S.P.; Kesava Rao, A.V.R. Root growth in Jatropha and its implications for drought adaptation. Biomass Bioenergy 2012, 39, 247–252.
53. Sadras, V.O.; Milroy, S.P. Soil-water thresholds for the responses of leaf expansion and gas exchange: A review. Field Crop Res. 1996, 47, 253–266.
54. Phocaides, A. Technical Handbook on Pressurized Irrigation Techniques; Food and Agriculture Organization of the United Nations: Rome, Italy, 2000.
55. Bouwer, H.; Chaney, R.L. Land treatment of wastewater. Adv. Agron. 1974, 26, 133–176.
56. Asano, T.; Pettygrove, G. Using reclaimed municipal wastewater for irrigation. Calif. Agric. 1987, 41, 15–18.
57. Feigin, A.; Bielorai, H.; Dag, Y.; Kipnis, T.; Giskin, M. The Nitrogen factor in the management of effluent-irrigated soils. Soil Sci. 1978, 125, 248–254.

CHAPTER 10

Comparison of Different Methods to Estimate Mean Daily Evapotranspiration from Weekly Data at Patna, India

Upadhyaya A*

Division of Land and Water Management, ICAR Research Complex for Eastern Region, Patna, India

ABSTRACT

Estimation of evapotranspiration is necessary for efficient water management and crop planning. Fifteen different methods of ET0 estimation were employed to compute daily reference evapotranspiration for the period 2010 to 2014. In the absence of reliable open pan evaporimeter data, FAO-56 Penman Monteith method was considered as one of the reliable method of ET0 estimation. The results showed that mean weekly evapotranspiration values obtained from Penman-Monteith method were very closer to FAO-56 Penman-Monteith method and values from all the other methods except FAO- 24 Pan, Christiansen Pan and Hargreaves methods generally predicted higher values of mean weekly daily ET0 in comparison to FAO-56 Penman-Monteith method. The analysis shows that mean weekly daily ET0 estimates of combination methods resulted better ET0 estimates than radiation, and temperature and evaporation methods.

Weekly ET0 values estimated by FAO-56 Penman-Monteith method were found to vary in the range of 1.3 mm/day to 6.7 mm/day. Average annual reference evapotranspiration was found as 1517.1 mm.

Keywords: Evapotranspiration; Penman-Monteith method; Pan evaporation; Crop water requirement

INTRODUCTION

Crop water requirement mainly depends upon crop sowing date, crop growth stages, crop duration and climatic conditions during the growing season of an individual crop. A good estimate of **EVAPOTRANSPIRATION** is essential for water balance irrigation scheduling and water resource planning and management. Since, direct measurement of ET_0 is difficult and time consuming, so most common practice is to estimate ET_0 from meteorological data available for the area employing empirical equations. There are numerous evapotranspiration equations developed and used by researchers but still there is confusion in selecting the reliable evapotranspiration estimation method for the area concerned. Some authors compared estimates of different ET_0estimation methods for different regions of the world. Al-Ghobari compared five different methods: FAO-Penman, Jensen- Haise, Blaney and Criddle, Pan Evaporation and calibrated FAO-Penman for four areas under local conditions of Saudi Arabia [1]. The results showed that estimated ET values were highly correlated with measured evapotranspiration values. George et al. used DSS model for estimation of reference evapotranspiration for three climatic conditions: Davis, Jagdalpur, and Kharagpur by different applicable methods [2]. Comparison of the estimates of different methods with the Penman- Monteith ET_0 estimates showed that Hargreaves, FAO- 24 Blaney Criddle and 1982 Kimberley- Penman methods ranked first, respectively for the Davis, Jagdalpur and Kharagpur stations. They recommended that DSS model is a user- friendly tool for estimating ET_0 under different data availability and climatic conditions. Itenfisu et al. made comparison between various methods of reference evapotranspiration with ASCE-Penman-Montieth equations in the United States [3]. Results showed that the ASCE standardized equation agreed best with full form of ASCE Penman-Montieth. The International Commission for Irrigation and Drainage and the Food and Agriculture Organization of the United Nations experts' consultation on revision of FAO methodologies for crop water requirements. Smith et al. recommended the use of FAO56-**PENMAN-MONTEITH METHOD** as the standard method to estimate ET_0[4]. Irmark et al. in north-central Florida, compared ET_0 values obtained from two equations derived from FAO56-PM (first equation was solar radiation (R_s) based and second net radiation (R_n) based) against FAO56-Penman-Monteith method and results were found correlated very well with the standard FAO56-Penman-Monteith method in all the locations [5]. Irmark et al. compared the estimates of 21 evapotranspiration methods with the estimate of FAO56- PM Method in humid climate of Florida [6]. The 1948 Penman method daily estimates were closest to the FAO56- PM method followed by 1963 Penman on the basis of standard error of estimate. Krishan et al. computed evapotranspiration rate for Pusa area in Bihar by various estimation methods [7]. They concluded that FAO- 24 Penman(c=1) is the closest whereas Hargreaves method is the farthest. The objective of this study is to compare the estimates of various methods with the FAO56-Penman Montieth Method estimates and evaluate the performance of these methods for Patna **CANAL** command.

METHODOLOGY

COLLECTION OF METEOROLOGICAL PARAMETERS

In order to estimate daily reference evapotranspiration, available daily meteorological data for the period from 2010 to April 2015 were collected from automatic weather station located at 25° 27' North latitude and 85° 10' East Longitude in Indian Council of Agricultural Research- Research Complex for Eastern Region, (ICAR-RCER) Patna. These **METEOROLOGICAL DATA** include minimum and maximum temperature, minimum and maximum relative humidity, wind velocity, and solar radiation. Mean weekly meteorological parameters are presented below in **TABLE 1**.

It may be observed from the Table that mean maximum temperature varied from 18.9°C (1st week) to 37.4°C (17th week) and mean minimum temperature varied from 7.3°C (2nd week) to 26.8°C (32nd week). Mean maximum relative humidity ranged from 63.2 percent (14th week) to 96.0 percent (1st week) whereas mean minimum relative **HUMIDITY** from 24.5 percent (14^{th}week) to 75.8 percent (32nd week). Wind velocity varied in the range of 0.4 m/s to 2 m/s. Mean solar radiation was observed to be minimum of 8.85 $MJm^{-2}day^{-1}$ in the 51st week and maximum of 30.95 $MJm^{-2}day^{-1}$ in the 22nd week, respectively.

Table 1: Mean meteorological parameters as observed at ICAR-RCER, Patna.

Week No.	Max. Temp. (°C)	Min. Temp. (°C)	Relative Humidity Max. (%)	Relative Humidity Min. (%)	Wind Speed (m/s)	Solar Radiation, MJ/m²/ day
1	18.9	8.2	96.0	57.9	0.7	12.34
2	20.1	7.3	95.4	57.4	0.5	13.08
3	20.2	7.8	94.7	60.6	0.6	12.98
4	22.3	9.2	91.9	53.3	0.9	13.88
5	21.9	10.4	91.0	53.0	0.7	14.20
6	23.5	11.4	90.4	54.3	0.7	14.23
7	25.6	12.1	89.7	47.7	0.8	15.96
8	26.9	13.4	90.6	49.6	0.7	15.32
9	28.0	14.1	84.8	43.3	0.8	16.99
10	29.6	13.6	81.0	32.3	0.8	18.98
11	30.4	15.3	76.3	33.7	0.8	19.33
12	32.8	16.1	73.4	27.6	0.7	21.34
13	34.7	18.5	72.2	26.1	0.8	23.16
14	36.6	20.1	63.2	24.5	1.1	24.29
15	36.4	21.5	64.1	28.7	1.3	23.20
16	36.7	22.6	73.5	31.5	1.2	24.80
17	37.4	23.3	68.8	35.7	1.4	24.70
18	35.6	24.3	76.5	46.5	2.0	28.78
19	34.8	24.9	72.5	46.0	1.3	28.03
20	34.8	24.9	76.0	48.4	1.4	28.83
21	35.3	25.6	77.9	49.4	1.5	29.94
22	35.2	25.7	79.5	49.2	1.2	30.95
23	35.2	25.9	78.0	50.7	1.4	27.32
24	33.9	26.0	85.1	62.3	1.4	26.85
25	32.5	25.5	88.4	69.6	1.5	21.45
26	32.9	26.3	87.5	66.0	1.3	22.56
27	33.0	26.6	89.3	70.1	1.1	23.73
28	32.7	26.5	86.4	70.6	1.3	20.18
29	32.6	26.7	89.8	73.4	1.3	22.20
30	33.0	26.5	86.2	62.7	1.1	22.55
31	32.5	25.0	89.0	67.2	1.1	26.47
32	32.9	26.8	90.8	75.8	1.2	24.57
33	31.5	26.2	91.8	73.0	1.4	21.18
34	31.9	26.6	91.0	73.1	1.4	22.71
35	31.6	25.9	90.7	74.8	1.3	22.17
36	31.4	26.3	89.1	73.6	1.1	21.57
37	31.8	26.0	90.2	71.4	1.2	23.64
38	31.3	25.5	90.3	70.3	1.3	25.06
39	32.0	24.9	87.6	70.1	1.0	25.63
40	32.4	24.6	87.3	70.0	1.0	22.96
41	32.8	24.1	89.1	62.4	0.9	22.87
42	32.1	23.2	89.6	62.5	0.9	20.58
43	31.2	20.8	88.1	51.3	0.9	19.65
44	31.4	20.7	91.2	56.4	0.7	20.30
45	30.9	17.9	91.6	52.3	0.4	19.10
46	29.7	15.7	91.2	49.3	0.4	16.53
47	28.8	14.5	90.4	48.4	0.4	14.02
48	26.8	11.6	89.3	49.9	0.5	13.02
49	25.8	10.8	91.4	54.0	0.5	10.78
50	24.4	10.1	91.3	60.7	0.4	9.14
51	23.1	9.9	93.8	59.0	0.5	8.85
52	22.9	10.1	94.3	49.2	0.5	10.65

COMPUTATION OF EVAPOTRANSPIRATION

Computation of crop water requirement needs estimation of reference evapotranspiration (ET_0). Direct measurement of ET_0 is difficult and time consuming, so most common practice is to estimate ET_0 from meteorological data available for the area employing empirical equations. Daily reference evapotranspiration (ET_0) values for the period from 2010 to 2014 were estimated using thirteen different methods. All the methods used in this study are cited and described in DSS-ET model of ET_0 estimation. The methods used were grouped according to their classification as (1) combination methods such as FAO-56. Penman-Monteith Allen et al. [8] Penman-Monteith Monteith, [9] Kimberley Penman Wright, [10] FAO-PPP-17 Penman Frere and Popov, [11] Penman VPD#1, Penman VPD#3 Kimberley Penman Wright and Jensen, [12] FAO-24 Penman c=1 Doorenbos and Pruitt [13,14], Businger-van-Bavel, [15,16] and CIMIS Penman Snyder and Pruitt [17] (2) **RADIATION METHODS** such as Turc [18] and Priestly-Taylor, [19] and (3) Temperature method such as Hargreaves [20]. The list of data requirement for each method is given below in **TABLE 2**.

Table 2: Input parameters for different methods of ETo estimation.

Methods	Temperature		Relative Humidity		Wind Velocity	Solar Radiation	Pan Evaporation
	Min.	Max.	Min.	Max.			
FAO-56Penman-Monteith	√	√	√	√	√	√	-
Penman-Monteith	√	√	√	√	√	√	-
1982Kimberley Penman	√	√	√	√	√	√	-
FAO-PPP	√	√	√	√	√	√	-
Penman 1963 VPD#1	√	√	√	√	√	√	-
Penman 1963 VPD#3	√	√	√	√	√	√	-
1972Kimberley Penman	√	√	√	√	√	√	-
FAO-24 Penman c=1	√	√	√	√	√	√	-
Hargreaves	√	√	-	-	-	-	-
Businger-van-Bavel	√	√	√	√	√	√	-
FAO-24 Pan	-	-	√	√	√	-	√
Christiansen Pan	√	√	√	√	√	-	√
Turc	√	√	√	√	-	√	-
Priestly-Taylor	√	√	√	√	√	√	-
CIMIS Penman	√	√	√	√	√	√	-

The ICID and the FAO expert consultation on revision of FAO methodologies for crop water requirements have recommended that FAO-56 Penman-Monteith method be used as the standard method to estimate ET_0. In this study also FAO-56 Penman-Monteith method was used as an index to estimate daily ET_0. The FAO-56 Penman-Monteith is a grass reference equation that was derived from the ASCE equations by fixing grass height=0.12 m for clipped grass using latent heat of vaporization (λ)=2.45 MJ kg^{-1}; bulk surface resistance of 70 sm-1, and albedo of 0.23. The equation to estimate daily ET_0 (mm/d) is given as:

$$ETo = \frac{0.408\,\Delta\,(R_n - G) + \gamma \frac{900}{T+273} U_2\,(e_s - e_a)}{\Delta + \gamma\,(1 + 0.34\,U_2)} \tag{1}$$

METHODS TO CALCULATE EVAPOTRANSPIRATION (ET_0)

ET methods recommended by ASCE are classified into four categories, namely, temperature, radiation, evaporation and combination methods. Description of each ET method is given below:

Hargreaves and Samani recommended estimation of solar radiation from extraterrestrial radiation and used temperature difference term [20]. Net solar radiation, relative humidity and wind parameters were not considered in the proposed equation mentioned below.

$$E_{to} = 0.0023\,R_A\,\sqrt{TD}\,(T_{mean} + 17.8) \tag{2}$$

Among the radiation methods, Turc proposed two equations of ETo estimation as a function of daily mean air temperature and total incoming solar radiation (Rs) for the mean relative humidity (RH) >50% and (RH) <50% as given below [18].

When RHmean > 50%

$$E_{to} = 0.013 \frac{T_{mean}}{(T_{mean} + 15)} (R_s + 50) \frac{1}{\lambda} \tag{3}$$

When RHmean ≤ 50%

$$E_{to} = 0.013 \frac{T_{mean}}{(T_{mean} + 15)} (R_s + 50) \frac{1}{\lambda} \left(1 + \frac{(50 - RH_{mean})}{70}\right) \tag{4}$$

Priestly and Taylor proposed an equation for surface area generally wet, which is a condition, required for potential evaporation [19]. The aerodynamic component was deleted and energy component was multiplied by a coefficient, α=1.26. The final equation is as below

$$E_p = \alpha \frac{1}{\lambda} \frac{\Delta}{(\Delta + \gamma)} (R_n - G) \tag{5}$$

Where Δ is expressed as

$$\Delta = \frac{4098\,e^0_{mean}}{(T_{mean} + 237.3)^2} \tag{6}$$

e0 mean (kPa) is expressed as

$$e^{0}_{mean} = 0.6108 \exp\left[\frac{17.27\,T_{mean}}{T_{mean} + 237.3}\right] \tag{7}$$

Christiansen and Hargreaves developed equation for estimating reference evapotranspiration from USWB class 'A' **PAN EVAPORATION** and several weather parameters.

$$E_{to} = 0.755\, E_{v}\, C_{T2}\, C_{w2}\, C_{H2}\, C_{s2} \tag{8}$$

C_{T2}, C_{w2}, C_{H2} and C_{s2} are estimated by following set of equations:

$$C_{T2}=0.862 + 0.179\,(T_{mean}/20) - 0.041((T_{mean}/20)^2 \tag{9}$$

$$C_{w2}=1.189 - 0.240\,(W/6.7) + 0.051(W/6.7)^2 \tag{10}$$

$$C_{H2}=0.499 + 0.620\,(RH_{mean}/60) - 0.119\,(RH_{mean}/60)^2 \tag{11}$$

$$C_{s2}=0.904 - 0.0080\,(S/0.80) + 0.088\,(S/0.80)^2 \tag{12}$$

Doorenbos and Pruitt suggested FAO-24 pan method in which, ET_o depends on two different conditions for estimating the pan coefficient values, one is for **VEGETATION** while other is for bare soil [13].

$$E_{to}=k_p\,E_{pan} \tag{13}$$

Frevert et al. has mentioned **POLYNOMIAL EQUATIONS** for estimating kp.

Penman in Vapor pressure deficit (VPD) #1, combined the two components i.e., radiation and aerodynamic components and defined general form of resulting equation for a **WELL-WATERED GRASS** reference as [21]:

$$E_{to} = \frac{1}{\lambda}\frac{\Delta}{(\Delta+\gamma)}(R_n - G) + \frac{1}{\lambda}\frac{\gamma}{(\Delta+\gamma)}\,6.43\,W_f^{P1}\,(e_z^0 - e_z)^{VPD\#1} \tag{14}$$

Here,

$$W_f^{P1}=1.0 + 0.537\,u_2$$

Businger, Monteith and van Bavel adopted a more theoretical vapour pressure function based on earlier work of resulting the following Businger

Bavel equation for a wet surface with zero resistance to vapour transfer [9,15,16].

$$E_{to} = \frac{1}{\lambda}\frac{\Delta}{(\Delta+\gamma)}(R_n - G) + \frac{1}{\lambda}\frac{\gamma}{(\Delta+\gamma)}\frac{0.622\rho\lambda k^2}{P}\frac{u_z}{[\ln((z-d)/z_0)]^2}(e_z^0 - e_z)^{VPD\#1} \tag{15}$$

Here 0.622 is the ratio of the molecular mass of **WATER** to the apparent molecular mass of dry air. Penman suggested to use VPD#3 in place of VPD#1 in his proposed equation, the resulting equation becomes [15]:

$$E_{to} = \frac{1}{\lambda}\frac{\Delta}{(\Delta+\gamma)}(R_n - G) + \frac{1}{\lambda}\frac{\gamma}{(\Delta+\gamma)} 6.43\, W_f^{P3}\,(e_z^0 - e_z)^{VPD\#3} \tag{16}$$

$$(e_z^0 - e_z)^{VPD\#3} = 0.5\,(e^0_{min} + e^0_{max}) - e_d^d$$

Penman-Monteith equation proposed by Monteith [9] was employed to estimate ETo for clipped crop surface condition considering crop height as 0.12 for humidity and temperature measurement at the height of 2 m. The equation used variable daily value of latent heat of vaporization and bulk surface resistance and aerodynamic resistance. The equation is available in many text books.

Wright and Jensen recommended more common wind function for the Penman method. The revised method is referred as 1972 Kimberly Penman method and is given below [12]:

$$E_{to} = \frac{1}{\lambda}\frac{\Delta}{(\Delta+\gamma)}(R_n - G) + \frac{1}{\lambda}\frac{\gamma}{(\Delta+\gamma)} 6.43\, W_f^{72KP}\,(e_z^0 - e_z)^{VPD\#3} \tag{17}$$

Here W_f 72KP=0.75 + 0.993 u_2.

Doorenbos and Pruitt [13,14] presented a modified Penman equation for estimating reference ET for grass which is known as FAO-24 corrected Penman [15]. If value of c is set to 1.0 in FAO 24 corrected Penman equation, then the resulting equation is called FAO- 24 Penman (c=1) method.

$$E_{to} = c\,[\frac{\Delta}{(\Delta+\gamma)}(R_n - G) + \frac{\gamma}{(\Delta+\gamma)} 2.7\, W_f^{F24CP}\,(e_z^0 - e_z)^{VPD\#1}] \tag{18}$$

Here W_f^{F24CP} is wind function for FAO Corrected Penman method and it can be calculated by W_f^{F24CP}=1 + 0.864 u_2 and c is adjustment factor and expressed as:

$c=0.68 + 0.0028\, RH_{max} + 0.018\, R_s - 0.068\, u_d + 0.013\,(u_d/u_n) + 0.0097\, u_d(u_d/u_n) + 0.430 \times 10^{-4}\, RH_{max} R_s u_d$. If c=1, it becomes FAO 24 Penman (c=1) method.

Frere and Popov suggested the different wind functions for various ranges of temperatures and used VPD#1 for estimating vapour pressure deficit. The equation can be written as below [11]:

$$E_{tu} = \frac{1}{\lambda}\frac{\Delta}{(\Delta+\gamma)}(R_n - G) + \frac{1}{\lambda}\frac{\gamma}{(\Delta+\gamma)}\, 6.43\, W_f^{FP17P}\,(e_z^0 - e_z)^{VPD\#1} \qquad (19)$$

Here Wf FP17P is wind function for FAO-PPP Penman method. The suggested wind functions depend upon temperature difference.

Wright presented variable wind function coefficients for Etr using fifth order polynomials with calendar day D as independent variable at Kimberly, Idaho, Kimberly Penman equation used variable value of wind function coefficients using fifth order polynomials with calendar day as independent variable [10]. Simplified equation for wind function is expressed as:

$$W_f^{82KP} = a_w + b_w\, u_2 \qquad (20)$$

Where,

$a_w=0.4 + 1.4 \exp\{-[(D-173)/58]^2\}$ and $b_w=0.605 + 0.345 \exp\{-[(D-243)/80]^2\}$

D is calendar Day (i.e., day of the year). For southern latitude use D' in place of D and take D'=(d-182) for D≥ 182 and D'=(D+182) for D <182. Keeping all other parameters same, the 1982 Kimberly Penman may be written as follows:

$$E_{tw} = \frac{1}{\lambda}\frac{\Delta}{(\Delta+\gamma)}(R_n - G) + \frac{1}{\lambda}\frac{\gamma}{(\Delta+\gamma)}\, 6.43\, W_f^{82KP}\,(e_z^0 - e_z)^{VPD\#3} \qquad (21)$$

RESULTS AND DISCUSSION

Daily reference evapotranspiration were computed by the above mentioned methods and converted into mean weekly ET_0(mm/day). The results are presented below in **FIGURE 1**.

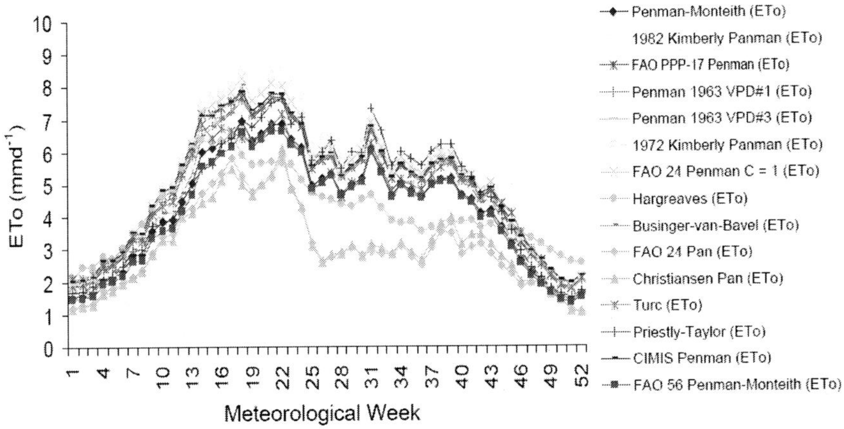

Figure 1: Variation of ET_0 values as estimated by various methods.

The weekly reference evapotranspiration (ET_0) estimates in year 2010, 2011, 2012 and 2013, and 2014 computed using FAO 56 Penman Monteith method varied from 1.4 mm/day (3rd week) to 8.0 mm/day (22nd week), 1.4 mm/day (50th and 51th weeks) to 7.2 mm/ day (22nd week), 1.1 mm/day (51st weeks) to 6.6 mm/day (18th week), and 1.3 mm/day (51st week) to 6.6 mm/day (16^{th}week), respectively. The average of five years weekly ET_0 values ranged from 1.3 mm/day (51st week) to 6.7 mm/day (21st and 22nd weeks), respectively. Annual reference evapotranspiration in these years were computed as 1535.4 mm, 1523.5 mm, 1485.2 mm, 1501.6 and 1539.9 mm, respectively, and average annual reference evapotranspiration was found as 1517.1 mm.

It may be observed from **FIGURE 1** that all the methods except Hargreaves, FAO 24 Pan and Christiansen Pan methods, follow the same trend of variation of ET_0 values in different weeks. ET_0 values computed by FAO 24 Pan and Christiansen Pan methods are always lower than ET_0 values computed by all other methods in all the weeks. ET_0 values computed by Hargreaves method are lower during 18th to 43rd week and higher during 44th week to 17th week compared to ET_0 values obtained from FAO 56 Penman-Monteith method. In rest of the methods the difference in ET_0 values obtained from Penman-Monteith and FAO 56 Penman-Monteith method is very less as compared to other methods.

The relationships between the ET_0 (mm/day) computed by fourteen various methods with ET_0 obtained from FAO56-Penman-Montetith method are given below in **FIGURE 2A TO 2N**.

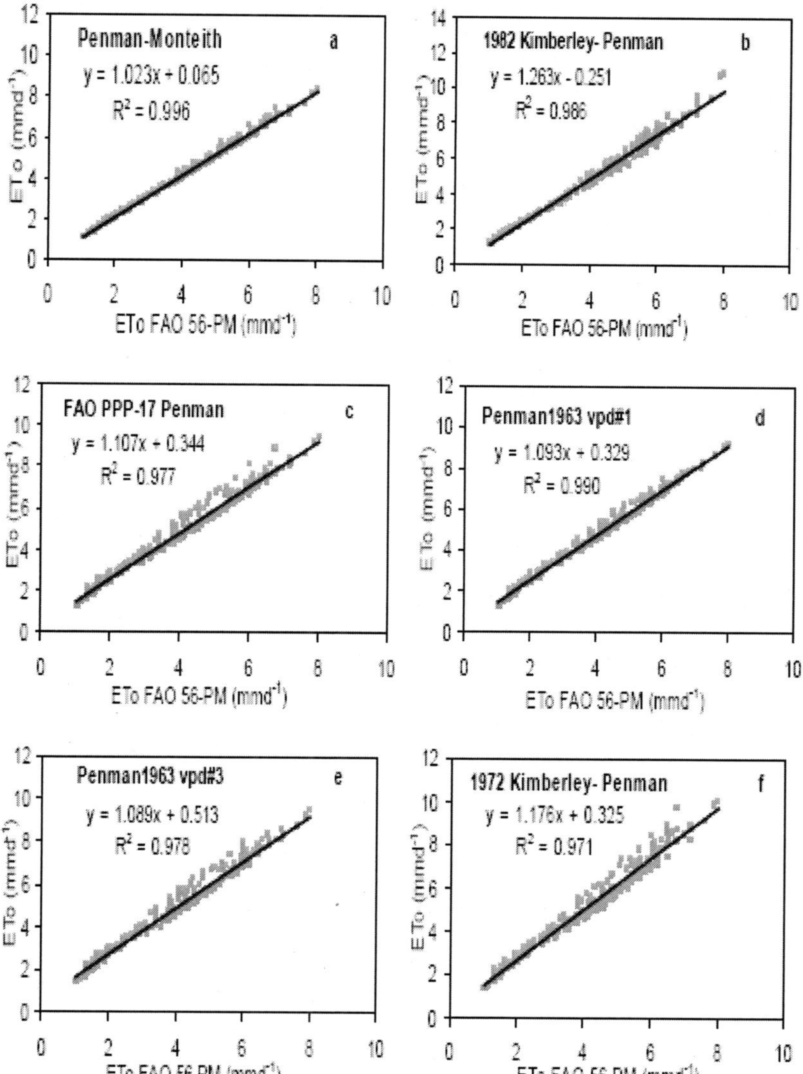
Penman-Monteith
a
y = 1.023x + 0.065
R^2 = 0.996
ETo (mmd^{-1})
ETo FAO 56-PM (mmd^{-1})
1982 Kimberley- Penman
b
y = 1.263x - 0.251
R^2 = 0.986
ETo (mmd^{-1})
ETo FAO 56-PM (mmd^{-1})
FAO PPP-17 Penman
c
y = 1.107x + 0.344
R^2 = 0.977
ETo (mmd^{-1})
ETo FAO 56-PM (mmd^{-1})
Penman1963 vpd#1
d
y = 1.093x + 0.329
R^2 = 0.990
ETo (mmd^{-1})
ETo FAO 56-PM (mmd^{-1})
Penman1963 vpd#3
e
y = 1.089x + 0.513
R^2 = 0.978
ETo (mmd^{-1})
ETo FAO 56-PM (mmd^{-1})
1972 Kimberley- Penman
f
y = 1.176x + 0.325
R^2 = 0.971
ETo (mmd^{-1})
ETo FAO 56-PM (mmd^{-1})

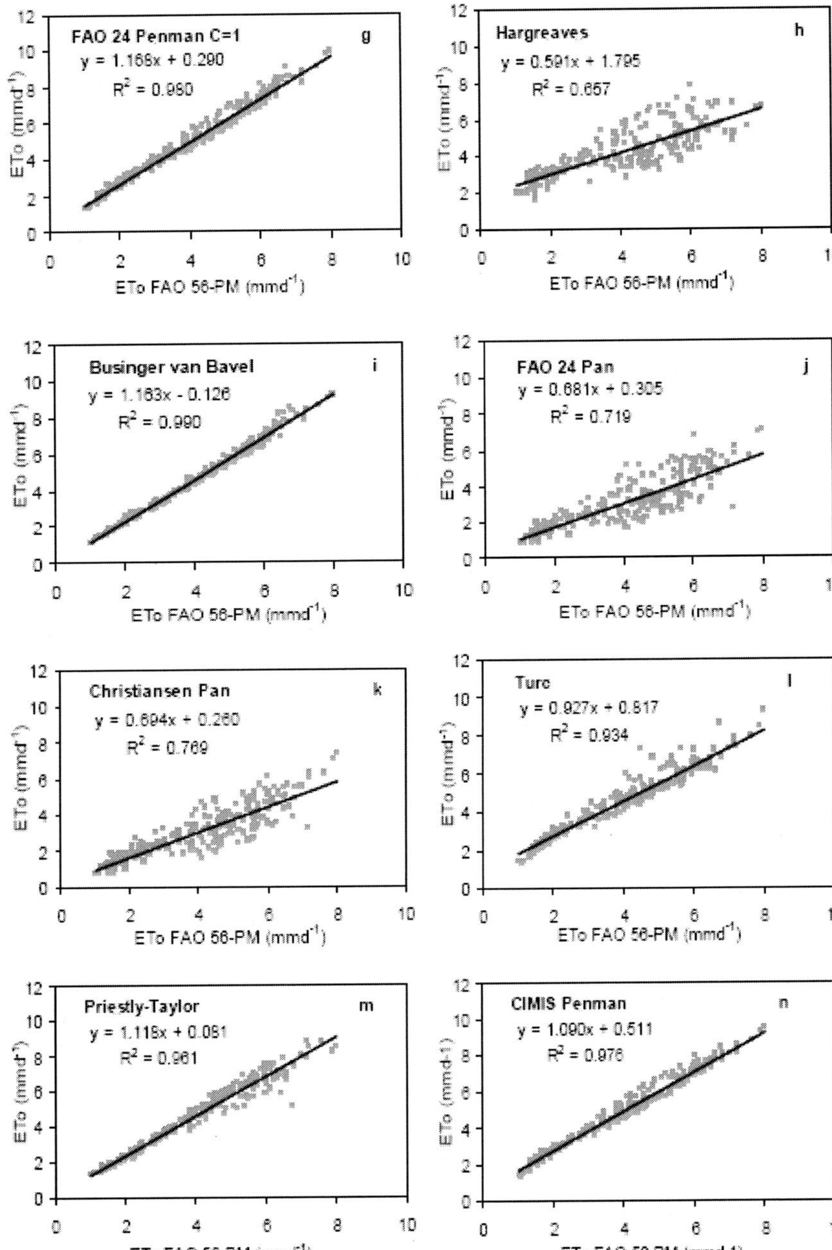

FIGURE 2: Variation of ET_0 values as estimated by various methods.

TABLE 3: Various statistical parameters and rank of ET_0 estimation methods.

Methods	Average ratio of ETo(method)/ ETo(FAO56-PM)	Slope	Intercept	Coefficient of determination	Rank
Penman-Monteith	1.04	1.023	0.065	0.996	1
1982 Kimberley Penman	1.20	1.263	-0.251	0.986	10
FAO-PPP	1.19	1.107	0.344	0.977	4
Penman1963 VPD#1	1.17	1.093	0.329	0.990	2
Penman1963 VPD#3	1.21	1.089	0.513	0.978	3
1972 Kimberley Penman	1.26	1.176	0.325	0.971	9
FAO-24 Penman C=1	1.24	1.168	0.290	0.980	8
Hargreaves	1.03	0.591	1.795	0.657	14
Businger-van-Bavel	1.13	1.163	-0.126	0.990	7
FAO-24 Pan	0.76	0.681	0.305	0.719	13
Christiansen Pan	0.76	0.694	0.260	0.769	12
Turc	1.13	0.927	0.817	0.934	11
Priestly-Taylor	1.14	1.118	0.081	0.961	6
CIMIS Penman	1.22	1.090	0.511	0.976	5

The ratio of Five years average of ET_0 (method) and ET_0 (FAO56- PM), slope (m), intercept (c) and coefficient of determination (r2) were computed in order to know the over and under estimation of all the methods with respect to the FAO-56 Penman-Monteith method are given below in **TABLE 3**.

In the above Table, the average ratios above and below 1.0 indicate over and under estimation of ET_0 compared to ET_0 values computed by FAO-56 Penman-Monteith method. The average ratios lie in the range of 0.76 to 1.26, which indicates that Christiansen Pan and FAO- 24 Pan under estimate, whereas other methods overestimate ET_0 values compared to ET_0 estimated by FAO-56 Penman-Monteith method [22]. Coefficient of determination represents the closeness of relationship of other ET_0estimation methods with FAO-56 Penman Monteith method and can be considered good indicator for deciding the rank. Penman Monteith with coefficient of determination of 0.996 was observed having 1st rank and Hargreaves with coefficient of determination 0.657 at 14th rank. The methods having rank between 1 and 5 were Penman Monteith, Penman1963 VPD#1, Penman1963 VPD#3, FAO-PPP, and CIMIS Penman, respectively and the methods having rank between 10 and 14 were 1982 Kimberley Penman, Turc, Christiansen Pan, FAO-24 Pan, and Hargreaves, respectively.

CONCLUSION

Fifteen different methods of ET_0 estimation were employed for climatic data of the period 2010 to 2014 and daily estimates of reference evapotranspiration were computed and converted into mean weekly daily values. The mean weekly daily ET_0 estimates from all the methods were compared against the ET_0 values obtained using FAO-56 Penman-Monteith method. On the basis of results, it can be said that all the methods produced significantly different evapotranspiration estimates than the FAO-56 Penman-Monteith method. Mean weekly daily evapotranspiration rates obtained from Penman-Monteith method were found very closer to FAO-56 Penman-Monteith method and values from all the other methods except Christiansen Pan, FAO- 24 Pan and Hargreaves methods, generally predicted higher values of mean weekly daily ET_0 in comparison to FAO-56 Penman-Monteith method. The performance of all the methods against FAO 56 Penman- Monteith was evaluated by assigning rank on the basis of coefficient of determination. The analysis shows that mean weekly daily ET_0 estimates of combination methods resulted better ET_0 estimates than radiation, temperature and evaporation methods. Penman-Monteith method produced best ET_0 estimate against FAO-56 Penman-Monteith estimated ET_0. Among the combination methods its performance followed by Penman VPD#1, PenmanVPD#3, FAO-PPP-17 Penman, CIMIS Penman, Businger-van-Bavel, FAO-24 Penman (c=1), 1972 Kimberley Penman method, 1982 Kimberley Penman. The performance of Turc radiation methods was less reliable than combination methods. Among the radiation methods Priestley-Taylor method performed much better than Turc method which may be due to consideration of wind parameter in its equation. The poorest estimates of temperature and evaporation methods were observed because these methods do not account for net radiation, vapor pressure deficit, solar radiation, or other important parameters, which have greater impact on ET_0 estimation.

REFERENCES

1. Al- Ghobari HM (2000) Estimation of reference evapotranspiration for Southern region of Saudi Arabia. Irrigation Science 19: 81-86.
2. George BA, Reddy BRS, Raghubanshi NS, Wallender WW (2002) Decision sport system for estimating reference evapotranspiration. Journal of Irrigation and Drainage engineering 128: 1-10.
3. Itenfisu D, Elliott RL, Allen RG, Walter IA (2003) Comparison of reference evapotranspiration calculations as part of the ASCE standardization effort. Journal of Irrigation and Drainage engineering 129: 440-448.
4. Smith M, Allen RG, Montieth JL, Perrier A, Pereira L, et al. (1991) Rep of the expert consultation on procedures for revision of FAO guidelines for prediction of crop water requirements. Food and Agricultural Organization of the United Nations, Rome.
5. Irmak S, Irmak A, Allen RG, Jones JW (2003a) Solar and net radiation-based equations to estimate reference evapotranspiration in humid climates. Journal of Irrigation and Drainage engineering 129: 336-347.
6. Irmak S, Allen RG, Whitty EB (2003b) Daily grass and alfalfa-reference evapotranspiration estimates and alfalfa-to-grass evapotranspiration ratios in Florida. Journal of Irrigation and Drainage engineering 129: 360-370.

7. Krishan R, Upadhyaya A, Roy LB (2015) Comparison of evapotranspiration rate for Pusa area of Bihar, by employing different methods through DSS. J Indian Water ResourSoc 35: 14-22.
8. Allen RG, Pereira LS, Raes D, Smith M (1998) Crop evapotranspiration: Guidelines for computing crop water requirements. Food and Agricultural Organization of the United Nations.
9. Monteith JL (1964) Evaporation and the environment. Proceedings of 19th Symposium of the Society for Experimental Biology. Cambridge University Press. pp: 205-234.
10. Wright JL (1982) New Evapotranspiration crop coefficients. Journal of the Irrigation and Drainage Divison 108: 57-74.
11. Frere M, Popov GF (1979) Agrometeorological crop monitoring and forecasting. FAO Plant Production and Protection Paper 17. FAO. Rome. pp: 38-43.
12. Wright JL, Jensen ME (1972) Peak water requirements of crops in Southern Idaho. Journal of Irrigation and Drainage Divison ASCE 96: 193-201.
13. Doorenbos J, Pruitt WO (1977) Guidelines for predicting crop water requirements. Irrigation and Drainage Paper No. 24. FAO. Rome.
14. Doorenbos J, Pruitt WO (1975) Guidelines for predicting crop water requirements. Irrigation and Drainage Paper No. 24. FAO. Rome.
15. Businger JA (1956) Some remarks on Penman's equations for the evapotranspiration. J AgricSci 4: 77.
16. Van Bavel CHM (1966) Potential evaporation: the combination concept and its experimental verification. Water resources Res 2: 455-467.
17. Snyder RL, Pruitt WO (1992) ET_0 User's Guide and Program Documentation. University of California, USA.
18. Turc L (1961) Estimation of irrigation water requirements, potential evapotranspiration: A simple climatic formula evolved up to date. Ann Agron12: 13-14.
19. Priestley CHB, Taylor RJ (1972) On the assessment of surface heat flux and evaporation using large-scale parameters. Mon Weather Rev 100: 81-92.
20. Hargreaves GL, Samani ZA (1985) Reference crop evapotranspiration from temperature. Applied engineering in Agriculture 1: 96-99.
21. Penamn HL (1963) Vegetation and hydrology. Tech Comm No 53, Commonwealth Bureau of Soils. England. p.125.
22. Penamn HL (1948) Natural evaporation from open water, bare soil and grass. Pro Roy Soc A 193: 120-146.

CHAPTER 11

Effect of Reference Conveyance Parameter Usage on Real Time Canal Performance: The Case of Fentale Irrigation Scheme in Ethiopia

Mebruk Mohammed[1,2], Adam Tefera[3]

[1]*Addis Ababa Institute of Technology, Addis Ababa University, Addis Ababa, Ethiopia*

[2]*Federal Policy Study and Research Center, Addis Ababa, Ethiopia*

[3]*Madawalabu University, Bale Robe, Ethiopia*

ABSTRACT

In designing a canal system, a major problem is to decide what conveyance parameter to apply in the calculations. Since basic knowledge on this subject is lacking, it is usually taken from literatures. Most of the irrigation projects in Ethiopia are found to work below their expectation. One of the main reasons is the conveyance parameters variation from the expected (design) value which ultimately affects the envisioned conveyance efficiency. To evaluate this variation, Fentale irrigation scheme was used as case study. The conveyance efficiency used at the design stage was 80%, which was within Food and Agricultural Organization of the united nations recommendations; while the field survey value was 17%. Such huge variation was due to the fact that the assigned conveyance parameter values (roughness coefficient, hydraulic radius and bed slope) no longer represent the current situation of the scheme. Such variation has resulted in increase in the depth and top width of the water surface which further resulted in 13% and 3% increase in wetted perimeter and top width of the canal, respectively. Thus this study suggests that conveyance parameters shall be derived from history of existing irrigation schemes in a country, rather than adopting it from standard literatures. As such construction quality, maintenance activities and technological transfer activities in a country shall be seen in deciding the conveyance parameters. The study also suggests that the ever increasing water shortage in an irrigation project could be managed by proper maintenance of the entire irrigation system.

Keywords: Conveyance Efficiency, Conveyance Parameters, Fentale Irrigation Scheme, Irrigation Canal Design

1. INTRODUCTION

Different irrigation schemes did not satisfy the net irrigation demand even though the available water is more than that, due to high losses from storage, conveyance, and application to irrigation plots. Identifying the various components with their level of losses and knowing what improvements can be made are essential in making most effective use of vital water resource. As available water resources become scarcer, more emphasis is given to efficient use of irrigation water for maximum economic return and water resources sustainability. This requires appropriate methods of measuring and evaluating how effectively water extracted from a water resource is used to produce crop yield [1] , which intern shall start from the planning and design stage of an irrigation project.

In planning and designing a canal system, a major problem is to decide what conveyance parameters to apply in the calculations. Since basic knowledge on this subject is lacking, it is common practice that this efficiency is usually assumed. Obviously, the parameters thus obtained are unlikely to suit the conditions of the canal in its future state. Because conveyance parameters future state is usually the "guess" factor in the design of canal system, engineers are facing the problem of uncertainty in their calculations. Due to this uncertainty, canals and their appurtenant hydraulic structures are being given a greater or lower capacity than would be necessary. Apart from harmful side effects, this way of doing things leads to investments that may be considerably higher than would otherwise be necessary [2] .

Stakeholders in Fentale Irrigation scheme complain that water stress is increasing from year to year as information obtained from staff members of supervisory bureaus. The same idea was supported by other studies ([3] [4] [5]). Water conveyance loss consists mainly of operation losses, evaporation and seepage into the soil from the sloping surfaces and bed of the canal. The most important of these is seepage. The seepage loss in the irrigation canals accounts for the major portion of water conveyance loss (98.37%) while approximately 0.3 per cent of the total stream is lost due to evaporation [6] . Evaporation loss in irrigation networks is generally not taken into consideration [7] [8] .

Seepage losses could be enhanced due to inappropriate conveyance parameter estimation which leads to smaller irrigation efficiency. Among the causes of irrigation inefficiency, conveyance efficiency is one of them. Main causes for low canal conveyance efficiency would be hydrologic, hydraulic and/or technological transfer. This research limits the hydrologic and hydraulic reasons as it would be hydrologic in the sense that the evaporation and seepage losses are under estimated. It would be hydraulic, if the designed hydraulic parameter values of the conveyance system might not reflect the real values. If the trained farmers in the water use association failed to operate the irrigation system, the effect of technological transfer in having low conveyance efficiency is an obvious consequence.

Appropriate value for conveyance parameters in planning and design will lead to improve equity in water distribution and minimize the gap between crop water requirement and actual water use. Therefore, the assessment of the potential cause of canal conveyance inefficiency is vital for sustainable functionality of irrigation schemes. Two main principal factors affecting conveyance efficiency (hydraulic and hydrologic factors) will be studied in this paper by taking a main canal as a case study. The envisioned (designed) and actual conveyance characters of the main canal in Fentale Irrigation scheme found in Awash Basin of Ethiopia will be assessed for possible hydraulic and hydrologic reasons of inefficiency.

2. MATERIALS AND METHOD

2.1. Location of the Study Area

In order to evaluate canal conveyance parameters, Fentale irrigation scheme in Awash River Basin was selected. Awash River basin, geographically located in the world's main rift system, covers a total area of 110,000 km^2 and is by far the most utilized river basin of Ethiopia which lends itself for usual water shortage complaints [9] . Fentale Irrigation Scheme is found at a distance of 200 km the capital of Ethiopia, Addis Ababa. Generally, the project is in low land agro-ecological zone characterized by low annual precipitation (~500 mm) and high temperature (10°C to 38.3°C) and. It falls in between 8°50'N and 9°04'N latitude and 39°41'E and 39°52'E longitude, with altitude ranging from 970 to 1020 m above mean sea level. The headwork is at an elevation of 1050 m amsl [9] .

The planned irrigation area is about 27,000 hectares (ha). Irrigation water for abstraction is raised by weir at the headwork of the project and conveyed through main canal of 49.3 km length and two Branch Canals (38 and 31 km length) with pertinent structures including Electro-mechanical equipment for Kawa command area [9] [10] . The whole canal system uses gravity driven flow except Kawa command area, which is irrigated by pump. Except the major conveyance canal, all the project distribution systems were subsurface pipes.

Fentale was designed and constructed to take maximum of 18 m^3/sec of water from the river by Diversion Weir with conveyance efficiency of 83% (see Table 1). The diverted irrigation water is conveyed through the 49.3 km Main Canal. At 34.650 km a Branch Canal (BC-I) bifurcates towards the Haro-arbo area and the other Branch Canal-II (BC-II) going towards Galcha (North Block) after 49.3 km. The designed discharge of branch Canal-I and Branch canal-II were 6.72 m^3/sec and 6.0 m^3/sec respectively [9] (Figure 1).

2.2. Data Collection

The first activity done was the reconnaissance survey in order to get the general insight of irrigation projects in the basin. During which water use associations, professional staffs and some farmers were consulted about water use practices, highlight of the status of the projects and inefficiencies encountered by different projects. Such consultation and reconnaissance survey results were used to identify the data type, location and time to collect these data. Frequent field observation at Fentale project was made to judge places where canal discharge measurement must be taken. After that, main canal discharge at carefully selected points of the project was taken by using the Current Meter in collaboration with the Hydrology Department of the Ministry of Water Resources and Energy of Ethiopia.

Table 1. Computed and designed Ec at different locations along the main canal.

Location at	Slope		Manning's n		Discharge		Ec	
	Existing	Design	Existing	Design	Existing	Design	Existing	Design
HR	0.001	0.0012	0.049	0.02	4.351	18	-	-
17.52 km	0.0012	0.001	0.073	0.03	3.355	17	0.77	0.94
26.9 km	0.001	0.001	0.048	0.02	1.239	15	0.37	0.88
42.7 km	0.0008	0.001	0.026	0.017	0.536 + 0.2**	7.2 + 6.72*	0.59	0.93
Canal conveyance efficiency							**0.17**	**0.77**

*The discharge carried by branch canal (BC-I) located at 34th km. **The discharge observed in the branch canal I (BC-I) during the study time.

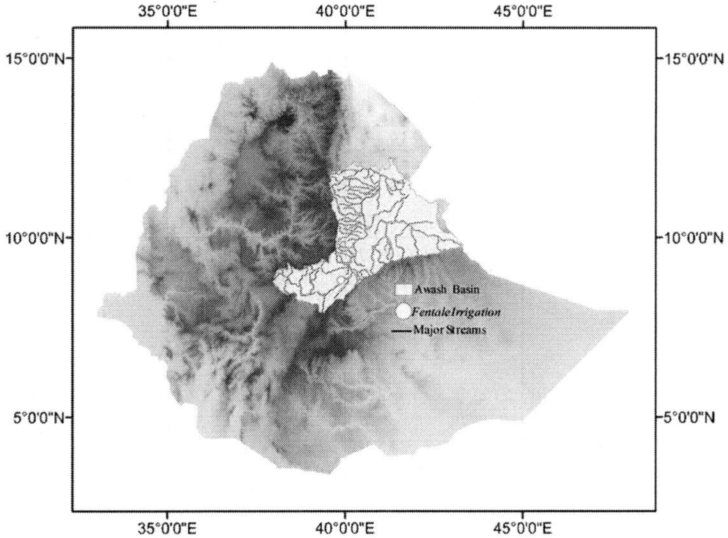

Figure 1. Location of the study area.

With this Current meter, canal discharge was measured at different depths across the width of the canal. The discharge measured by this equipment was obtained by dividing the number of clicks by the duration in which the number of clicks was counted. To bring about high precision in measurement, the duration in which the number of clicks counted was made to be greater than forty seconds, which is in accordance with [11] . The points of measurement taken were at straight reaches of the canal. Besides the canal slope and cross-sectional survey was also conducted along and across the location where discharge measurement was taken. The discharge is measured through careful measurement of velocity, slope, wetted area and perimeter of the canal.

While measuring the velocity, the canal section was first divided in to 0.5 m - 1 m interval cross-sectional lengths. Then on the basis of the existing depth of flow at each interval, velocities were measured at one or two depth locations. Such velocity data were collected at four locations (see Table 2) along the main canal of Fentale irrigation scheme.

Slope of the canal were measured through surveying the canal bed profile for 200 m upstream and 150 m downstream of the canal section at which the velocity is measured. Wetted area and perimeter were also obtained through careful survey of the canal cross section.

Table 2. Observed data location naming and canal section description.

Main canal reach	Designed Q (m^3/sec)	Canal shape*	Canal-section taken at	Canal bed elevation (m amsl)
From the head regulator (HR) to about 17.52 km	18	Trapezoidal 1.5 H × 1 V and 1.0 H × 1.0 V (short distance for stretch covered by sound rock)	HR	1050.0
From 17.52 km to 26.9 km	17		17.52 km	1032.48
From 26.9 km to 42.7 km	15		26.9 km	1023.1
From 42.7 km to 49.3 km	7.2	Trapezoidal 1.5 H × 1 V except rectangular** for stretches covered by sound rocks	42.7 km	1007.3

*Canal stretches of permeable soil formation; geo-membrane lining with a protection of boulders covered with geo-mesh was adopted. **The particularity of this part of the canal is that, it was constructed to serve as storage by widening the canal section and providing head regulating weir at the end.

2.3. Data Analysis

Irrigation water is normally conveyed from a water source to the farm or field through natural drainage ways, constructed earthen or lined canals, or pipelines. Many conveyance systems have transmission losses, meaning that water delivered to the farm or field is usually less than the water diverted from the source. Water losses in the conveyance system include canal seepage, canal spills, evaporation losses from canals, and leaks in pipelines. These losses have to be taken into account when calculating the gross irrigation requirements of an irrigation project. This can be done through the use of an efficiency factor called conveyance efficiency, which has to be estimated at the planning stage. The higher the conveyance efficiency, the lesser the water loss, thus the larger the area that can be irrigated from a given finite water source, and the less the leaching of nutrients and damage to the soil and the more environmentally friendly the irrigation system will become. Canal conveyance efficiency, for the purpose of this study being defined as the ratio of the volume of water delivered at the downstream end to the volume of water released at the upstream end of the canal section, can be derived by the use of the following equation [2] :

$$Ec = \frac{V_d + V_2}{V_c + V_1} \times 100 \quad (1)$$

where

Ec = conveyance efficiency;

V_c = volume of water at upstream canal end (m^3);

V_1 = inflow, if any, from other sources within the section (m^3);

V_d = volume of water delivered to the farmer (at the downstream end of the section) (m^3);

V_2 = non-irrigation deliveries from the conveyance system (m^3).

Conveyance efficiency could also be analyzed on the basis of measured discharge data, which is the volume of water per unit time. In this study discharge was measured at different locations over the length of the canal in order to calculate the Ec_i at different sections. The overall conveyance efficiency of the canal is then the product of Ec_i calculated at different sections over the length of the canal.

Many factors affect the conveyance efficiency of a canal, among which, hydraulic and hydrologic causes are presented in this paper. To evaluate the effect of change in the conveyance efficiency the study and design documents of Fentale irrigation scheme was thoroughly assessed. The study and design documents of the project have used manning equation for canal design. This is acceptable, as the canals could be treated as uniform flow where in manning's equation is popular [12] [13] [14] [15] . In using the manning's equation for conveyance efficiency evaluation, conveyance parameters like manning's roughness coefficient, canal slope and canal hydraulic radius were scrutinized. Thus observed values of manning's roughness coefficient, wetted perimeter and bed slope shall be compared with the designed values. Observed values of these conveyance characters were obtained through careful measurement of velocity, canal slope, wetted area and wetted perimeters of the canal at different locations and use these in manning equation for manning's roughness coefficient estimation as follows [16] .

$$Q = \frac{1}{n} S^{1/2} AR^{2/3} \Rightarrow n = \frac{S^{1/2} AR^{2/3}}{Q} \quad (2)$$

where,

Q is discharge of the canal (m^3/s) ? Measured in a field;

R = (A/P) is Hydraulic radius (m) ? Computed from field observed values;

A = Wetted cross-sectional area (m^2) ? Surveyed in filed;

P = Wetted perimeter (m) ? Surveyed in field;

S = bed slope (-) ? Surveyed in field;

n = Manning's roughness coefficient (-) ? Computed from field observed values.

In evaluating the hydrologic effects, the resulting canal dimension change effect on the possible hydrologic water loss reason were analyzed. Thus the possible evaporation and seepage loss along the canal length is evaluated and compared with the expected (designed) one.

3. RESULTS AND DISCUSSIONS

The first reconnaissance study to the site has consulted the potential stakeholders of Fenatale irrigation scheme (District Agricultural office, Irrigation Water users Association, and Oromia irrigation development authority). From the consultation, in addition to the canal cross section near by the head regulator, it was decided to observe canal discharges at three different locations along the main canal. The same consultation is used to decide the time and location on which non-irrigation delivery along these canals is minimal (V_2 in Equation (1) is zero). Besides the main canal do not have any other inflow along the reach selected (V_1 in Equation (1) is zero). Accordingly the data were collected during such defined time and locations. The locations of the measurements are shown in Table 2.

Fentale was designed and constructed to take maximum of 18 m^3/sec of water from the river by Diversion Weir. The main canal off taking from the diversion weir and serving the command area was designed and constructed for 24-hour continuous flow [9] . Nevertheless, during site observation (January, 2013), the discharge released from the weir was measured to be 4.513 m^3/s and the total command area developed was estimated to be 8000 ha.

The four locations naming, design capacity and observed discharges during field study were shown in Table 2 and the observed canal section profile with the corresponding velocity is depicted in Figures 2-5 below in which the scale bars show the velocity in m/s unit. For better depiction of the flow velocity, Figure 4 and Figure 5 are drawn by making the scale in the depth direction exaggerated.

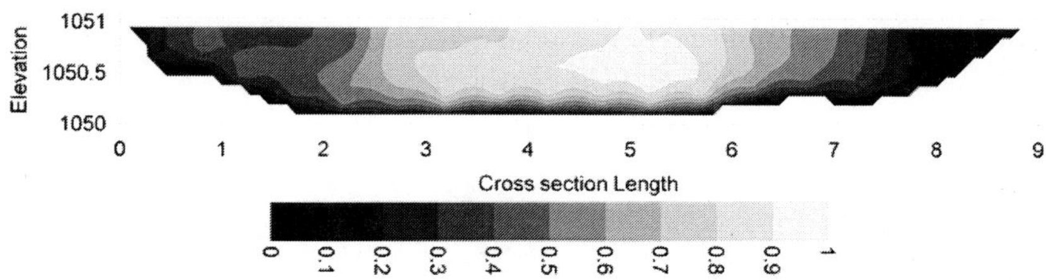

Figure 2. Observed canal section and velocity distribution at head regulator.

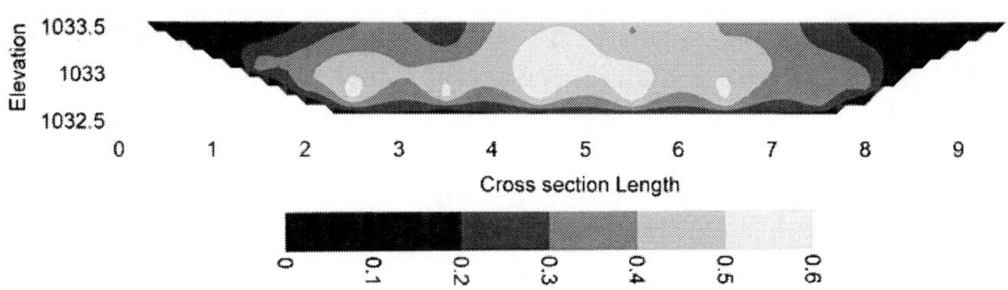

Figure 3. Observed canal section and velocity distribution at 17.52 km.

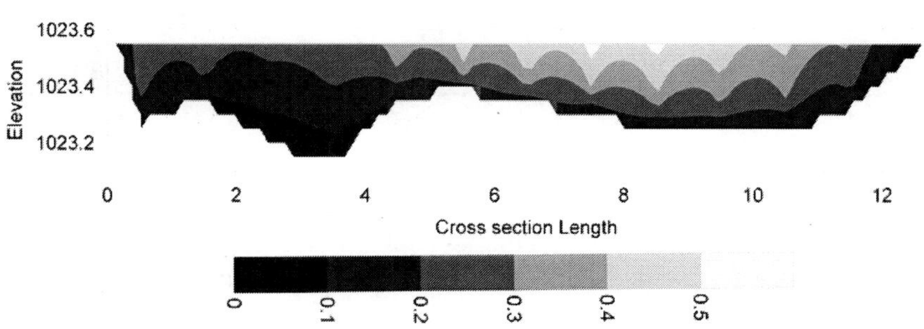

Figure 4. Observed canal section and velocity distribution at 26.9 km.

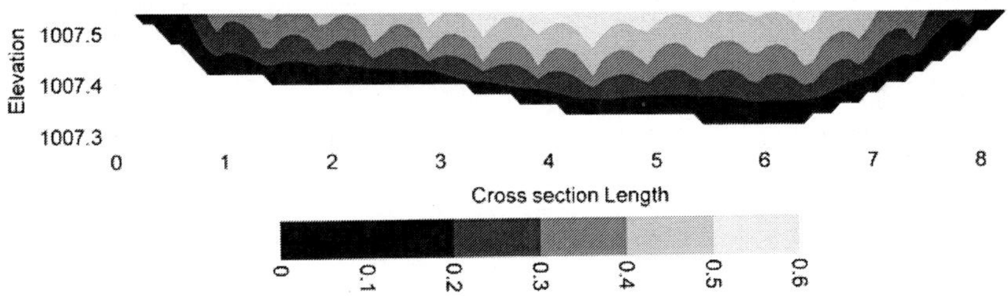

Figure 5. Observed canal section and velocity distribution at 42.7 km.

Conveyance Parameters

Hydraulic parameters variation from what has been estimated during design has a great effect on the conveyance efficiency (Ec) of canals. The conveyance parameters variation is seen by carefully analyzing the manning's roughness coefficient, wetted perimeter of the canal and top width of the canal water surface. The values of the existing manning's roughness coefficient for different sections along the main canal were between 0.026 - 0.073 (see Table 3); however, at the same locations the designed values range from 0.017 - 0.03. The increase in the value of roughness coefficient is mainly due to the canal section material change which ultimately decreases the discharge carrying capacity of the canal.

Table 3. Conveyance parameters at the four sections along the main canal.

Location at	S Bed slope*	A (m^2)*	P (m)*	R (m) (A/P)	Q (m^3/s)*	Manning's n**
HR	0.0012	7.409	9.682	0.636	4.351	0.049
17.525 km	0.001	8.706	10.428	0.683	3.355	0.073
26.9 km	0.001	4.118	13.201	0.29	1.239	0.048
42.7 km	0.001	1.445	8.431	0.16	0.536	0.026

*Measured at field level. **Computed by making use of Equation (2).

The conveyance efficiency of the main canal is also compared in Table 1 for the existing and designed condition. Conveyance efficiency is affected by several factors among which are size irrigated area, size of rotational unit, number and types of crops grown, type conveyance system and the technical and managerial facilities for water control [14] . Conveyance efficiency could be as high as 0.9, if continuous supply with no substantial change in flow occur. According to Food and Agricultural Organization of the united nations (FAO) [14] the lower limit for conveyance efficiency is

0.65, which presumed to exist if rotational supply in large schemes (>10,000 ha) and small schemes (<1000 ha) with respective problematic communication and less effective management.

The design Ec value for Fentale Irrigation project is within the recommended range, as far as the FAO recommendations are accepted. However, the existing conveyance efficiency could not add up to the expectations. The low efficiency indicates possible assumption errors in the design or construction of infrastructure. This variation complicates operation and maintenance of the irrigation system. The difference between the planned and actual efficiency is partly due to a too optimistic estimate of the post-con- struction efficiency the main canal.

It could be seen here that the values of Ec are by far different from one another. The decrease in efficiency is larger in the main canal reach after the 17.52 km length. This is mainly due to high silt deposition and wide coverage of the canal section with grass and weed as observed during the field survey.

To evaluate the possible water loss through seepage and evaporation due to the altered conveyance parameters, the wetted perimeter and top width variation of the canal were also computed for the design discharges by taking the bed width and side slopes of the design parameters. Accordingly Table 4 shows the results, from the table it is possible to deduce that the design dimensions will no longer carry the discharges, which exacerbate the seepage and the evaporation losses. The same table show that the design water depth has increased by 0.66 m, 0.43 m, 0.47 m, and 0.41 m at the head regulator at 17.52 km, at 26.9 km and at 42.7 km, respectively. This increase in water level will bring about overflow of canal water which is again another cause for conveyance Efficiency variation.

Table 4. Top width (TW) and wetted parameter (P) increase on the main canal.

Location	Design Q	Existing				Designed			% increase	
		Slope	y (m)	TW (m)	P (m)	y (m)	TW (m)	P (m)	TW	P
HR	18	0.0012	1.79	10.2	15.5	1.13	9.7	13.1	5	18
17.52 km	17	0.001	1.65	11.0	15.9	1.22	10.7	14.3	3	11
26.9 km	15	0.001	1.33	13.9	17.8	0.86	13.6	16.1	2	10
42.7 km	7.2	0.001	1.1	9.13	12.4	0.69	8.86	10.9	3	14
		Average increase in percent							3	13

4. CONCLUSIONS

The expected (designed) and the actual value of conveyance parameters obtained for the main canal of Fentale irrigation scheme show great difference; 77% during design against the 17% existing conveyance efficiency obtained during the study time. Such huge variation in conveyance efficiency resulted from variation in the hydraulic parameters like manning's roughness coefficient and canal slope; the former being the main contributor to the decrease in conveyance efficiency. Such variation will result in 42 - 66 cm water level rise, if the canal is expected to carry the design discharge. This rise in the water level has resulted in 3% increase in the top width of the canal water surface and 13% increase in wetted perimeter across the canal section. This in turn results in loses like evaporation from the top water surface and seepage through the increased wetted perimeter. Besides the increase in water depth could bring about over bank flow along the main canal section.

Such differences are basically either be improper conveyance parameter estimation during design and planning phase or improper operation and maintenance of the canal. The study clearly shows that, irrespective of the cause, there is a lot of water which is lost in the main canal. This is also true for other canal systems (secondary and lower order canals) which were not seen in this study, which makes the total loss a much larger threat to the overall benefit of the project. Thus this study suggests that conveyance parameters shall be derived from experience of existing irrigation systems, rather than adopting it from standard literatures. As such construction quality and maintenance activities shall be seen in deciding the conveyance efficiency estimation. The study also suggests that the ever increasing water shortage in an irrigation project could be managed by proper maintenance of the entire irrigation system.

ACKNOWLEDGEMENTS

The authors are grateful to Addis Ababa Institute of technology, Federal Policy study and research center for their strong support of the research publication. We would also like to acknowledge Ministry of water irrigation and Electricity for giving us the current meter for the discharge measurement. Last but not least the authors are grateful to the anonymous reviewers for their valuable comments that improve this publication.

CITE THIS PAPER

Mohammed, M. and Tefera, A. (2017) Effect of Reference Conveyance Parameter Usage on Real Time Canal Performance: The Case of Fentale Irrigation Scheme in Ethiopia. Computational Water, Energy, and Environmental Engineering, 6, 79-88.

REFERENCES

1. McCornic, P.G., Kamara, A.B. and Tadesse, G. (2003) Integrated Water and Land Research and Capacity Building Priorities for Ethiopia.
2. Bos, M.G. and Nugteren, J. (1990) On Irrigation Efficiencies. ILRI Publication.
3. Abdu, B. and Bekele, S. (2007) Analysis of Irrigation Systems Using Comparative Performance Indicators: A Case Study of Two Large-Scale Irrigation Systems in the Upper Awash Basin. International Water Management Institute, Addis Ababa.
4. Yesuf, K. (2004) Assessment of Small-Scale Irrigation Using Comparative Performance Indicators on Two Selected Schemes in Upper Awash River Valley. Unpublished MSc Thesis, Alemaya University.
5. OWWDSE (2009) Fentale Irrigation Based Integrated Project: Socioeconomic Study. Unpublished, Finfinne.
6. Badenhorst, J.W., De Lange, M., Mokwena, M.E. and Rutherford, R.J. (2002) Water Conservation and Water Demand Management in Agriculture: Development of Water Management Plans by Irrigation Water Suppliers in South Africa. ICID Eighteenth Congress Best Paper for 6th Hassan Ismail Award Paper, Montreal.
7. Xie, M., Kuffner, U. and Le Moigne, G. (1993) Using Water Efficiently: Technological Options. World Bank Technical Paper Number 205, Washington DC.
8. Australian National Committee on Irrigation and Drainage (ANCID) (2000) Open Channel Seepage and Control—Literature Review of Channel Seepage Identification and Measurement. Vol. 1.1, Victoria, Australia.
9. OWWDSE (2009) Fentale Irrigation Based Integrated Project: Engineering Report. Unpublished, Fifinne.
10. OWWDSE and Synergics Hydro (India) (2009) Fentale Irrigation Scheme Operation and Management Study Report. Unpublished, Finfinne.
11. Turnipseed, D.P. and Sauer, V.B. (2010) Discharge Measurements at Gaging Stations. USGS, Virginia.
12. Michael, A.M. (2009) Irrigation Theory and Practice. Vikas Publication House, New Delhi.
13. Bos, M.G., Wolters, W., Drovandi, A. and Morabito, J.A. (1991) The Viejo Retamo Secondary Canal-Performance Evaluation (Case Study). Irrigation and Draniage Systems, Mendiza.
14. Food and Agricultural Organization of the United Nations (FAO) (1992) Irrigation Scheme Operation and Maintenance: Irrigation Water Management. Manual No. 10, Rome.
15. FAO (1989) Guidelines for Designing and Evaluating Surface Irrigation Systems. Irrigation and Drainage Paper No. 45.
16. VenTe Chow (1959) Open-Channel Hydraulics. McGraw-Hill, New York.

CHAPTER 12

Optimizing Preplant Irrigation for Maize under Limited Water in the High Plains

I. Kisekka[a], A. Schlegel [b], L. Ma[c], P.H. Gowda [d], P.V.V. Prasad[e]

[a] *Kansas State University, Southwest Research-Extension Center, 4500 E. Mary St., Garden City, KS 67846, United States*

[b] *Kansas State University, Southwest Research-Extension Center, 1474 State Highway 96, Tribune, KS 67879, United States*

[c] *USDA ARS, Plains Area, Agricultural Systems Research Unit, 2150 Centre Ave., Fort Collins, CO 80526, United States*

[d] *USDA-ARS Grazing Lands Research Laboratory, El Reno, OK 73036, United States*

[e] *Kansas State University, Department of Agronomy, Manhattan, KS 66506, United States*

ABSTRACT

Due to inadequate irrigation capacity, some farmers in the United States High Plains apply preplant irrigation to buffer the crop between irrigation events during the cropping season. The purpose of the study was to determine preplant irrigation amount and irrigation capacity combinations that optimize yield, water productivity, and precipitation use efficiency (PUE) and minimize soil water evaporation losses prior to planting. The CERES-Maize model embedded in the RZWQM2 model in combination with long-term climatic data from 1986 to 2014 for southwest Kansas were used for this research.

Experimental data from 2006 to 2009 was used to calibrate and validate the model. Model performance was satisfactory with high index of agreement (IA > 0.88). Relative root mean square error (RRMSE) ranged between 4.5% and 27%. Under very limited irrigation capacity (2.5 mm/day), applying 75–100 mm of preplant irrigation produced median yields that were 10–17% higher than not applying preplant irrigation. However, even at limited irrigation capacity the benefit of preplant irrigation were only realized if the seasonal yield potential was in the range of 6000 to 10,000 kg/ha corresponding to years with normal seasonal rainfall. Irrigation capacity had a stronger effect on maize grain yield compared to preplant irrigation amount. Preplant irrigation increased ET and transpiration under 2.5 mm/day irrigation capacity. Preplant irrigation amount did not have a substantial impact on water productivity at high and moderate irrigation capacity but had second order dominant effect under limited irrigation capacity. At low irrigation capacity (2.5 mm/day) increasing preplant irrigation increased median PUE up to 18% although the effect was second order dominant. Negligible water losses through deep percolation from 2.4 m soil profile were simulated. Increasing preplant irrigation resulted in significantly higher soil water evaporation losses prior to planting at all irrigation capacities. Overall preplant irrigation is beneficial under very limited irrigation capacity but is not necessary under sufficient irrigation capacity in most years. The decision to apply preplant irrigation should be evaluated and implemented carefully in combination with other agricultural water management technologies and strategies such as soil water monitoring, drip irrigation, and residue management.

Keywords: Preplant irrigation; Irrigation capacity; RZWQM; DSSAT-CSM; CERES-Maize; Maize yield; Evapotranspiration; Soil water evaporation

1. INTRODUCTION

In the U. S. south and central High Plains, groundwater levels in the Ogallala aquifer are declining due to water withdrawals for irrigation exceeding average annual recharge (McGuire, 2004). This results in diminished well capacities that eventually became incapable of meeting full crop water needs during the summer growing season. For this reason, some irrigators on fine and medium textured soils apply preplant irrigation from groundwater to ensure that the soil profile has adequate water before planting to help their low irrigation capacity systems to better keep up with the crop water needs during the growing season. Preplant irrigation (also known as preseason, dormant season, off-season, or winter irrigation) is a water management strategy in which water is applied prior to or several months before planting. This practice is common to the U.S. South and Central High Plains (including areas in western Kansas, eastern Colorado and the panhandles of Texas and Oklahoma) as previously reported by Stone et al. (1994). Surveys conducted in western Kansas in the later part of the 20th century indicated over 60% of respondents used some form of preplant irrigation (Kromm and White, 1990). Recent droughts of 2011 and 2012 coupled with declining well capacities have stimulated interest again in preplant irrigation.

The major question for the producers is: when should preplant irrigation be used or when is it beneficial? Also, what is the best way to implement preplant irrigation in order to minimize nonproductive water losses that make this practice inefficient? Optimum preplant irrigation strategies target minimizing soil water evaporation, enhancing precipitation capture and storage and eliminating deep drainage. Preplant irrigation water is lost through two major processes; deep drainage and soil water evaporation. In wet years preplant irrigation might also use up storage that would have been used by spring rainfall which reduces rainfall storage efficiency. Stone et al. (2008) simulated efficiency of preplant irrigation and reported that available water in the soil at the time of the preplant irrigation application had the greatest effect on storage efficiency, which decreased dramatically when available water was greater than 60%. Stone et al. (2008) also reported that preplant irrigations made in the early spring were more efficient compared to irrigation made in the fall. Soil water during the off season is influenced by soil water at harvest which in turn is influenced by irrigation scheduling during the growing season. Given the significant effect of soil water on efficiency of preplant irrigation, a need exists for dynamic site specific decision support tools that can be used for predicting soil water over time and space and to assess need for and effect of preplant irrigation on soil water evaporation, transpiration, deep drainage, crop yield, and water productivity. Such a decision support tool can be developed from whole system agricultural models.

Whole system models integrate the physical, biological and chemical processes of an agricultural system and are very useful for extrapolating field research to different soils, climate and management, technology transfer and decision making (Ahuja et al., 2000). The need for such models and decision support tools will increase as farmers and other stakeholders demand quick transfer of research results in an integrated and usable form for site specific management. Calibrated and validated cropping systems models are an example of whole system agricultural decision support tools. Performance of these decision support tools can be enhanced by integrating them with field measurements such as soil water feedbacks. Examples of cropping systems models that can be used in assessing need and potential benefits of preplant irrigation include RZWQM2 (Ahuja et al., 2000), DSSAT-

CSM (Hoogenboom et al., 2015), WOFOST (Van Diepen et al., 1989), APSIM (Keating et al., 2003), and AquaCrop (Raes et al., 2009) Agricultural systems models have been successfully applied in assessing irrigation management in the U.S. Great Plains (Saseendran *et al.*, 2008; DeJonge *et al.*, 2012; Ma *et al.*, 2012; Kisekka *et al.*, 2016a ; Kisekka *et al.*, 2016b).

In this paper, the RZWQM2 (Root Zone Water Quality Model) was selected for the following reasons: 1) process oriented, dynamic, and simulates the impact of agricultural management practices such as tillage, residue management, irrigation, and fertility on soil water, crop production, and water quality (Ahuja et al., 2000), 2) DSSAT v4.0 has been embedded into RZWQM2 which provides a suite of detailed biophysical crop models for simulating crop growth and development (Ma et al., 2006), 3) RZWQM2 has advanced features for assessing limited irrigation strategies such as water allocation limits by period, irrigation scheduling based on ET or soil water deficit, and 4) an automatic optimization algorithm called PEST (Doherty, 2009) has also been embedded into RZWQM2 to facilitate parameter estimation which allows for reproducible and objective model calibrations.

The purpose of the study was to determine preplant irrigation amount and irrigation capacity combinations that optimize yield, water productivity, and precipitation use efficiency (PUE) and minimize soil water evaporation losses prior to planting. The research involved combining short term experimental data with long-term historic climatic data (1986–2014), and crop simulation modeling to determine optimum preplant irrigation water management in the United States High Plains.

2. MATERIALS AND METHODS

2.1. Experimental data

A field experiment was conducted at the Kansas State University Southwest Research-Extension Center near Tribune Kansas (38.5°N, 101.7°W, and 1086 m above sea level) from 2006 to 2009. The soil at the study site is a deep well drained Ulysses silt loam (fine-silty, mixed, mesic Aridic Haplustolls). The climate of the study site is semi-arid with mean annual precipitation of 440 mm and mean annual evapotranspiration of 1943 mm (1986–2014). The growing season precipitation from 2006 to 2009 is given in Table 1. The experimental design was a factorial with preplant irrigation (0 and 76 mm), irrigation capacity (2.5, 3.8 and 5.0 mm/day) and seeding rate (80,000 plants/ha) replicated four times. Experimental plots were 36 m long by 18 m wide with irrigation as the main plot factor and seeding rate as subplots (3 × 36 m) with maize row spacing of 76.2 cm. Standard practices for maize production in the region were followed including preplant application of 269 kg N ha^{-1} as urea-ammonium nitrate (the model developed based on this experimental data was not calibrated for nitrogen). The full season maize hybrids used were 33B53 in 2006, Pioneer 33B54 in 2007, Pioneer 34B99 in 2008, and Pioneer 34B94 in 2009. Maize was planted on 01/May/2006, 10/May/2007, 26/Apr/2008, and 06/May/2009. Two middle rows of each plot were mechanically harvested. For model calibration and validation only experimental data corresponding to the 80,000 plants/ha seeding rates was used. Total irrigation applications are summarized in Table 1. Details and results from the field study are reported in Schlegel et al. (2012).

Table 1. Preplant and growing season irrigation and growing season precipitation for a field study from 2006 to 2009 at Kansas State University Southwest Research-Extension Center near Tribune, Kansas (Schlegel et al., 2012).

Year	Preplant irrigation (mm)	Growing season irrigation (mm)			Growing season precipitation (mm)
		2.5[a] (mm/day)	3.8 (mm/day)	5 (mm/day)	
2006	76	243	320	483	176
2007	76	183	257	397	205
2008	76	209	278	375	238
2009	76	225	299	453	364

[a] Irrigation capacity.

Irrigation was done using a linear move system modified to apply irrigation based on different treatment combinations. Preplant irrigation was applied in two events of 38 mm several days apart in early spring (April). In-season irrigations were initiated from late May to early September. In-season irrigation was scheduled as follows 35 mm weekly, 25 mm weekly and 35 mm bi-weekly for the 5.0, 3.8 and 2.5 mm/day irrigation capacities respectively. In case of sufficient rainfall, a scheduled irrigation was postponed. Soil water was measured using neutron attenuation with a field calibrated neutron probe. Seasonal crop water use (ETc) was estimated from a soil water balance as reported in Schlegel et al. (2012).

2.2. Model description

The RZWQM2 model (version 3.0) with embedded DSSAT-CSM CERES-Maize 4.0 crop growth module was used in this study (Ma et al., 2012). In RZWQM2, unsaturated soil water flow and redistribution was modeled using a one dimensional Richards's equation. The potential evaporation and transpiration demand of the atmosphere were computed using the Shuttleworth–Wallace ET model (Ahuja et al., 2000). The Shuttleworth–Wallace ET model in RZWQM2 is an extension of the Penman-Monteith ET model but the former takes into account incomplete canopy cover and plant height in potential evaporation and transpiration estimations. Actual root water uptake described by the sink term in Richards's equation was calculated numerically using the Nimah and Hanks (1973) procedure. Actual root water uptake, potential evaporation and potential transpiration were used in computing water stress factors used in CERES-Maize to modulate plant growth process like leaf growth as soil water gets depleted. The CERES-Maize model embedded in RZWQM2 is a radiation-based mechanistic crop model that predicted maize growth and development based on weather (precipitation, solar radiation, maximum and minimum temperature, to a lesser extent photoperiod); 6 cultivar/genetic coefficients (P1, P2, P5, G2, G3, and PHINT [as defined in Table 2]); and management practices (plant population, row spacing, planting date, and irrigation management) as described in Jones et al. (2003). More details about the RZWQM can be found in Ahuja et al. (2000).

Table 2. Soil and plant parameters for simulating soil water flow and maize growth and development in RZWQM-DSSAT at the Kansas State University Southwest Research-Extension Center near Tribune, Kansas.

Name	Description	Layer	Initial value	Lower bound	Upper bound	PEST[a] optimized value
		1	1.32	1	2	1.37
		2	1.32	1	2	1.32
		3	1.32	1	2	1.28
		4	1.32	1	2	1.28
	Bulk Density (g/cm3)	5	1.32	1	2	1.32
		6	1.32	1	2	1.32
		7	1.32	1	2	1.29
		8	1.32	1	2	1.32
Ks	Saturated Hydraulic Conductivity (cmh^{-1})	1	1.57	1	20	1.56
		2	3.24	1	20	3.23
		3	2.53	1	20	2.52
		4	1.00	1	20	1.00
		5	1.98	1	20	1.98
		6	3.51	1	20	3.49
		7	1.44	1	20	1.43
		8	1.82	1	20	1.82
$\theta_{1/3bar}$	Volumetric Soil Water Content at 1/3 bar ($cm^3 cm^{-3}$)	1	0.33	0.28	0.38	0.34
		2	0.34	0.29	0.39	0.34
		3	0.30	0.26	0.35	0.29
		4	0.30	0.25	0.34	0.29
		5	0.33	0.28	0.38	0.32
		6	0.34	0.29	0.39	0.33
		7	0.24	0.20	0.28	0.24
		8	0.23	0.20	0.27	0.23
θ_{15bar}	Volumetric Soil Water Content at 1/3 bar ($cm^3 cm^{-3}$)	1	0.16	0.13	0.18	0.16
		2	0.09	0.08	0.11	0.09
		3	0.15	0.13	0.18	0.16
		4	0.13	0.11	0.15	0.13
		5	0.14	0.12	0.16	0.14
		6	0.11	0.10	0.13	0.11
		7	0.14	0.12	0.16	0.14
		8	0.16	0.13	0.18	0.16
P1	Thermal time from seedling emergence to the end of Juvenile (degree days)		200.00	100.00	450.00	205.0
P2	Extent to which development is delayed due to increase in photoperiod		0.70	0.01	2.0	0.70
P5	Thermal time from silking to physiological maturity (Degree Days)		900.00	600	1000	867.0
G2	Maximum possible number of kernels per plant		516	440	1000	531.0
G3	Grain filling rate during the linear grain filling stage ($mgday^{-1}$)		12	5	16	12.5
PHINT	Phyllochron interval (degree days)		38.9	25	55	41.0

[a] Parameter Estimation algorithm.

2.3. Calibration and validation

Meteorological data was obtained from a weather station located near the study site and data accessed through the High Plains Regional Climate Center (http://www.hprcc.unl.edu/). The soil profile was set to 2.4 m comprising of eight layers 30 cm thick. Soil texture measured at the site was 12% sand, 63% silt, 25% clay within typical ranges of a Ulysses silt loam soil. Initial values and range of saturated hydraulic conductivity were obtained from the NRCS's Web Soil Survey for Ulysses silt loam soil. Initial values of cultivar coefficients were taken from one of the maize hybrids (7000 maize IB0031 PIO 3475) in the DSSAT-CSM database that produced close predictions to measured data during initial manual calibration. In this study, the treatment that received highest irrigation (75 mm preplant irrigation and 5 mm/day irrigation capacity) for each of the four years 2006–2009 was used for model calibration. The rest of the treatments (irrigation capacity of 3.8 and 2.5 mm/day with and without preplant irrigation) were used for model validation using data from 2006 to 2009 (Schlegel et al., 2012). One third of the data was used for calibration and the remaining two thirds were used for validation.

Following initial manual calibration of cultivar coefficients, all soil and plant parameters (P1, P2, P5, G2, G3, and PHINT), soil properties (bulk density, vertical *Ks*), and soil water capacity ($\theta_{1/3}$ and θ_{15}) were fitted using a gradient-based parameter optimization algorithm (PEST) embedded in the RZWQM2 model (Doherty, 2009). Default values for parameters related to carbon and nitrogen cycle processes were used. Parameter bounds in Table 2 were used to constrain automatic parameter estimation by PEST to ranges that were physically meaningfully for the study site based on expert knowledge. Parameters were log transformed to improve linearity to ensure the objective function was efficiently minimized. PEST Regularization was also used to minimize model overfitting. Following PEST optimization, the model was executed with uncertainty (with the best 26 parameter sets) to determine the parameter set that produced the least RMSE following similar procedures as reported in (Malone et al., 2010).

The model was executed continuously from January 1st, 2006 to December 31st, 2009 for each treatment. Observed grain yield, and seasonal crop water use (ET) were used to fit model parameters. Precise dates of critical growth stages were not recorded during the study but tasseling and physiological maturity are generally observed to occur in mid-July and late September respectively (Kisekka *et al.*, 2016a ; Kisekka *et al.*, 2016b). Initial soil water at planting was measured during the study. Model validation was done by comparing measured grain yield, seasonal ET and soil water to predicted values. The statistical goodness-of-fit indicators used in this study included Relative Root Mean Square Error (RRMSE), index of agreement (IA) and Relative Error.

2.4. Model application

The calibrated and validated model was applied to assess the effect of five preplant irrigation amounts (0, 25, 50, 75, 100%) applied in late spring between April 1st to 30th of each year on grain yield, ET, soil water evaporation, and precipitation use efficiency. This analysis was repeated for three irrigation capacities (2.5, 3.8, and 6.4 mm/day) over 29 years of historical weather data from January 1st, 1986 to December 31st, 2014. The model was continuously executed for the 29 years. Initial soil water on January 1st, 1986 was set to field capacity in the top 1.2 m of the soil profile and 60% of field capacity in the deeper profile up to 2.4 m. Planting was set to occur during the first week of May each year at a seeding rating of 80,000 seeds/ha. Nitrogen fertilizer was applied at a rate of 269 kg N ha^{-1} as urea-ammonium nitrate each year as pre-plant. Preplant and in-season irrigation were triggered whenever the soil water reached a defined threshold (60% of plant available water) and each irrigation application was limited to 25 mm. Frequency of irrigation was limited by irrigation capacity. Simulated maize yield was adjusted to 15.5% nominal moisture content to match nominal moisture content of measured yields.

2.5. Determining optimum preplant irrigation management strategies

The most efficient/optimum preplant irrigation management strategy was determined by evaluating two criteria including the mean-Gini dominance (MGD), and Stochastic Dominance Analysis (SD) (Tsuji *et al.*, 1998; Kisekka *et al.*, 2016a ; Kisekka *et al.*, 2016b). The MGD decision criteria states that for example if two risky preplant irrigation management strategies A and B have the following characteristics, as shown in equation 1, then A is the more efficient of the two strategies:

$$E(A) \geq E(B)$$
$$\text{and} E(A) - G(A) \geq E(B) - G(B) \tag{1}$$

where E(.) is the expected value and G(.) is Gini-mean difference of distributions A and B.

The MGD decision criteria approach assumes that the decision maker is averse to risk and helps to avoid risky strategies. The Gini-mean difference in equation 1 is a function of Gini coefficient a measure of statistical dispersion. Gini coefficient is computed as half of the Gini-mean difference expressed as in equation 2 for unordered data following Dixon et al. (1987):

$$G = \frac{\sum_{i=1}^{n}\sum_{j=1}^{n}|x_i - x_j|}{2n^2\mu} \quad (2)$$

where |xi – xj| is the absolute difference of a randomly selected pair of values of a random variable, and μ is the arithmetic mean. For stochastic dominance analysis as an example, preplant irrigation management strategy A dominates strategy B if cumulative probability density function (CDF) of yield or water productivity gains from A lie to the right of the CDF of B over the entire probability interval 0–1.

3. RESULTS AND DISCUSSION

3.1. Model calibration and validation

Calibrated parameters for soil physical and hydraulic properties and maize cultivar coefficients are shown in Table 2. These values are within range of those reported by Stone et al. (2011) for Ulysses silt loam soils near Tribune Kansas. Bulk density values in Table 2 are within range of 1.24–1.46 g/cm^3 reported by USDA NRCS in western Kansas (http://websoilsurvey.sc.egov.usda.gov/App/WebSoilSurvey.aspx). Optimized saturated hydraulic conductivity values in Table 2 were also within range of 0.4–5 cm/h that was measured in the USDA NRCS soil surveys. Other studies that have used inverse modeling and measured soil water to characterize soil physical properties include (Šimůnek and Nimmo, 2005; Muñoz-Carpena *et al.*, 2008 ; Kisekka *et al.*, 2015).

Maize cultivar coefficients are within range of those reported by Kisekka *et al.*, 2016a ; Kisekka *et al.*, 2016b for maize hybrids grown in the US Central High Plains. Simulated grain yield from 2006 to 2009 for the treatment with preplant irrigation of 75 mm and irrigation capacity of 6.4 mm/day (well-watered treatment) were in agreement with measured yield (Figure 1). Although phenology data was not recorded, it was observed that anthesis occurred in mid-July (personal communication) which was in agreement with simulated phenology, which indicated that silking occurred on 07/10/2006, 07/19/2007, 07/12/2008 and 07/13/2009 respectively. Physiological maturity was generally observed to occur in late September although exact dates were not recorded, the corresponding model simulations were September 01, 2006, September 07, 2007, September 10, 2008 and September 12, 2009.

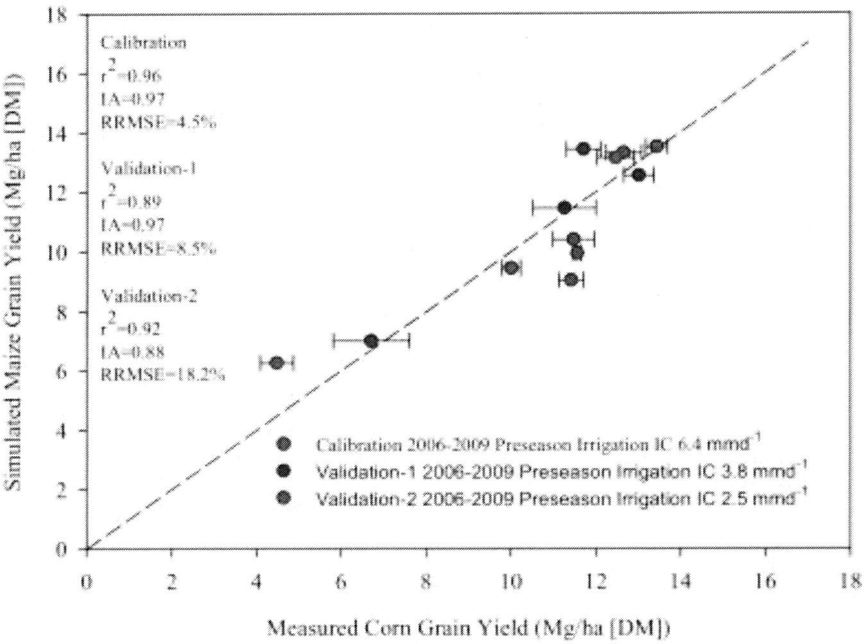

Figure 1. Comparing simulated to measured maize grain yield in a preplant irrigation study that was conducted at the Kansas State University Southwest Research-Extension Center near Tribune, Kansas, IC refers to irrigation capacity, IA is index of agreement and RRMSE is the relative root mean square.

Seasonal ET was well simulated for all treatments with predicted values within 4–8% of the measured seasonal ET as shown by the goodness-fit-indicators in Figure 2 and Table 3 with the exception of the 75 mm preplant irrigation and 6.4 irrigation capacity treatment in 2009. The ability of the model to predict grain yield and seasonal ET of maize with acceptable accuracy implies that this model could be applied for various assessments including generation of production functions, optimization of preplant irrigation management and strategic irrigation scheduling under deficit irrigation.

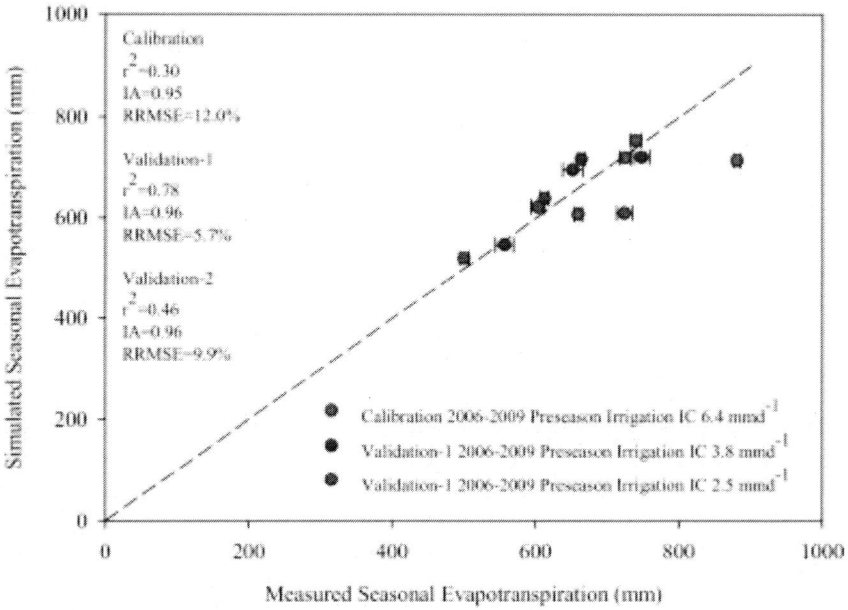

Figure 2. Comparing simulated to measured seasonal evapotranspiration in a preplant irrigation study that was conducted at the Kansas State University Southwest Research-Extension Center near Tribune, Kansas, IC refers to irrigation capacity, IA is index of agreement and RRMSE is the relative root mean square.

Table 3. Maize grain yield and seasonal evapotranspiration validation goodness-of-fit indicators for treatments that did not receive 75 mm of preplant irrigation during a field study conducted at the Kansas State University Southwest Research-Extension Center near Tribune, Kansas and simulations were from CERES-maize in RZWQM2.

Irrigation capacity (mm/day) with no-preplant irrigation	Grain yield			Seasonal evapotranspiration		
	[a]r^2	[b]IA	[c]RRMSE (%)	r^2	IA	RRMSE (%)
6.4	0.64	0.96	9.0	0.32	0.95	10.4
3.8	0.67	0.84	19.9	0.36	0.93	13.0
2.5	0.44	0.60	32.0	0.48	0.94	12.0

[a] Coefficient of determination.
[b] Willmont Index of Agreement.
[c] Relative root mean square error.

Table 4 Mean seasonal transpiration and average number of seasonal irrigation events at different amounts of preplant irrigation simulated using 29 years of historical weather data (1986–2014) and the DSSAT-CSM embedded in the RZWQM2 at Kansas State University Southwest Research-Extension Center near Tribune.

Preplant irrigation amount (mm)	Seasonal Transpiration (mm) Irrigation capacity (mm/day)			Average number of seasonal irrigation events[1] Irrigation capacity (mm/day)		
	6.4	3.8	2.5	6.4	3.8	2.5
0	497 a	497 a	401 a	19	16	12
25	492 a	492 a	408 b	19	15	12
50	492 a	492 a	418 b	18	15	11
75	486 a	486 a	421 c	18	15	11
100	495 a	495 a	430 c	17	14	11

[1] Each irrigation event was 25 mm.

Total soil water storage was also well simulated for both the calibration and validation treatments as shown in Figure 3. However, the model had a tendency to under predict measured total profile soil water. Possible reasons for the discrepancy include uncertainty in conceptualization of unsaturated flow, measurement uncertainty due to heterogeneity in soil conditions e.g., the sensing volume of the neutron attenuation technique decreases as soil water content increases. Ma et al. (2016) also noted the challenge of spatial soil heterogeneity of soil properties in agricultural systems modeling and reported that best results were obtained when crop parameters and soil field capacity were fit simultaneously.

3.2. Optimizing preplant irrigation amount and irrigation capacity combinations

In order to determine preplant irrigation amount and irrigation capacity combinations that optimize maize grain yields, water productivity, and precipitation use efficiency (PUE) and minimize soil water evaporation losses the mean-Gini dominance (MGD), and Stochastic Dominance Analysis (SD) analysis techniques were used.

3.2.1. Maize grain yield

At very limited irrigation capacity of 2.5 mm/day, applying 50–100 mm of preplant irrigation produced median yields that were 10–17% higher than not applying preplant irrigation. Increases in yield were small under 3.8 and 6.4 irrigation capacity. It can be seen in Figure 4 that the upper whiskers of the boxplots were larger under 2.5 mm/day compared to higher irrigation capacities, implying increased variability in high yields as irrigation capacity decreases. At very low irrigation capacity, the crop relies heavily on plant available water at planting that is made available to the crop to between rainfall and irrigation events during the growing season. Under limited irrigation capacity, irrigation might be triggered based on soil

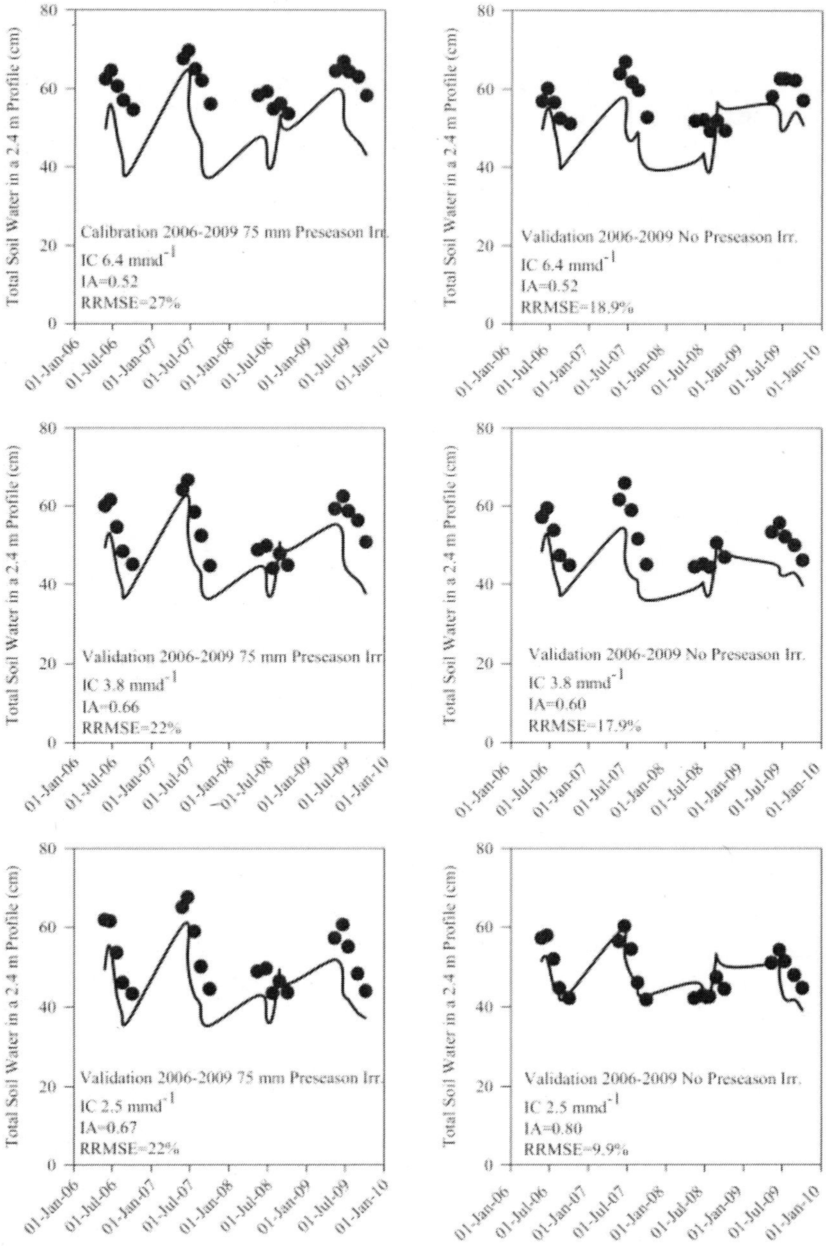

Figure 3. Simulated and measured total soil profile soil water in a preplant irrigation study that was conducted at the Kansas State University Southwest Research-Extension Center near Tribune, Kansas, IC refers to irrigation capacity.

water status but the actual irrigation events may not occur. This is particularly common with center pivot systems in the High Plains that take several days to complete one revolution due to diminished well capacity from the Ogallala aquifer.

At low (2.5 mm/day) to moderate (3.8 mm/day) irrigation capacity, applying 100 mm of preplant irrigation was the optimum strategy (Table 5). At high irrigation capacity of 6.4 mm/day, applying 75 mm of preplant irrigation produced the highest expected mean grain yield and highest difference between expected grain yield and Gini mean difference as shown in Table 5.

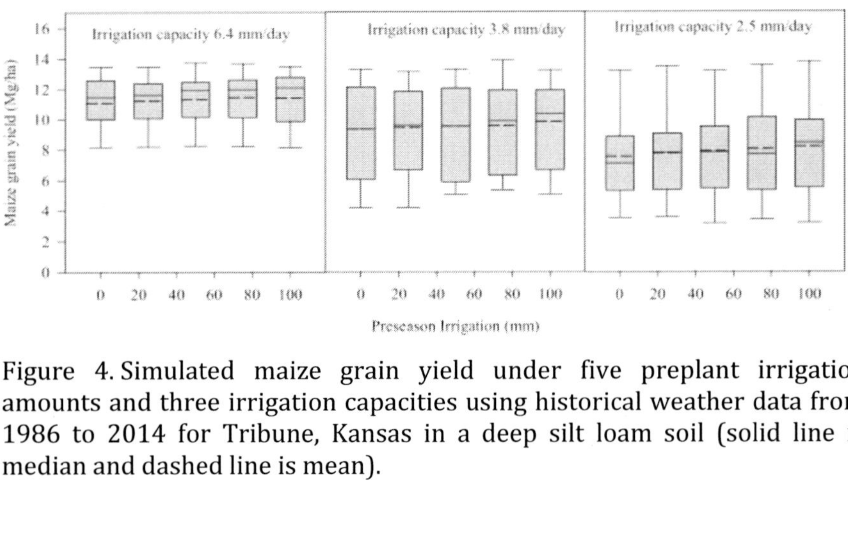

Figure 4. Simulated maize grain yield under five preplant irrigation amounts and three irrigation capacities using historical weather data from 1986 to 2014 for Tribune, Kansas in a deep silt loam soil (solid line is median and dashed line is mean).

Table 5. Maize grain yield (kg/ha) for different preplant irrigation amounts and irrigation capacities averaged over 29 years (1986–2014) of weather data at Kansas State University Southwest Research-Extension Center near Tribune, Kansas.

Preplant irrigation (mm)	Irrigation capacity					
	2.5 mm d^{-1}		3.8 mm d^{-1}		5.0 mm d^{-1}	
	$E(x)$	$E(x)-G(x)$	$E(x)$	$E(x)-G(x)$	$E(x)$	$E(x)-G(x)$
	Grain Yield (kg/ha)					
0	7464	3851	9306	5353	11083	8842
25	7675	4020	9407	5659	11225	8950
50	7814	4056	9491	5787	11321	9141
75	7948	4123	9526	5836	11450	9293
100	8098	4229	9798	6228	11424	9174

Stochastic dominance analysis results are shown in Figure 5 and they confirm results from the MGD analysis. It is shown in Figure 5 that applying 100 mm of preplant irrigation was dominant at the lowest irrigation capacity. However, the dominance of the 100 mm preplant irrigation scenario was second order since its cumulative density function (CDF) was not to the right of all the other CDFs over the entire range of possible yields. Under limited irrigation capacity (2.5 mm/day), preplant irrigation produced higher yields if the yield potential in a given cropping season was in the range of 6000 to 10,000 kg/ha. This corresponded to years with normal growing season (May to October) rainfall of 336 mm. In very wet years and very dry years there was no substantial differences in yields between different preplant irrigation amounts and irrigation capacity combinations. For irrigation capacity of 3.8 and 6.4 mm/day, the benefit from preplant irrigation were not consistent meaning preplant irrigation might not be necessary in most years. It can also be seen from Figure 5 that there is a 90% chance that increasing irrigation capacity would increase yields since the CDFs of the highest irrigation capacity were to the right of those with lower irrigation capacity. If a producer has a constrained water supply, reducing the total number of irrigated land also termed as concentrating water (Howell et al., 2012) to increase irrigation capacity would increase median yield and also reduce year to year variability in grain yield (Klocke et al., 2015). However, as noted by Howell et al. (2012) concentrating water could reduce income potential in wet years.

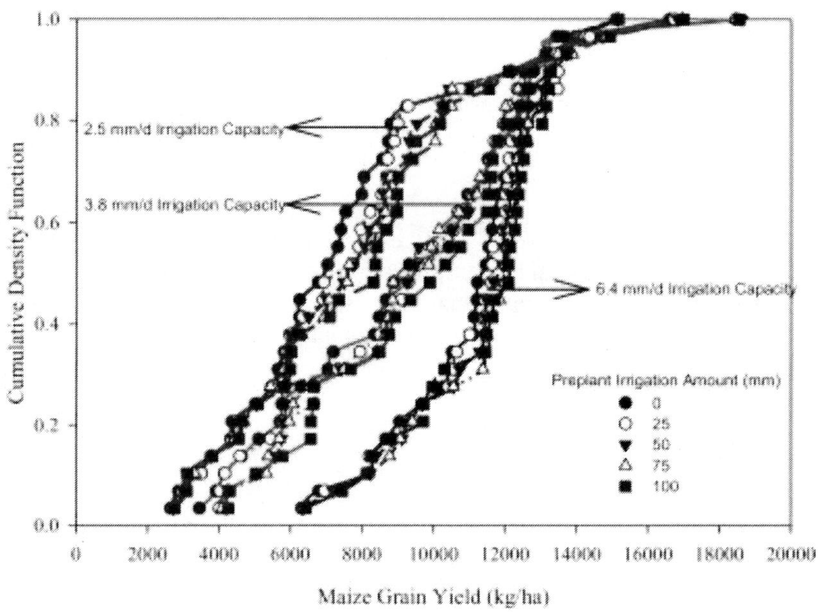

Figure 5. Cumulative probability density function of maize grain yield for different amounts of preplant irrigation and irrigation capacity simulated using CERES-Maize in RZWQM for climatic data from 1986 to 2014 for Tribune, Kansas.

3.2.2. Evapotranspiration and water productivity

Figure 6 also shows that the median ET for all levels of preplant irrigation was higher than the 625 mm on average required to meet full maize water requirements in western Kansas under irrigation capacities of 6.4 and 3.8 mm/day (Stone and Schlegel, 2006). For limited irrigation capacity of 2.5 mm/day, at least 75 mm to 100 mm of preplant irrigation was required to attain median seasonal ET of 625 mm/day or higher. The goal in water limited cropping systems is to maximize ET since it is directly related to biomass production and grain yield (Stewart and Peterson, 2014). Under very limited water scenarios, late spring preplant irrigation can increase seasonal ET but the goal should be to partition more water to transpiration while reducing soil water evaporation to improve water productivity. Figure 6 also shows that reducing irrigation capacity reduced seasonal ET due to reduced root uptake and transpiration.

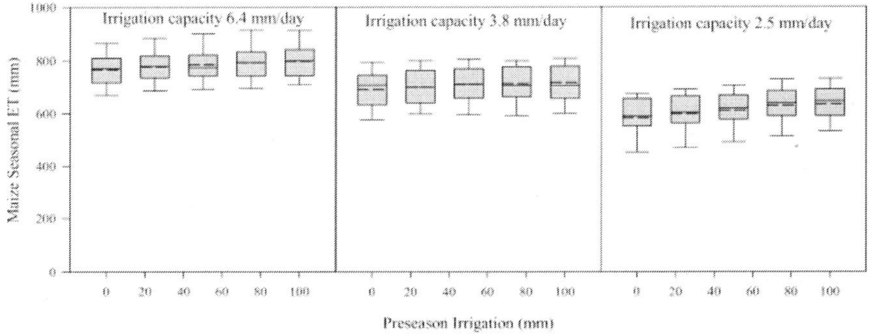

Figure 6. Simulated seasonal maize evapotranspiration (solid line in boxplot is median and dashed line in mean) under five preplant irrigation amounts and three irrigation capacities using historical weather data from 1986 to 2014 for Tribune, Kansas in a deep silt loam soil.

Transpiration did not vary substantially at different amounts of preplant irrigation for irrigation capacities of 6.4 and 3.8 mm/day. This implies that extra water could have been lost to soil water evaporation or deep drainage. But results from this study indicated that deep drainage from the 2.4 m soil profile was negligible therefore most of the water was lost to soil water evaporation. However, for very limited irrigation capacity of 2.5 mm/day, median transpiration increased with increase in preplant irrigation amount (Table 4). The results suggest that preplant irrigation is beneficial under very limited irrigation capacities but is not necessary under sufficient irrigation capacity (≥3.8 mm/day).

Table 6. Maize water productivity (kg/ha-mm) for different preplant irrigation amounts and irrigation capacities averaged over 29 years (1986–2014) of weather data at Kansas State University Southwest Research-Extension Center near Tribune, Kansas.

Preplant irrigation (mm)	Irrigation capacity					
	2.5 mm d^{-1}		3.8 mm d^{-1}		5.0 mm d^{-1}	
	$E(x)$	$E(x)$-$G(x)$	$E(x)$	$E(x)$-$G(x)$	$E(x)$	$E(x)$-$G(x)$
	Grain Yield (kg/ha-mm)					
0	12.48	7.45	13.22	8.54	14.45	11.69
25	12.52	7.59	13.25	8.58	14.44	11.38
50	12.49	7.39	13.19	8.77	14.44	11.59
75	12.34	7.20	13.20	8.70	14.47	11.64
100	12.47	7.36	13.52	9.24	14.35	11.28

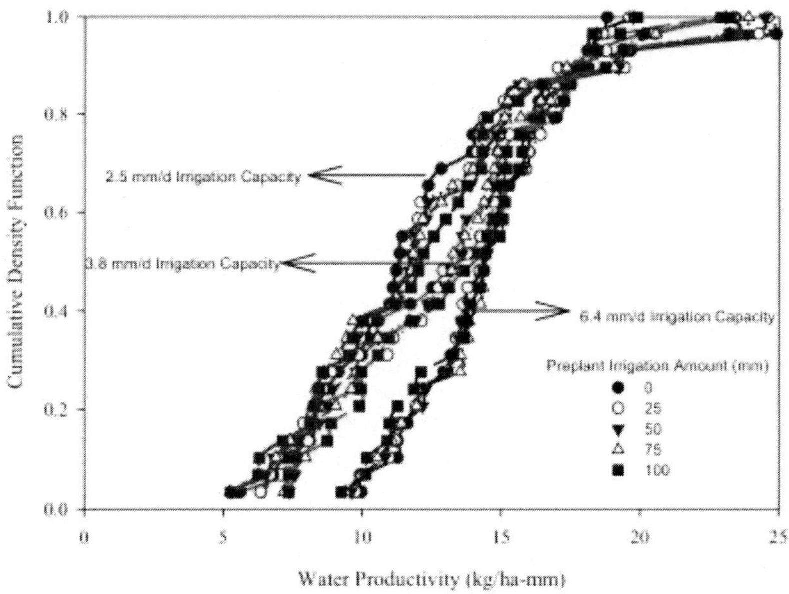

Figure 7. Cumulative probability density function of maize water productivity for different amounts of preplant irrigation and irrigation capacity simulated using CERES-Maize in RZWQM for climatic data from 1986 to 2014 for Tribune, Kansas.

Optimum preplant irrigation amounts were also analyzed based on water productivity as an indicator of how well water applied before planting was used for grain production. Table 6 shows that preplant irrigation amount did not have a substantial impact on water productivity at high and moderate irrigation capacity of 6.4 and 3.8 mm/day respectively. Stochastic dominance analysis revealed that similar to yield, there is a narrow range (11–15 kg/ha-mm) of water productivity where applying 100 mm preplant irrigation improved water productivity under the 2.5 mm/day irrigation capacity. Results in Figure 7 also indicate that irrigation capacity has a more dominant effect on water productivity compared to preplant irrigation amounts.

3.2.3. Precipitation use efficiency

Optimum combinations of preplant irrigation amounts and irrigation capacity were also evaluated based on their ability to enhance precipitation capture and storage expressed in form of precipitation use efficiency (PUE). PUE was calculated as a ratio of grain yield to precipitation. The precipitation was summed from harvest of previous year's crop to physical maturity of the crop in the current year (i.e., from October to September). Figure 8 shows that preplant irrigation had strong effect on PUE at the lowest irrigation capacity. At low irrigation capacity (2.5 mm/day) increasing preplant irrigation increased median PUE up to 18%. This is due to the fact that applying 75–100 mm of preplant irrigation resulted in increased yield compared to the scenario with 0 mm of preplant irrigation at 2.5 mm/day irrigation capacity. A concern with preplant irrigation is to ensure that it does not reduce storage for precipitation, this study shows that if preplant irrigation is applied in late spring in combination with soil water monitoring (irrigation was triggered at 60% plant available in this study) it did not reduce PUE in fact it enhanced PUE at very limited irrigation capacity. It is worth noting from Figure 8 that at moderate irrigation capacity of 3.8 mm/day preplant irrigation amounts of 100 mm improved PUE at lower levels. No effect of preplant irrigation on PUE at high irrigation capacity of 6.4 mm/day. It can also be seen in Figure 8 that reducing irrigation capacity by irrigating larger land area compared to available water supply reduced PUE. We recommend that producers match size of irrigated land to available water resources in order to guarantee sufficient irrigation capacity that enhances PUE.

3.2.4. Soil water evaporation losses prior to planting

Water losses due to soil water evaporation prior to planting increased with preplant irrigation amount at all irrigation capacities as shown in Figure 9. Scenario with 0 mm preplant irrigation had the least soil water evaporation loss prior to planting. Increase in soil water evaporation with preplant irrigation is expected as there would be more water available for the evaporation process. Under sufficient irrigation capacity (≥3.8 mm/day), we recommend no preplant irrigation to minimize soil water evaporation losses. Under limited well capacity we recommend applying at least 75 mm of preplant irrigation in late spring (April) prior to planting. Also, this analysis indicated negligible losses through deep percolation from 2.4 m soil profile indicating the biggest potential water loss from preplant irrigation would be soil water evaporation.

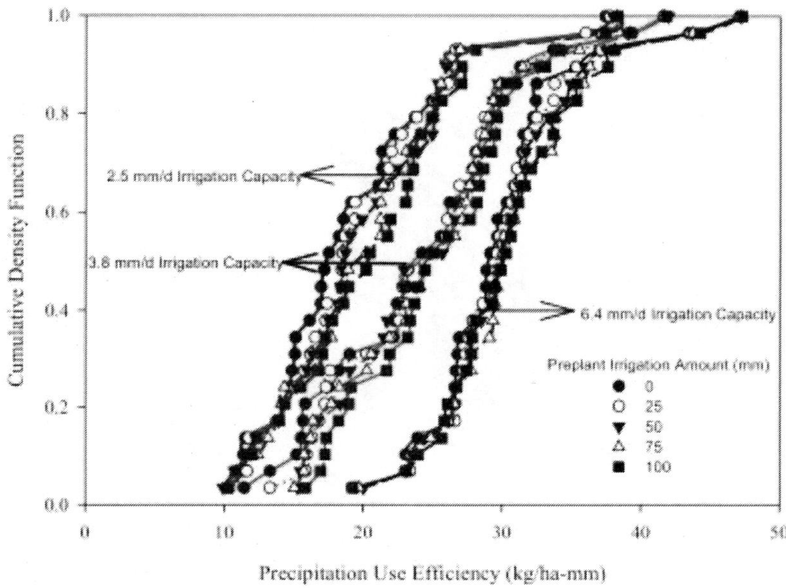

Figure 8. Cumulative probability density function of precipitation use efficiency (PUE) for different amounts of preplant irrigation and irrigation capacity simulated using CERES-maize in RZWQM for climatic data from 1986 to 2014 for Tribune, Kansas.

Different irrigation application methods might influence storage efficiency of preplant irrigation. For example, subsurface drip irrigation (SDI) would reduce soil water evaporation losses compared to preplant irrigation done with a sprinkler irrigation system. Mobile Drip Irrigation (MDI) might also reduce soil water evaporation due to reduced surface wetting (Kisekka *et al.*, 2016a ; Kisekka *et al.*, 2016b). Lamm (2002) noted that SDI could be beneficial for applying preplant irrigation to increase seasonal capacity by allowing irrigation to be applied under near freezing temperatures which would not be possible with center pivot irrigation systems. Residue management might also improve the efficiency of preplant irrigation. Irrigators need to integrate technologies (e.g., soil water sensors) in implementation of preplant irrigation that help reduce water losses due to soil water evaporation and deep drainage but enhances precipitation capture and storage.

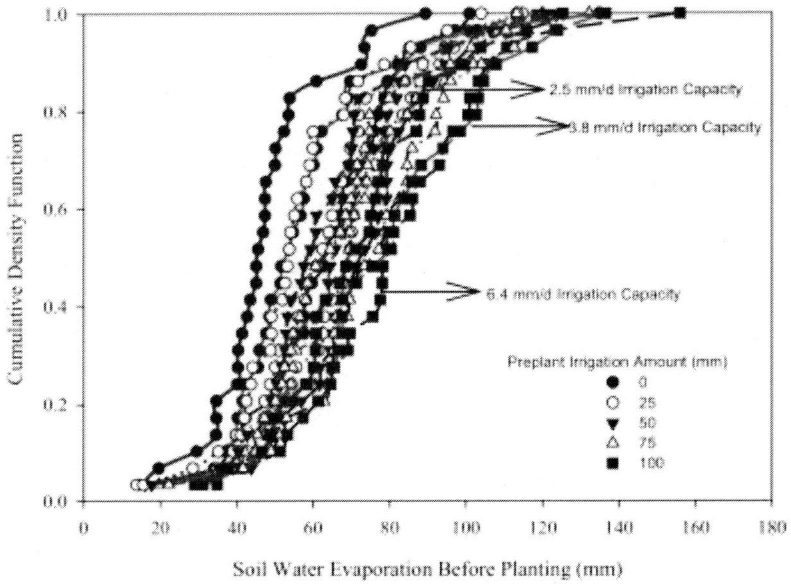

Figure 9. Cumulative probability density function of soil water evaporation prior to planting in late spring for different amounts of preplant irrigation and irrigation capacity simulated using CERES-maize in RZWQM for climatic data from 1986 to 2014 for Tribune, Kansas.

Although most of the earlier experimental research (Musick *et al.*, 1971; Stone *et al.*, 1983; Stone *et al.*, 1987; Stone *et al.*, 1994 ; Stone *et al.*, 2008) showed that preplant irrigation was an inefficient use of water in most years, recent findings by Schlegel et al. (2012) have showed that under very limited well capacity (2.5 mm/day) preplant irrigation increased yield and was profitable. In addition, Ma et al. (2012) noted that preplant irrigation was necessary to ensure good germination in dry years. Musick and Lamm (1990) in their review of preplant irrigation in the South and Central High Plains listed conditions where preplant irrigation would have the best and least benefit, in general these conditions underscored the need to make preplant irrigation decisions based on knowledge of soil water and local rainfall patterns. The approach used in this study helps to build on earlier work by extrapolating short term experimental results to long term climatic conditions to understand effect of preplant irrigation on water limited maize production in the high plains under a wide range of growing conditions.

The decision to apply preplant irrigation should be evaluated carefully since it depends on irrigation capacity and cropping season yield potential. Optimum preplant irrigation for systems with 2.5 mm/day irrigation capacity is 100 mm. No preplant irrigation for systems with 3.8 and 6.4 irrigation capacity is required in most years. Overall, a producer would obtain higher yields, water productivity and PUE by increasing their irrigation capacity by reducing irrigated areas to match limited water supply.

4. CONCLUSIONS

A simulation study was conducted to assess the effect of preplant irrigation amounts and irrigation capacities on maize production in western Kansas region of the United States High Plains. The calibrated and validated CERES-maize model embedded in the RZWQM2 adequately predicted yield, ET, and soil water. The calibrated and validated model was applied to assess effect of five preplant irrigation amounts (0, 25, 50, 75 and 100 mm) and three irrigation capacities (2.5, 3.8 and 6.4 mm/day) on maize yield. Under very limited irrigation capacity of 2.5 mm/day, applying 75–100 mm of preplant irrigation produced median yields that were 10–17% higher than not applying preplant irrigation. However, even at limited irrigation capacity the benefit of preplant irrigation were only realized if the seasonal yield potential was in the range of 6000 to 10,000 kg/ha corresponding to years with normal seasonal rainfall. Irrigation capacity had a stronger effect on maize grain yield compared to preplant irrigation amount. Preplant irrigation increased ET and transpiration under 2.5 mm/day irrigation capacity. Preplant irrigation amount did not have a substantial impact on water productivity at high and moderate irrigation capacities but had second order dominant effect under limited irrigation capacity. At low irrigation capacity (2.5 mm/day) increasing preplant irrigation increased median PUE up to 18% although the effect was second order dominant. No effect on PUE at high irrigation capacity. Water losses due to soil water evaporation prior to planting increased with preplant irrigation amount at all irrigation capacities. Under very limited irrigation capacity (≤2.5 mm/day) we recommend applying 75–100 mm of preplant irrigation in late spring (April). No preplant irrigation is recommended at high irrigation capacity to minimize soil water evaporation losses. Deep percolation losses from 2.4 m soil profile were negligible. The decision to apply preplant irrigation should be evaluated and implemented carefully in combination with other agricultural water management technologies and strategies such as soil water monitoring, drip irrigation, and residue management to mitigate negative effects of soil water evaporation and deep drainage.

ACKNOWLEDGEMENTS

The authors would like to thank the following for providing funding for this research USDA Ogallala Aquifer Project, The Foundation for Food and Agricultural Research (Award # 430871), and USDA Project No. 2016-68007-25066, through the NIFA Water for Agriculture Challenge Area.

This is contribution number 17-181-J from the Kansas Agricultural Experiment Station. Authors thank the anonymous reviewers for their constructive comments.

REFERENCES

1. Ahuja, L.R., Rojas, K.W., Hanson, H.D., Shafer, M.J., Ma, L., 2000. Root zone water quality model. In: Modeling Management Effects on Water Quality and Crop Production. Water resources publications, LLC, CO, USA.
2. DeJonge, K.C., Ascough, J.C., Andales, A.A., Hansen, N.C., Garcia, L.A., Arabi, M., 2012. Improving evapotranspiration simulations in the CERES-Maize model under limited irrigation. Agric. Water Manage. 115, 92–103,
3. Dixon, P.M., Weiner, J., Mitchell-Olds, T., Woodley, R., 1987. Bootstrapping the Gini coefficient of inequality. Ecology 68 (5), 1548–1551.
4. Doherty, J., 2009. PEST: Model Independent Parameter Estimation. Watermark Numer. Comput., Corinda, Queensland, Australia, Available at http://www.pesthomepage.org.
5. Hoogenboom, G., Jones, J.W., Wilkens, P.W., Porter, C.H., Boote, K.J., Hunt, L.A., Singh, U., Lizaso, J.I., White, J.W., Uryasev, O., Ogoshi, R., Koo, J., Shelia, V., Tsuji, G.Y., 2015. Decision Support System for Agrotechnology Transfer (DSSAT) Version 4.6. www.DSSAT.net.
6. Howell, T.A., O'shaughnessy, S.A., Evett, S.R., 2012. Integrating multiple irrigation technologies for overall improvement in irrigation management. In: Proceedings of the 2012 Central Plains Irrigation Conference, Central Plains Irrigation conference proceedings, Colby, Kansas, pp. 170–186.
7. Jones, J.W., Hoogenboom, G., Porter, C.H., Boote, K.J., Batchelor, W.D., Hunt, L.A., Wilkens, P.W., Singh, U., Gijsman, A.J., Ritchie, J.T., 2003. DSSAT cropping system model. Eur. J. Agron. 18, 235–265.
8. Keating, B.A., Carberry, P.S., Hammer, G.L., Probert, M.E., Robertson, M.J., Holzworth, D., Huth, N.I., Hargreaves, J.N.G., Meinke, H., Hochman, Z., McLean, G., Verburg, K., Snow, V., Dimes, J.P., Silburn, M., Wang, E., Brown, S., Bristow, K.L., Asseng, S., Chapman, S., McCown, R.L., Freebairn, D.M., Smith, C.J., 2003. An overview of APSIM, a model designed for farming systems simulation. Eur. J. Agron. 18, 267–288.
9. Kisekka, I., Migliaccio, K.W., Munoz-Carpena, ˜ R., Khare, Y., Li, Y.C., 2015. Modeling soil water dynamics considering measurement uncertainty. Hydrol. Processes 29, 692–711.
10. Kisekka, I., Nguyen, G., Aguilar, J., Rogers, D., 2016a. Mobile drip irrigation evaluation in maize. Kans. Agric. Exp. Stat. Res. Rep. 2 (7),
11. Kisekka, I., Aguilar, J., Rogers, D., Klocke, N., 2016b. Assessing deficit irrigation strategies for maize using simulation. Trans. ASABE 59, 303–317.
12. Klocke, N., Currie, R., Kisekka, I., Stone, L., 2015. Maize and grain sorghum response to limited irrigation, Drought, and Hail. Appl. Eng. Agric. 30, 915–924.

13. Kromm, D.E., White, S.E., 1990. Conserving Water in the High Plains. Kansas State University, Department of Geography, Manhattan, Kans. Lamm, F.R., 2002. Advantages and disadvantages. In: International Meeting on Advances in Drip/Micro Irrigation, Puerto de La Cruz, Tenerife, Canary Islands, December 2–5, p. 2002.
14. Ma, L., Hoogenboom, G., Ahuja, L.R., Ascough II, J.C., Saseendran, S.A., 2006. Evaluation of the RZWQM-CERES-Maize hybrid model for maize production. Agric. Syst. 87, 274–295.
15. Ma, L., Trout, T.J., Ahuja, L.R., Bausch, W.C., Saseendran, S.A., Malone, R.W., Nielsen, D.C., 2012. Calibrating RZWQM2 model for maize responses to deficit irrigation. Agric. Water Manage. 103, 140–149.
16. Ma, L., Ahuja, L.J., Trout, T., Nolan, B.T., Malone, R.W., 2016. Simulating maize yield and bomass with spatial variability of soil field capacity. Agron. J.,
17. Malone, R.W., Jaynes, D.B., Ma, L., Nola, B.T., Meek, D.W., Karle, D.L., 2010. Soil-test N recommendations augmented with PEST optimized RZWQM simulations. J. Environ. Qual. 39, 1711–1723.
18. McGuire, V.L., 2004. Water-level changes in the High Plains Aquifer, predevelopment to 2003 and 2002–2003. In: Fact Sheet FS-2004-3097. U.S. Geological Survey, Lincoln, Neb. Munoz-Carpena, ˜ R., Ritter, A., Bosch, D.D., Schaffer, B., Potter, T.L., 2008. Summer cover crop impacts on soil percolation and nitrogen leaching from a winter maize field. Agric. Water Manage. 95, 633–644.
19. Musick, J.T., Lamm, F.R., 1990. Preplant irrigation in the central and southern High Plains: a review. Trans. ASAE 33, 1834–1842.
20. Musick, T.J., Sletten, W.H., Dusek, D.A., 1971. Preplant irrigation of grain sorghum in the southern high plains. Trans. ASAE 14, 93–97.
21. Nimah, M.N., Hanks, R.J., 1973. Model for estimating soil water, plant, and atmospheric interrelations: I. Description and sensitivity. Soil Sci. Soc. Am. Proc. 37, 522–527,
22. Raes, D., Steduto, P., Hsiao, T.C., Fereres, E., 2009. AquaCrop- the FAO crop model to simulate yield response to water: II. Main algorithms and soft ware description. Agron. J. 101, 438–447.
23. Saseendran, S.A., Ahuja, L.R., Nielsen, D.C., Trout, T.J., Ma, L., 2008. Use of crop simulation models to evaluate limited irrigation management options for corn in a semiarid environment. Water Resour. Res. 44, W00E02,
24. Schlegel, A.J., Stone, L.R., Dumler, T.J., Lamm, F.R., 2012. Managing diminished irrigation capacity with preseason irrigation and plant density for crop production. Trans. ASABE 55, 525–531.
25. Simˇ unek, ˚ J., Nimmo, J.R., 2005. Estimating soil hydraulic parameters from transient flow experiments in a centrifuge using parameter optimization technique. Water Resour. Res. 41,
26. Stewart, B.A., Peterson, G.A., 2014. Managing green water in dryland agriculture. Agron. J. 106, 1–10,
27. Stone, L.R., Schlegel, A.J., 2006. Crop water use in limited-irrigation environments. In: Proc. 2006 Central Plains Irrigation Conf., Colby, Kans.: Central Plains Irrigation Association, pp. 173–184.
28. Stone, L.R., Carlson, C.G., Hanson, T.L., Gwin Jr., R.E., Gallagher, P., Horton, M.L., 1983. Amount of profile water in early spring resulting from increased profile water in fall. SSSA J. 47, 305–309.
29. Stone, L.R., Gwin Jr., R.E., Gallagher, P.J., Hattendorf, M.J., 1987. Dormant-season irrigation Grain yield, water use, and water loss. Agron. J. 79, 632–636.
30. Stone, L.R., Schlegel, A.J., Lamm, F.R., Spurgeon, W.E., 1994. Storage efficiency of preplant irrigation. J. Soil Water Conserv. 49, 72–76.
31. Stone, L.R., Lamm, F.R., Schlegel, A.J., Klocke, N.L., 2008. Storage efficiency of off-season irrigation. Agron. J. 100, 1185–1192.
32. Stone, L.R., Klocke, N.L., Schlegel, A.J., Lamm, F.R., Tomsicek, D.J., 2011. Equations for drainage component of the field water balance. Appl. Eng. Agric. 27, 345–350.

33. Tsuji, G.Y., Hoogenboom, G., Thornton, P.K. (Eds.), 1998. Understanding Options for Agricultural Production. Systems Approaches for Sustainable Agricultural Development, vol. 7. Kluwer Academic, Dordrecht, The Netherlands, http://dx.
34. Van Diepen, C.A., Wolf, J., van Keulen, H., 1989. WOFOST: a simulation model of crop production. Soil Use Manage. 5 (1), 16–24.

CHAPTER 13

Assessment of Wastewater Quality of Drains for Irrigation

Almas Hamid[1], Mahrukh Zeb[1], Asim Mehmood[2], Sana Akhtar[1*], Samia Saif[2]

[1]Department of Environmental Sciences, Kinnaird College for Women, Lahore, Pakistan; [2]Global Environmental Laboratories, Lahore, Pakistan.

ABSTRACT

Rapid industrialization and overpopulation have stimulated increase in waste water both domestic and industrial. Due to lack of inefficient sewerage system and absence of treatment plants, the wastewater is discharged into drainage systems causing environmental and health implications. At various points, the wastewater of these drains is used for irrigation purposes. This study characterizes the effluent of major drains of Lahore city and assesses its suitability for irrigation. Wastewater samples were collected from drains in winter and spring seasons and analyzed for various parameters of special concern to irrigation. TSS, oil and grease, BOD, COD, chloride, TDS, sodium, SAR, bicarbonate ions, EC, iron, Copper and Nickel were found to be varied in concentration in different drains when compared with NEQS (National Environment Quality Standards) and FAO guidelines for irrigation in both seasons. Regular monitoring and proper treatment of wastewater before discharging into the drains can reduce the pollution in these drains.

Keywords: Industrialization; Wastewater Quality; Drains; Irrigation; Winter and Spring Season; FAO Guidelines

1. INTRODUCTION

Pakistan is the sixth most populous country of the world with current population of approximately 183.06 million according to latest survey of 2013. It is expected that population will reach 242.1 million by 2030 at the existing growth rate [1]. Water and food demand will be directly affected which, in turn, will increase the effluent discharges from domestic, industrial and agricultural sectors [2].

Being an agricultural country, the water demands of Pakistan are very high. Due to the water scarcity issues, the untreated effluents from the cities are being used for irrigation purposes in many parts of the country including Lahore [3]. Lack of adequate resources is resulting in the usage of wastewater for irrigation without any treatment. Consequently, urban and peri-urban farmers use marginal or untreated wastewater to cultivate a variety of crops. In Pakistan, the percentage of the wastewater, which is treated before use in irrigation is only 2 per cent [4].

The Lahore's drainage system comprises 8 major and 75 tributary drains having carrying capacity of 7474 Cusecs. The length of total surface drains is 212 km, with 4 pumping stations and pumping capacity of 663 Cusecs. The major drains of Lahore which include Shahdara drain, Mian Mir drain, Iqbal town drain, Garden town drain, Sattu Katla drain, Charar drain, Chota Ravi drain and Hudiara drain discharge their waste directly into River Ravi, within a reach of about 65 km [5].

Originally, the purpose of these drains was to collect the flood water and agricultural excess water. But due to increase in population and rapid industrialization of Lahore, the drains are now mainly used to collect the industrial and municipal effluents of diverse composition resulting in increasing pollution problems in the irrigation outlets [5].

271 industrial units in Lahore district, mainly constituting of textile, dairy, paints & plastic, chemical, food processing, pulp and paper processing, poultry, pesticides, leather, tanneries and pharmaceuticals are disposing of around 281.6 cusec untreated waste into the irrigation canals and drains [6]. Moreover, 3136 cusecs effluent is added by municipalities making a total of 3418 cusecs .It has been reported that in Lahore, very few industries, like 3 out of some 100 industries, using hazardous chemicals, treat their wastewater effectively. The untreated effluent discharges have created multiple environmental and social challenges for the city [5].

Afzal et al. [7] conducted a study to investigate the degree of pollution, in the Hudiara drain, caused by disposal of untreated sewage and industrial waste of Pakistan and India. The results showed high values of biochemical oxygen demand (BOD), total organic carbon (TOC) chemical oxygen demand (COD), and trace metals in the samples taken from the Indo-Pak border. This study demonstrated that the major pollutants in the water of Hudiara drain were suspended solids (SS), fecal coliform (FC) and chemical oxygen demand (COD).

The discharge of untreated industrial and municipal effluents into the irrigation system can pose a serious threat to the quality of soil & crops and to the health of human beings [6]. According to estimation, untreated effluents from the Lahore city are being used to irrigate approximately 800 to 1000 hectares area. Most trace elements and other heavy metals mount up in soils eventually contaminating or rendering soils non-productive [3].

The change in physicochemical properties of the soil is the chronic impact of the usage of wastewater for irrigation eventually increasing the amount of heavy metals in the soil. Long term use of wastewater of drains, can lead to accumulation of high amounts of trace elements in soils and can enter the food chain through absorption via plants.

Cd can be easily absorbed and accumulated in plants and animals in very high quantities [8]. Sharma et al. [9] carried out study in three industrially different cities viz., Ludhiana, Jalandhar and Malerkotla to observe the level of Cd buildup in soils and plants being irrigated with untreated sewage water. Cd level was high at all sites in those plants which were receiving wastewater for irrigation as compared to those plants getting tubewell irrigation. Similarly, Beta vulgaris (palak) has been highly contaminated with Cd, Pb and Ni due to wastewater irrigation in suburban areas of Varanasi, India causing serious health risks for the human beings.

Sahu et al. [10] assessed the wastewater quality of loamy drain receiving effluent from tanneries and other industries and studied their impacts on soil and vegetables irrigated with this water. The results showed that most of the parameters were not complying with the prescribed limits for safe disposal of effluents into the surface water. Samples of soil and vegetables from the land irrigated with drain water were also collected and analyzed for Cu, Zn, Ni, Cr, Pb and Cd. The different metals were in diverse concentrations, which may create chronic health hazards to human and cattle through food chain in the long run.

Other effects posed on groundwater quality, wild life & health impacts on farmers have also been assessed. For instance, J. H. J. Ensink et al., [11] carried out a cross sectional study in the city of Faisalabad, Pakistan, to assess the risk of Giardia duodenalis (Giardia) infection in farmers using untreated wastewater in agriculture. A strong correlation was found between Giardia infections in those households who were using wastewater farming as compared to the ones using regular irrigation water. Cattle grazing on freshly irrigated grasslands with untreated wastewater reportedly suffered from disease (cysticerosis) in Melbourne, Australia and Denmark [4].

Severe contamination in Ravi River has destroyed 42 species of fish. The bird life has also been acutely affected and has migrated to other areas [2]. An extensive and seasonal study was carried out by Akhtar and Nawaz on the effluent quality of six drains of Lahore and their impact on the aquatic life of River Ravi, into which these drains were discharging their effluents. The results when compared with USEPA guidelines for fish and aquatic life proved that River Ravi water quality was not by any means suitable for the propagation of fish during dry weather and canal closure period [12].

Due to the varied nature of industrial manufacturing processes, effluents vary widely in composition and often contain toxic materials to cause serious problems towards human health, plants, crops, animals and aquatic life. The present study was conducted to characterize the physical and chemical characteristics of effluents of the major drains of Lahore city in order to determine the pollutants being discharged into the drains which ultimately found their way into the receiving water bodies, mainly rivers, and to evaluate that whether the surface water quality was suitable for the purpose of irrigation or not, for which it was being extensively utilized [13].

2. MATERIALS AND METHODS

2.1. Selection of the Study Area

Eight major drains of Lahore city were selected for the present study and the samples were collected from each site namely Shahdara drain, Mian Mir drain, Iqbal town drain, Garden town drain, Sattu Katla drain ,Charar drain, Hudiara drain, Chota Ravi drain [12]. **Figure 1** shows the locations of study area.

2.2. Sampling Procedure

The study was carried out from January to April. During this period, four samples were taken from each drain; twice in January and then in March, in pre-sterilized 1.5 L sampling bottles according to standard sampling procedures to ensure the integrity of the collected samples and validity of test results. All the bottles were labeled by the name of the site and date & time at which the sample was taken. All of the samples were preserved in an icebox at or near 4°C immediately after collection. Nitric acid and Sulfuric acid were added to the sampling bottles in quantities sufficient to

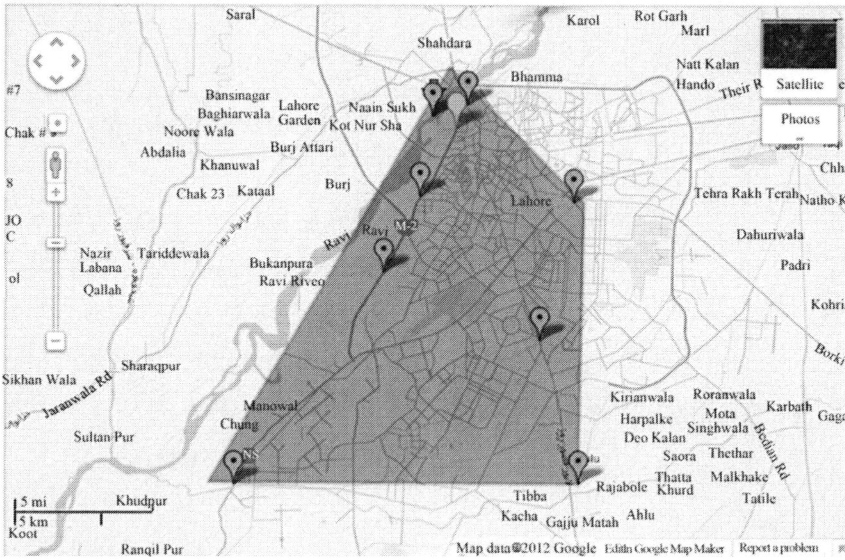

Figure 1. Sampling points of major Drains of Lahore city (View of the Terrain).

meet the requirements of preservation and to lower the pH of the samples to just about 2. After collection, all the samples were transferred carefully to laboratory for further analysis. A total of 16 samples were collected in two months from these drains.

2.3. Selection of Parameters

Each sample of the wastewater was tested for various Physio-Chemical parameters as shown in **Table 1**. Some of these parameters like pH, Sodium, EC, SAR, RSC, Bicarbonate, Chloride, TDS, Fe, Cu, Ni and Cr were analyzed to test the water quality of these drains for the purpose of irrigation. The results were compared with National Environmental Quality Standards (NEQS) and FAO guidelines for irrigation to establish a baseline condition of pollution level in the drains. **Table 1** shows the instruments & procedures used for the analysis of physiochemical parameters and heavy metals in the waste-water.

Table 1. Instruments/Procedures used for the analysis of effluents of the drains.

Sr. No.	Parameter	Instrument/Procedure
1.	Temperature	Thermohunter PT-2LD
2.	pH	Eco tester pH 1/USEPA 8156
3.	Color	Spectrophotometer/USEPA 8025
4.	TSS	Filtration/USEPA 8158 & 8164
5.	TDS	Evaporation/USEPA 8163
6.	Electrical conductivity	Conductivity Meter DIST 3 New
7.	COD	Spectrophotometer
8.	BOD	BOD track
9.	DO	DO meter
10.	Oil and grease	Gravimetric method/USEPA 10056
11.	Chloride	Digital titrator/USEPA 8206
12.	Sulfate	Spectrophotometer/USEPA 8051
13.	Sodium	Flame Photometer 1381- E
14.	Calcium & Magnesium	Titration/USEPA 8222
15.	CO_3^- & HCO_3^-	Digital titration/USEPA 10244
16.	SAR	Na/√ (Ca + Mg)/2
17.	RSC	$(CO_3 + HCO_3)$ - (Ca + Mg)
18.	Iron	Atomic Absorption Spectroscopy/ USEPA 2361
19.	Heavy Metals (Copper, Chromium and Nickel)	Atomic Absorption Spectroscopy/ USEPA 2201

3. RESULTS AND DISCUSSION

The focus of the present study was to quantitatively characterize the pollution load in the major drains of the Lahore city in the month of January and March and to assess the suitability of drain water for irrigation as it is being frequently used in irrigation.

Samples of wastewater collected in the month of January and March from the major drains of Lahore city were analyzed for parameters including pH, Temperature, TDS, TSS, Colour, COD, BOD, DO, Oil and Grease, EC, Cl, SO_4^{2-}, Na, Ca, Mg, CO^{2-}, HCO_3^-, RSC and SAR, Fe, Cu, Cr and Ni . The results of the analysis of these physical and chemical parameters are shown in Tables 2-7.

The pH measurement of the wastewater samples of the drains showed that the water of Shahdara drain, Iqbal town drain and Sattu katla drain was acidic in the winter season while in the spring season, the water of all the drains was alkaline which could be the reason of more flow of water and increase in domestic effluents. The causes of acidity and alkalinity may be runoff, sewage, geology and high nutrient levels. However, these results show that value of pH was within the acceptable range according to National Environmental Quality Standards and FAO guidelines for irrigation. Temperature is a very important parameter of the wastewater to be measured. At high temperatures, the water loses its dissolved oxygen. The value of temperature was within the recommended limit for no risk according to NEQS. FAO guidelines have no suggested value for temperature [14].

Table 2. Results of Physio-chemical parameters of wastewater of the drains in winters (January) and its comparison with NEQS.

Sr. No.	Parameter	Unit	NEQS Value	Shahdara Drain	Mian Mir Drain	Iqbal Town Drain	Garden Town Drain	Sattu Katla Drain	Charar Drain	Hudiara Drain	Chota Ravi Drain
1.	pH	-	6 - 9	6.8	7.9	6.8	7.8	6.9	7.8	7.5	7
2.	Temperature	°C	40	20	21	19	22	20	22	18	15
3.	TSS	mg/L	200	230	90	182	178	134	206	201	145
4.	TDS	mg/L	3500	1170	790	780	790	810	800	2800	940
5.	Color	TCU	*NGVS	156	79	504	350	104	342	618	771
6.	COD	mg/L	150	362	299	503	349	295	385	641	340
7.	BOD	mg/L	80	170	132	245	165	123	182	265	156
8.	DO	mg/L	NGVS	0.6	0.8	0.9	0.7	0.9	0.5	1	0.8
9.	Oil and Grease	mg/L	10	32	33	67	33	33	33	24	16
10.	Chloride	mg/L	1000	256	61	85	64	110	60	692	123
11.	Sulphate	mg/L	600	272	105	98	116	153	89	306	102
12.	Calcium	mg/L	NGVS	71	94	60	54	84	90	98	65
13.	Magnesium	mg/L	NGVS	49	44	61	39	50	65	70	45
14.	Carbonate	mg/L	NGVS	13	17	9	10	12	15	20	8

Table 3. Results of analysis of Physio-chemical parameters of the wastewater of the Drains in winters (January) and its comparison with FAO guidelines for irrigation.

Sr. No.	Parameter	Unit	*FAO Guidelines for irrigation – Degree of restriction on use			Shahdara Drain	Mian Mir Drain	Iqbal Town Drain	Garden Town Drain	SattuKatla Drain	Charar Drain	Hudiara Drain	Chota Ravi Drain
			None	*Slight to moderate*	*Severe*								
1.	pH	-	Normal range 6.5 - 8			6.8	7.9	6.8	7.8	6.9	7.8	7.5	7
2.	TDS	mg/L	<450	450 - 2000	> 2000	1170	790	780	790	810	800	2800	940
3.	Electrical Conductivity	dS/m	<0.7	0.7 - 3.0	> 3.0	5.6	5.0	3.9	4.6	4.4	5.9	6.5	5.6
4.	Sodium	me/L	<3	3 – 9	> 9	46.74	40.57	47.39	41.30	46.57	38.70	56.09	43.04
5.	Chloride	me/L	<4	4 – 10	> 10	7.21	1.72	2.39	1.80	3.10	1.69	19.49	3.46
6.	Bicarbonate	me/L	<1.5	1.5 - 8.5	> 8.5	20.20	19.28	20.98	16.89	18.56	17.70	21.15	18.38
7.	SAR	-	>0.7	0.7 - 0.2	< 0.2	33.8	22.5	28.4	28.2	26.4	19.7	27.5	32.5
8.	RSC	me/L		1.25		-ve	-ve	-ve	-ve	-ve	-ve	-ve	-ve

Table 4. Results of analysis of heavy metals in the wastewater of the drains in winters (January) and its comparison with NEQS and FAO guidelines for irrigation.

Sr. No.	Parameter	Unit	NEQS value	**FAO guidelines for irrigation	Shahdara Drain	Mian Mir Drain	Iqbal Town Drain	Garden Town Drain	Sattu Katla Drain	Charar Drain	Hudiara Drain	Chota Ravi Drain
1.	Iron	mg/L	8.0	5.0	5.3	1.3	3.3	2.0	5.5	1.7	1.9	2.1
2.	Copper	mg/L	1.0	0.20	0.74	0.08	0.50	0.009	3.75	0.07	0.40	0.03
3.	Nickel	mg/L	1.0	0.20	0.40	0.12	0.11	0.13	2.65	0.23	0.14	0.23
4.	Chromium	mg/L	1.0	0.10	*ND	ND	ND	ND	1.37	ND	ND	ND

NGVS = No guideline value suggested; ND = Not detected; NEQS Values: (Pakistan, Environmental Protection Agency EPA, 1997); **FAO Guidelines for irrigation: (Pescod, 1992) *Guidelines for Interpretation of Water Quality for Irrigation; **Threshold Levels Of Trace Elements for Crop Production (phytotoxic threshold levels of some selected trace elements in municipal wastewater).

The analysis of effluents showed high concentration of oil and grease in these drains in both seasons i.e. winter and spring. The values were violating the acceptable range in all the drains as specified by NEQS as shown in **Figure 2**. The sources of oil and grease in wastewater can be traced to domestic, commercial and industrial wastewater. The most common industries releasing oil and grease are textile industries, laundries and industries of wool scouring [15]. Large amounts of oil and grease into the water bodies increase the BOD while also causing foul odors by trapping plants and garbage and subsequently flies and mosquitoes are attracted to this type of water causing diseases. Oil will form a layer on the surface of water and may inhibit sunlight through it and a decrease in the dissolved oxygen. This will eventually affect the receiving water bodies and aquatic life within it [10]. FAO guidelines for irrigation doe not specify a standard value for oil and grease.

Table 5. Results of analysis of Physio-chemical parameters of wastewater of the drains in spring (March) and its comparison with NEQS.

Sr.No.	Parameter	Units	NEQS Value	Shahdara Drain	Mian Mir Drain	Iqbal Town Drain	Garden Town Drain	Sattu Katla Drain	Charar Drain	Hudiara Drain	Chota Ravi Drain
1.	pH	-	6 - 9	7.2	7.5	7.1	7.6	7.0	7.7	7.3	7.0
2.	Temperature	°C	40	25	26	24	25	26	28	24	25
3.	TSS	mg/L	200	170	85	146	150	122	147	160	118
4.	TDS	mg/L	3500	1080	670	630	741	790	760	2230	870
5.	Color	TCU	NGVS	105	60	301–	341	97	296	318	506
6.	Oil and Grease	mg/L	10	16	20	38	17	24	24	19	12
7.	COD	mg/L	150	336	330	414	289	243	350	436	318
8.	BOD	mg/L	80	152	124	196	146	118	162	204	135
9.	DO	mg/L	NGVS	0.7	0.7	0.8	0.6	0.8	0.6	0.8	0.6
10.	Chloride	mg/L	1000	143	52	68	58	88	46	504	95
11.	Sulphate	mg/L	600	242	97	72	98	139	63	264	89
12.	Calcium	mg/L	NGVS	65	82	55	51	75	82	85	58
13.	Magnesium	mg/L	NGVS	45	38	53	35	45	60	65	40
14.	Carbonate	mg/L	NGVS	11	14	5	7	9	13	16	6

Table 6. Results of analysis of Physio-chemical parameters of the wastewater of the Drains in spring (March) and its comparison with FAO guidelines for irrigation.

Sr. No.	Parameter	Unit	FAO Guidelines for irrigation – Degree of restriction on use			Shahdara Drain	Mian Mir Drain	Iqbal Town Drain	Garden Town Drain	Sattu Katla Drain	Charar Drain	Hudiara Drain	Chota Ravi Drain
			None	*Slight to moderate*	*Severe*								
1.	pH	-		Normal range 6.5-8		7.2	7.5	7.1	7.6	7.0	7.7	7.3	7.0
2.	TDS	mg/L	<450	450 - 2000	>2000	1080	670	630	741	790	760	2230	870
3.	Electrical conductivity	dS/m	<0.7	0.7 - 3.0	>3.0	5.2	4.9	4.5	4.3	4.2	5.3	6.1	5.0
4.	Sodium	me/L	<3	3 - 9	>9	43.3	40	44.34	39.5	42.6	35.2	50	36.5
5.	Chloride	me/L	<4	4 - 10	>10	4	1.46	3	1.63	2.5	1.3	14.2	2.7
6.	Bicarbonate	me/L	<1.5	1.5 - 8.5	>8.5	20	19.6	20.2	16.1	17.9	16.7	20.3	17.8
7.	SAR	-	>0.7	0.7 - 0.2	<0.2	26.5	29.8	33.3	27.4	31.3	23.4	32.0	29.5
8.	RSC	me/L		1.25		-ve	-ve	ve	-ve	ve	-ve	ve	-ve

COD values are always higher than BOD values and maybe much higher when significant amounts of biological resistant organic matter are present. The highest COD levels in the month of January and March found in Hudiara drain were 641 mg/L and 436 mg/L whereas the lowest COD values in the month of January and March were 295 mg/L and 243 mg/L respectively as shown in **Figure 3**.

Table 7. Results of analysis of heavy metals in the wastewater of the drains in spring (March) and its comparison with NEQS and FAO guidelines for irrigation.

Sr. No.	Parameter	Unit	NEQS value	FAO guidelines for irrigation	Shahdara Drain	Mian Mir Drain	Iqbal Town Drain	Garden Town Drain	Sattu Katla Drain	Charar Drain	Hudiara Drain	Chota Ravi Drain
1.	Iron	mg/L	8.0	5.0	5.0	1.2	3.0	1.9	5.3	1.5	1.6	2.0
2.	Copper	mg/L	1.0	0.20	0.72	0.04	0.47	0.008	3.79	0.04	0.39	0.01
3.	Nickel	mg/L	1.0	0.20	0.39	0.11	0.10	0.12	2.62	0.21	0.13	0.21
4.	Chromium	mg/L	1.0	0.10	*ND	ND	ND	ND	1.35	ND	ND	ND

NGVS = No guideline value suggested; ND = Not detected; NEQS Values: (Environmental Protection Agency EPA, 1997); FAO Guidelines for irrigation: (Pescod, 1992) *Guidelines for Interpretation Of Water Quality For Irrigation. **Threshold Levels Of Trace Elements for Crop Production (phytotoxic threshold levels of some selected trace elements in municipal wastewater).

The highest BOD value noted in the month of January and March were 265 mg/L and 204 mg/L as recorded at Hudiara Drain whereas the lowest BOD values were found in Sattu katla drain for both the months and the values were 123 mg/L and 118 mg/L respectively as shown in **Figure 4**. These Figures show that Hudiara drain has the highest BOD value and Sattu katla drain has lowest BOD value. When compared with NEQS (80 mg/L) all the BOD values were found above the limits.

DO in all of the wastewater samples collected from the eight drains during the sampling period was very low in the range 0.5 - 1 mg/L, less than standard DO level. The wastewater coming from sewage treatment plants mostly has organic materials that are decomposed by microorganisms, and uses oxygen to carry out their activities [10].

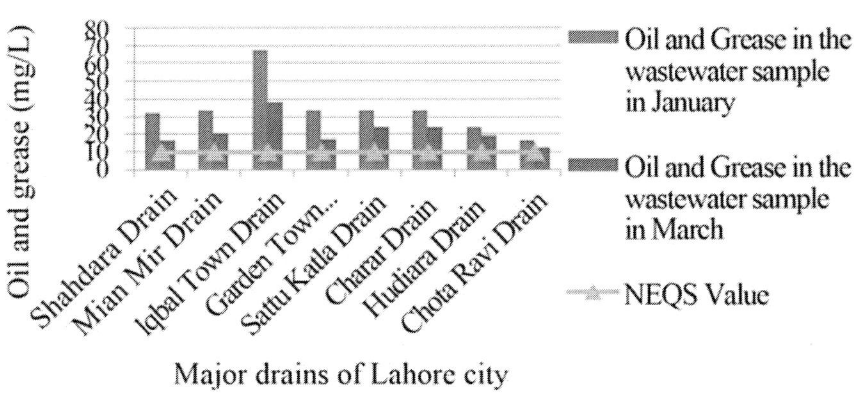

Figure 2. Oil and Grease in the wastewater samples in the winter and spring seasons.

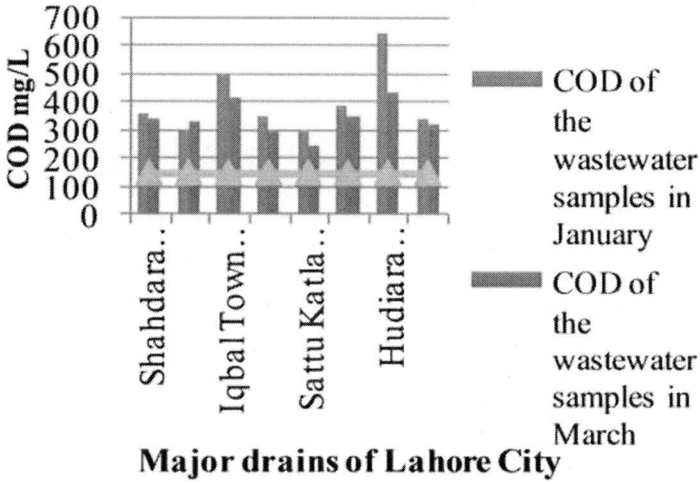

Figure 3. COD of wastewater samples in the winter and spring seasons.

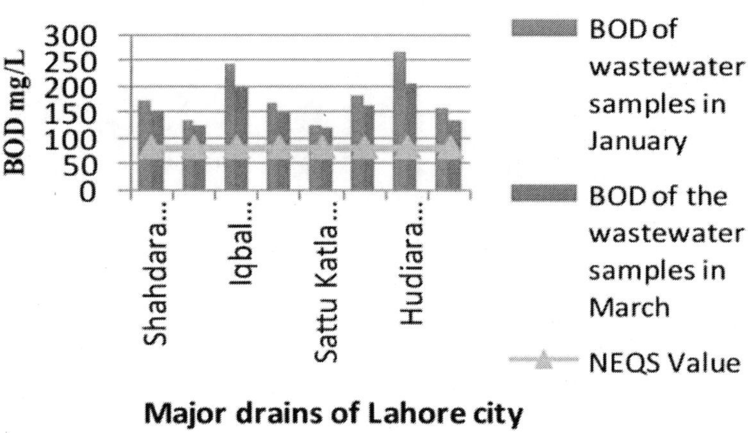

Figure 4. BOD of wastewater samples in the winter and spring seasons.

The concentration of Sulphate was in the range of 89 - 306 mg/L in January and in March it ranges from 63 to 246 mg/L which is within the NEQs. FAO guidelines for irrigation have no standard value for sulphate. Calcium is a major constituent of the surface water. The concentration of Ca in the wastewater of Hudiara, Charar and Mian Mir drain was high in both winter and spring season as compared to others. The concentration of magnesium was also high in Iqbal town drain, Hudiara and Charar drain in winter and spring season. There is no guideline value for calcium and magnesium in NEQS and FAO standards.

TDS refer to the total concentration of dissolved substances in the water. TDS is important parameter among agricultural water quality parameters. TDS plays important role in the plant growth, crop yield and quality of product [16]. The amount of TDS was again high in Hudiara drain as compared to others. Though TDS in the effluents was acceptable according to NEQS, but according to FAO guidelines for irrigation the water of the Hudiara drain is not suitable for irrigation purposes. The degree of restriction on the use of wastewater of the Hudiara drain is severe (>2000 mg/L), while for the rest of the drains it is from slight to moderate.

Water having more salt contents carries more electrical current. The electrical conductivity increases approx. 2 percent with every °C increase in temperature. It gives an indication of the amount of inorganic materials in the water including calcium, nitrogen, bicarbonate and others [16]. There is no standard value given for EC in NEQS. The results of the analysis of the wastewater showed that the degree of restriction on the use of wastewater is severe because all the drains have high EC than the standard value of FAO as shown in **Figure 5**.

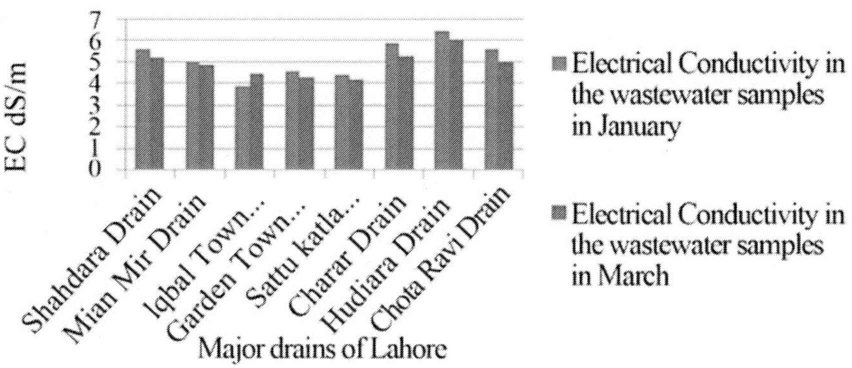

Figure 5. Electrical conductivity in the wastewater samples in winter and spring seasons.

It is important to measure chloride so that quality of the wastewater can be assessed for use in irrigation. The major source of chloride in the irrigation outlet can be treated effluents and the municipal sewage [7]. Chloride was present in the range of 60 - 692 mg/L in the samples of wastewater in the month of January while in the month of March it was in the range of 46 - 504 mg/L. When compared to NEQS, the value of chloride was within the acceptable range. But FAO guidelines for irrigation suggest that chloride concentration was above recommended limit for hudiara drain and the degree of restriction on the use of their wastewater is severe (>10 me/L). So it can be said that the water of Hudiara drain is unsuitable for irrigation purposes because it may have a harmful affect on the vegetables and crops grown by utilizing this water [17,18].

The comparison of the results of sodium in the wastewater of the drains with FAO guidelines for irrigation showed that the amount of sodium was very high in all the drains. The degree of restriction on the use of the water of these drains is severe because their concentration is very high (>9.0 me/L). High amount of sodium in the irrigation water causes toxicity problems like impaired growth, reduction in the yield of crops, change in the morphology of the plants and even its death [19].

The analysis of the bicarbonate ions in the wastewater of all the drains showed that the amount of these ions was very high in all the drains as compared to FAO guidelines for irrigation. The degree of restriction on the use of wastewater of these drains is severe because its value is very high and hence this water is not apposite for the purpose of irrigation.

Sodium adsorption ratio (SAR) is an indication of the probable pressure on the soil properties caused by sodium. SAR states the concentration of sodium in water relative to calcium and magnesium. When the amount of sodium in the irrigation water exceeds than allowable limit, it disperses the soil resulting in low penetration rate of air and water into the soil. It is difficult to harvest on the dispersed soil. Soils can have excess sodium because of the wastewater irrigation that is why it is important to measure SAR [19]. The results for SAR showed that the water of these drains was not suitable for irrigation as the value of SAR was very high in both winter and spring seasons. The degree of restriction on the use of wastewater of these drains is severe.

Assessment of sodium permeability hazard is referred to as Residual sodium carbonate (RSC). It is measured by taking the bicarbonate/carbonate and calcium/magnesium concentrations in irrigation water into account. RSC is important parameter for irrigation because in it the relative concentrations of bicarbonate and carbonate are compared with the concentrations of calcium, magnesium, and sodium [19]. The results showed that the negative value of RSC when compared to the FAO guideline for irrigation (1.25 me/L) is fit for irrigation.

Iron is one of the most important micronutrient for the humans and plants. Plants especially leafy crops absorb iron in high amounts. There is confirmation of the iron toxicity in Indus basin aquifers [17]. The results of the chemical analysis showed that the amount of iron was within the acceptable range as compared to NEQS. According to FAO guidelines for irrigation Shahdara and Sattu Katla drain is not suitable for irrigation due to high Iron content (>5 mg/L) as shown in **Figure 6**.

Industrial activities, atmospheric deposition and livestock manure are the reasons of contamination of soil with Cu. **Figure 7** shows the results of the analysis of Cu in the wastewater. The amount of Cu in all the drains falls within the acceptable range suggested by NEQS except in Sattu Katla drain which was very high than the standard value. Comparison of the Cu values with FAO guideline for irrigation showed that the water of Shahdara, Hudiara, Iqbal town and Sattu Katla drain is not fit for cultivation of crops.

Nickel is mostly present in those areas where domestic waste is combined with industrial wastewater [20]. Nickel is readily present in those agricultural lands where domestic wastewater is used for irrigation. High amount of Ni is toxic for the growth of plant. The results of the analysis of the wastewater of the drains of Lahore showed that the amount of Ni in all the drains was within the acceptable range of NEQS as given in **Figure 8**. But comparison of these values with FAO guideline for irrigation depicted that the quality of the water of Shahdara, Sattu Katla, Charar and Chota Ravi drain is not suitable for the irrigation of the crops.

The most dominant sources of Cr are phosphate fertil izers and metallurgic industries which cause atmospheric deposition. It is also released by tanning, ink manufacture, metal plating, dyes, wood preserving, textile and ceramic industries [20]. The results of the analysis of the wastewater showed that Cr was not detected into any drain except Sattu Katla drain, in concentration that exceeds both NEQs and FAO standards.

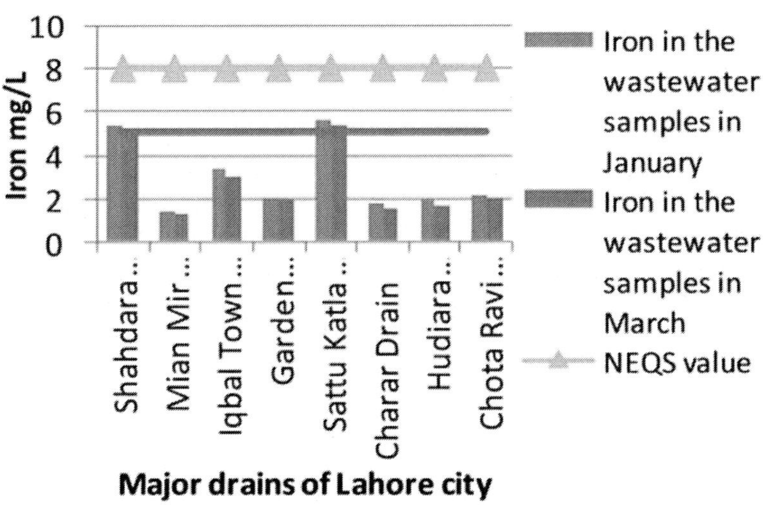

Figure 6. Iron in the wastewater samples in winter and spring seasons.

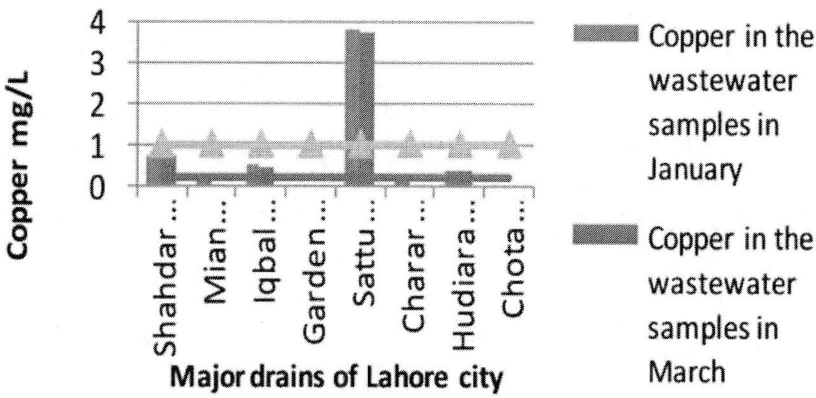

Figure 7. Copper in the wastewater samples in winter and spring seasons

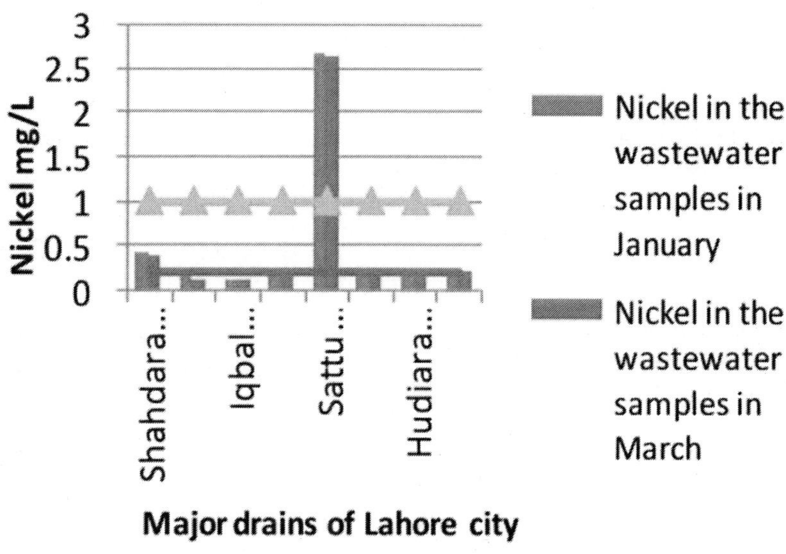

Figure 8. Nickel in the wastewater samples in winter.

4. CONCLUSION

The major drains of Lahore city were originally storm water drains. But due to increase in population and rapid industrialization in the city of Lahore, the drains are now mainly used to collect the industrial and municipal effluents of diverse nature. The characterization of the wastewater of these drains showed that many parameters were higher than the prescribed limit as compared to NEQS and FAO guidelines for irrigation. These results provide strong evidence that the water of these drains is not suitable for irrigation and other domestic purposes. Therefore, there is an urgent need to take essential steps to treat industrial wastewater before discharging into these drains so that pollution load can be minimised and the water can be used for beneficial purposes.

5. RECOMMENDATIONS

The findings of the study make it necessary for the government and other related authorities of wastewater management to take following actions:

- The wastewater of these drains needs to be monitored on a regular basis.
- The industrial and municipal effluents should be treated before discharging into the drains and the drain discharge should also be treated before use for irrigation. The implementation and follow-up of the environmental laws should be made compulsory by all the industries.
- A detailed study should be conducted to determine heavy metals in the vegetables and soil irrigated by the water of these drains.
- Livestock should not be taken to these drains for drinking and bathing. It can contaminate their meat and milk through accumulation of heavy metals and harmful chemicals.
- The pumping of groundwater near these drains must be shunned to lessen the hazard of waterborne diseases.

Almas Hamid, Mahrukh Zeb, Asim Mehmood, Sana Akhtar, Samia Saif • All of the drains should be covered in order to reduce the visual impacts on the society, and drains should be lined with concrete so that the risk of seepage of toxic chemicals and heavy metals into the groundwater could be evaded.

REFERENCES

1. GOP, "Population Census Organization, Government of Pakistan," 2012. http://www.census.gov.pk/
2. Economic Adviser's Wing, Finance Division, Government of Pakistan. "Highlights: Pakistan Economic Survey 2010- 2011," 2011, pp. 1-11.

3. E. Naeem, "An Investigation of the Characteristics of Effluent Mixing in Stream," University of Engineering and Technology, Taxila, 2009.
4. D. Pay, A. S. Christopher, S. L. Raschid, R. Mark and B. Akiça, "Wastewater Irrigation and Health; Assessing and Mitigating Risk in Low-Income Countries," In: NonPathogenic Trade-Offs of Wastewater Irrigation, Earthscan, London, 2010, pp. 101-126.
5. WASA, "Presentation on the Initiative Taken in WASA Lahore Regarding Benchmarking of UWSS," Water and Sanitation Agency, Lahore, 2007.
6. IPD, "An Atlas: Surface Water Industrial and Municipal Pollution in Punjab," Irrigation and Power Department, Directorate of Land Reclamation Punjab, 2008, pp. 1-132.
7. S. Afzal, I. Ahmad, M. Younas, M. D. Zahid, K. M. H. Atique, A. Ijaz and K. Ali, "Study of Water Quality of Hudiara Drain, India-Pakistan," Environment International, Vol. 26, No. 1-2, 2000, pp. 87-96.
8. A. M. Kizer and H. Zhang, "Understanding Your Irrigation Water Test Report," Oklahoma State University, Oklahoma, 2001.
9. K. R. Sharma, M. Agrawal and F. Marshall, "Heavy Metal Contamination of Soil and Vegetables in Suburban Areas of Varanasi, India," Ecotoxicology and Environmental Safety, Vol. 66 No. 2, 2007, pp. 258-266.
10. K. R. Sahu, S. Katiyar, J. Tiwari and C. G. Kisku, "Assessment of Drain Water Receiving Effluent from Tanneries and Its Impact on Soil and Plants with Particular Emphasis on Bioaccumulation of Heavy Metals," Journal of Environmental Biology, Vol. 28 No. 3, 2007, pp. 685- 690.
11. J. H. J. Ensink, W. van der Hoek and W. Simmons Robert, "Livelihoods from Wastewater: Water Reuse in Faisalabad, Pakistan," IWA Publishing, London, 2008, pp. 387- 400.
12. S. Akhtar and N. Mohammad, "Impact of Water Quality on Aquatic Life in River Ravi, Pakistan," Journal of Nature Environment and Pollution Technology, Vol. 11, No. 2, 2012, pp. 219-224.
13. M. Tariq, M. Ali, and Z. Shah, "Characteristics of Industrial Effluents and Their Possible Impacts on Quality of Underground Water," Soil & Environment, Vol. 25, No. 1, 2006, pp. 64-69.
14. E. M. Ali, "Impact of Drain Water on Water Quality and Eutrophication Status of Lake Burullus, Egypt, a Southern Mediterranean Lagoon," African Journal of Aquatic Science, Vol. 36, No. 3, 2011, pp. 267-277.
15. WWF, "Pakistan's Waters at Risk: Water & Health Related Issues in Pakistan & Key Recommendations: A Special Report," WWF, Lahore, 2007, pp. 1-25.
16. SFWF, "TDS and pH," Organization of Safe Drinking Water, Canada, 2002, pp. 1-6.
17. M. Ibrahim and S. Salamon, "Chemical Composition of Faisalabad City Sewage Effluent: II Irrigation Quality," Journal of Agriculture Research, Vol. 30, No. 1, 1992, pp. 391-340.
18. A. M. Kahlown, M. Ashraf, M. Hussain, A. H. Salam and Z. A. Bhatti, "Impact Assessment of Sewerage and Industrial Effluents on Water Resources, Soil, Crops and Human Health in Faisalabad," Pakistan Council of Research in Water Resources, Lahore, 2006, pp. 1-105.
19. M. B. Pescod, "Wastewater Treatment and Use in Agriculture-FAO Irrigation and Drainage Paper," Wastewater Characteristics and Effluent Quality Parameters, Food and Agriculture Organization of the United Nations, Rome, 1992.
20. I. Thornton, D. Butler, P. Docx, M. Hession, C. Makropoulos, M. McMullen, M. Nieuwenhuijsen, A. Pitman, R. Rautiu, R. Sawyer, S. Smith and D. White, "Pollutants in Urban Wastewater and Sewage Sludge," Final Report, European Union, London, 2001, pp. 1-244.

CHAPTER 14

Evaluating the Usefulness of Continuous Leaf Turgor Pressure Measurements for the Assessment of Persimmon Tree Water Status

M. A. Martínez-Gimeno[1] · M. Castiella[2] · S. Rüger[2] · D. S. Intrigliolo[1,3] · C. Ballester[3,4]

[1.] *Center for Applied Biology and Soil Sciences (CEBAS)MurciaSpain*

[2.] *YARA ZIM Plant Technology GmbHHennigsdorfGermany*

[3.] *Valencian Institute for Agricultural Research (IVIA)ValenciaSpain*

[4.] *Centre for Regional and Rural Futures (CeRRF)Deakin UniversityGriffithAustralia*

ABSTRACT

Continuous plant water status monitoring is crucial in order to improve irrigation management. The noninvasive Yara ZIM-probe was assessed for detecting plant water stress in Persimmon trees (*Diospyros kaki* L.f.). The probe measures the pressure transfer function (P_p) through a patch of an intact leaf, which is inversely correlated with the turgor pressure. This technology was evaluated in two parallel experiments involving either distinct watering regimes or rootstocks with different drought tolerance [*Diospyros lotus* (L) and *Diospyros virginiana* (V)]. Concomitant measurements of midday stem water potential (Ψ_{stem}) and trunk diameter variations were taken throughout the experiments. P_p was highly correlated with Ψ_{stem}. Persimmon leaves exhibited the inversed P_p curve phenomena under water stress, which enabled the association of a particular range of Ψ_{stem} to each of the three leaf turgor states defined. Persimmon trees with no sign of initial or total inversion ensured Ψ_{stem} above −0.8 MPa, values considered of a well-watered Persimmon tree. Yara ZIM-probe readings as well as Ψ_{stem} and trunk diameter variation measurements pointed L as a more sensitive rootstock to drought than V. In conclusion, results showed that the Yara ZIM-probe can be used to continuously monitor water status in Persimmon trees although further research would be needed to ensure their feasibility for scheduling irrigation.

INTRODUCTION

Water is becoming a limiting factor for crop production in much of the world (IPCC 2014). That is the case of the Mediterranean ecosystems where the scarce rainfall is not enough to cover the high crop water requirements during most of the season, and the water resources available are limited. Optimizing irrigation must be then a priority in order to ensure the sustainability of the agricultural systems.

Water-saving irrigation strategies such as the partial rootzone drying or regulated deficit irrigation (RDI) have been studied in depth in experimental and commercial orchards with successful results (Ruiz-Sanchez et al. 2010). A common conclusion in all the studies related to the implementation of water-saving irrigation strategies is that water stress monitoring is crucial in order to avoid an undesirable impact on yield. A series of plant-based water stress indicators can be found in the market to continuously monitor the plant water status throughout the season. Stem dendrometers, porometers, sap flow probes or the measurement of canopy temperature among others have been studied during the last decades in woody crops to automatically monitor the plant water status in an attempt to substitute the stem water potential measurement, which is the accepted method as reference despite being a destructive and time- and labor-consuming technique (Ballester et al. 2013a; Fernández and Cuevas 2010; Fernández 2014; Jones et al. 2009; Ortuño et al. 2010). The use of these methods in the field, however, is still a constraint for growers due to different reasons such as difficulty of installation, maintenance requirements and the need of processing a large amount of data (Fernández 2014).

Monitoring the leaf turgor pressure with a highly sensitive pressure sensor clamped to a patch of a leaf (Yara ZIM-probe, Zimmermann et al. 2008) has been reported as an easier to install and easier to use technique than the above-mentioned methods with a great potential to be used at field by growers (Fernández 2014).

The noninvasive Yara ZIM-probe is a magnetic-based probe that measures the pressure (P_p) transfer function through a patch of an intact leaf. This P_p has been shown to be inversely correlated with the turgor pressure (Zimmermann et al. 2008, 2009; Westhoff et al. 2009). The usefulness of the Yara ZIM-probe to detect plant water stress has been studied on several horticultural and fruit crops, providing evidence that it can detect changes in turgor pressure caused by variations in the microclimate or in the soil water availability (Westhoff et al. 2008; Zimmermman et al. 2009; Rüger et al. 2010, 2011; Ehrenberger et al. 2012; Fernández et al. 2011; Rodriguez-Dominguez et al. 2012). Moreover, some studies suggest that its use has the potential for efficiently select genotypes tolerant to water stress environments (Kant et al. 2014). Different approaches such as the assessment of the daily or nightly maximum P_p value, the turgor recovery phase during the afternoon or the reverse of the P_p curve, however, must be followed to determine the degree of stress reached by the plants depending on their physiological characteristics and their tolerance or sensitivity to drought stress, which makes necessary the assessment of this technology for each particular case.

Persimmon (*Diospyros kaki* L.f.) cultivation in Spain has steadily increased in the last decades from a cropped area of 2.000 ha in 2002, to an approximately 13.000 ha in 2014 (Perucho 2015). This notable increase in production is mainly due to the current replacement of citrus with Persimmon trees, particularly with the *cv.* 'Rojo Brillante,' which fruit reaches a higher value in the market than oranges, thanks to a postharvest treatment with high CO_2concentrations that removes its astringency (Arnal and Río 2003). Studies performed in Valencia, Spain, with the *cv.* 'Rojo Brillante' showed that RDI strategies may lead growers to obtain water savings of 20 % without any reduction in yield, increasing then the water use efficiency (Buesa et al. 2013). Results also showed that fruit weight reduction in RDI trees caused a decrease in the fruit commercial value, pointing out that further research would be needed to define a successful RDI strategy for this crop. The Yara ZIM-probe technology could be a suitable tool to be used in this crop for irrigation scheduling in order to properly manage water stress and avoid any fruit size reduction that could affect the economic return obtained by farmers.

The objectives of the present study were: (1) to assess the feasibility of the Yara ZIM-probe for detecting plant water stress in Persimmon trees; (2) to explore the relationships between $P_p$and Ψ_{stem} for their possible use as a tool for irrigation scheduling in this crop; and (3) finally, to test the sensitivity of the Yara ZIM-probe in two rootstocks of differing drought tolerance.

MATERIALS AND METHODS

Plot and irrigation treatments

The experiment was performed during 2014 in a commercial orchard planted with Persimmon trees *cv.* 'Rojo Brillante' grafted onto two contrasting rootstocks in vigor, *Diospyros lotus* (L) and *Diospyros virginiana* (V) at a spacing of 5 m × 2.5 m. The plot was located in Liria (40°N, elevation 300 m), Valencia, Spain, where the climate is typically Mediterranean and the soil is sandy loam with 32 % by weight stones and an effective depth of 0.8 m. Soil density ranged from 1.35 to 1.45 t m^{-3} and was considered of low fertility (0.66 organic matter and 0.05 % total N).

Irrigation was applied with two drip lines leaving ten pressure compensated emitters (4 L h^{-1}) per tree. Water had an average electrical conductivity (EC) of 1.1 dS m^{-1} and an average Cl concentration of 122 mg L^{-1}. Irrigation was applied according to the estimated crop evapotranspiration ($ET_c = ET_o \times K_c$). Reference evapotranspiration, ET_o, (Allen et al. 1998) was obtained from a weather station located near the orchard, which also measured the solar radiation (Pyranometer CMP3, Kipp & Zonen, Delft, The Netherlands). Crop coefficient applied ranged from 0.2 in March to 0.9 at full canopy growth. As a part of an outgoing experiment, trees had been irrigated at 100 and 125 % ET_c.

The Yara ZIM-probes were tested on trees grafted on both rootstocks and irrigated at both rates. In a first experiment, trees grafted onto L and irrigated at 100 % ET_c (L-WW) during the whole experiment (total amount applied 332.5 mm) were compared with trees grafted onto the same rootstock (L-DS) in which irrigation was withheld during two periods from August 18th to August 23rd [day of the year (DOY) 230–235] and from August 29th to September 3rd (DOY 241–246). L-DS trees were irrigated as L-WW trees between the two drought cycles. A second experiment was then set up with trees irrigated at 125 % ET_c (total amount received 366.6 mm) in order to compare the behavior of trees grafted onto L (L-125-DS) and V (V-125-DS) rootstocks when subjected to drought cycles. In these trees, irrigation was withheld from August 25th to September 3rd (DOY 237–246).

The orchard was divided into three blocks of three rows each where the treatments were applied. All the treatments consisted of 21 trees (seven trees per row) with the five trees of the mid-row (perimeter trees were avoid) used for the measurements. The L-125-DS and V-125-DS treatments, in which four trees were used for the plant water status measurements, were carried out in the block of the middle while the L-WW and L-DS treatments in the remainder blocks.

Leaf turgor monitoring

The basic principle of the Yara ZIM-probe is described by Zimmermann et al. (2008) and the principle of the magnetic leaf patch clamp pressure probe by Westhoff et al. (2009). Briefly, the Yara ZIM-probe consists of two magnets that exert an external pressure to a patch of a leaf covering an area of 87 mm^2. One of the magnets contains a highly sensitive pressure sensor able to detect pressure variations up to 300 kPa. The sensor measures the difference in pressure between the magnets and the leaf turgor, named patch pressure (P_p), and therefore provides information about relative changes in leaf turgor at real time. The distance between the magnets can be regulated in order to set up the most suitable initial P_p, which ranges between 10 and 60 kPa.

All the Yara ZIM-probes were previously tested under laboratory conditions to ensure that ambient temperature (T_a) did not have any influence on their readings.

Selected trees from each treatment were equipped with two Yara ZIM-probes each (8–10 Yara ZIM-probes/treatment) on 13th and 14th of May, 2014. The Yara ZIM-probes were installed in mature leaves located in the east side of the canopies. In order to distinguish between relative changes in leaf turgor caused by water stress and those caused by microclimate variations (Zimmermann et al. 2013a), relative humidity (RH) and T_a sensors (Yara ZIM Plant Technology GmbH (Hennigsdorf, Germany) were also installed in the orchard. All probes (leaf turgor, T_a and RH) were connected by cable to transmitters which sent the data wirelessly every 5 min over a distance of up to 1500 m to a central controller (Yara ZIM Plant Technology GmbH Hennigsdorf, Germany). The controller contains a GPRS modem which is linked to an Internet server where data are stored and available for real-time inspection and download.

Other plant water status determinations

Measurements of Ψ_{stem} were carried out during the experiment to determine the plant water status in all the trees equipped with the Yara ZIM-probes. Measurements were performed at solar noon with a Scholander pressure chamber (PMS Instrument Company, mod. 600, OR, USA) using 2–4 mature leaves per tree previously bagged with aluminum foil for at least 1 h before the measurements to avoid transpiration (Turner 1981).

The four selected trees from L-125-DS and V-125-DS treatments (grafted onto the two contrasting rootstocks) were equipped with linear variable differential transformers (LVDT, Schlumberger mod. DF-2.5) to monitor trunk diameter variations during the drought stress cycles. Each sensor was fixed to the main trunk of the tree by a metal frame of Invar (a metal alloy with a minimal thermal expansion), located about 25 cm from the ground. All the transformers were previously calibrated in the laboratory by means of a precision micrometer (Verdtech SA, Spain). The trunk diameter variations were used to calculate the maximum daily trunk shrinkage (MDS) by obtaining the difference between the maximum (MXTD) and minimum (MNTD) diameter in a day. At the beginning of the experiment, trees grafted onto L and V had, respectively, an average trunk perimeter of 0.25 ± 0.004 and 0.23 ± 0.020 m. Data were automatically recorded every 30 s using a data logger (model CR10X) connected to an AM16/32 multiplexer programmed to report mean values every 30 min.

Data analysis

Data were analyzed using Statgraphics X64, Origin 2015 (Microcal Software Inc., Northampton, MA) and SigmaPlot 11.0. The relationship between P_p and the others water status indicators used during the measurement was explored by ANOVA and the least significant difference (LSD) procedure. Both methods take into account that values of $P < 0.05$ are considered to be statistically significant. Data shown are mean ± standard deviation.

RESULTS

Meteorological conditions

Total ET_0 and rainfall registered during the experiment (from April 1st to September 10th) were of 787 and 63 mm, respectively, which can be considered typical values for the area of study. Daily means of T_a ranged from 12.2 to 28.57 °C with a maximum temperature recorded in August of 40.5 °C. Daily means of RH ranged from 28.9 to 76.2 % with a minimum value reached in July (9.2 %).

Experiment 1: L-WW versus L-DS

Stem water potential measurements

The L-WW treatment in which water restrictions were not applied had Ψ_{stem} values around −0.60 MPa during the whole experiment (mean Ψ_{stem} of −0.62 ± 0.01 MPa; Figure 1). L-DS trees had similar Ψ_{stem} values to L-WW trees at the beginning of the experiment. Once water restrictions began, Ψ_{stem} dropped steadily to a minimum value of −1.73 and −1.93 MPa during the first and second drought cycles, respectively (Figure 1).

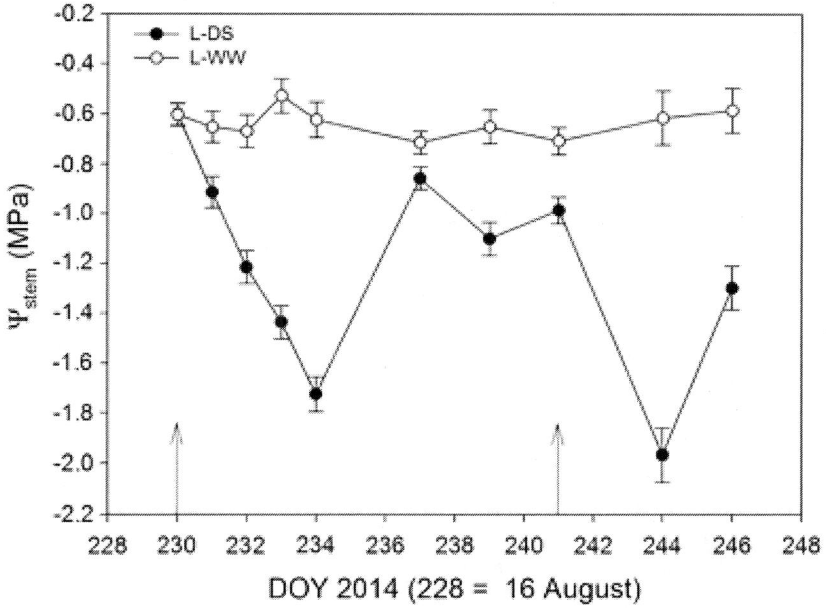

Figure 1. Stem water potential (Ψ_{stem}) evolution in control (L-WW) and non-irrigated trees (L-DS) during the drought cycles. Each point is the average of 10–20 leaves (5 trees per treatment). *Vertical bars* represent the ±LSD intervals, and *arrows* indicate the starting of the drought cycles

Leaf turgor monitoring with the Yara ZIM-probes

A daily continuous increase in maximum (recorded at midday) and minimum (recorded at night) P_p was observed in all of the Yara ZIM-probe sensors at the beginning of the experiment just after clamping. Some probes showed a steadily P_p increase similar to the increase in air temperature recorded in May and stabilized after a couple of weeks. Other probes (about 35 %), however, exhibited a sharp increase in P_p reaching the maximum values detected by the sensor (250–300 kPa) in weeks (Figure 2). These probes were reclamped again (several times in some occasions) until the P_p readings were more stable and treatments began (end of June).

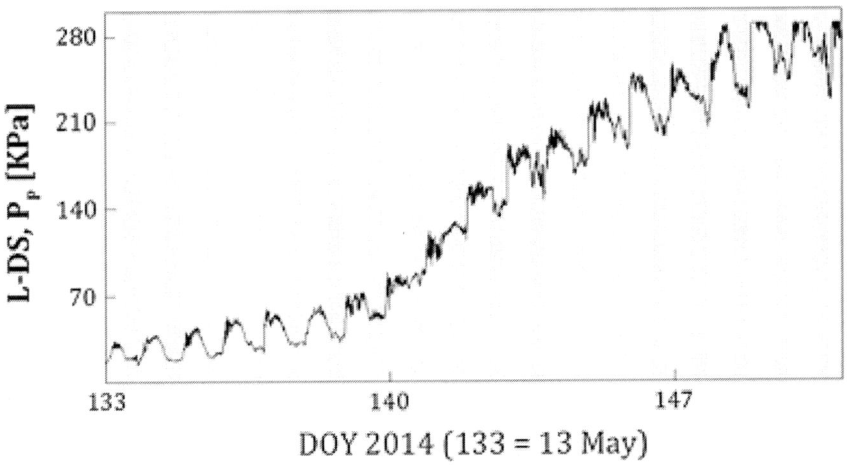

Figure 2. Patch pressure (P_p) increase after clamping at the first stages of the leaf growth. Sensors were reclamped when P_p reached

Three different daily P_p curve shapes were obtained depending on the range of stress reached by the trees. Each of these shapes was associated to a leaf turgor state as reported in Ehrenberger et al. (2012); thus, state I was related to a daily P_p shape with peaking values at noon and minimum values recorded at night; state II was related to a half inverted curve with a sharp decrease of P_p at noon; and state III was related to a complete inversion of the P_p curve with minimum values recorded during the day and maximum values during the night. Figure 3depicts the Ψ_{stem} and P_p evolution of a representative L-WW and L-DS tree during the experiment as well as the classification of the P_p shape curves in the above-mentioned states. The P_p curve in L-WW trees was all the time in state I, while L-DS trees showed a progressive change in the P_p curve shape from state I to state II and III as Ψ_{stem} decreased during the drought cycles. Similarly, the P_p curve shape in L-DS trees changed from state III to state II and I when irrigation was resumed as Ψ_{stem} recovered to similar values to the L-WW treatment. When data from all the Yara ZIM-probes of each treatment were analyzed, the P_p curve in L-WW trees was in the state I during the whole experiment. In L-DS trees, on the other hand, the P_p curve remained in the state I just 9.7 % of the time while it was 34.7 and 55.6 % of the time in the state II and III, respectively.

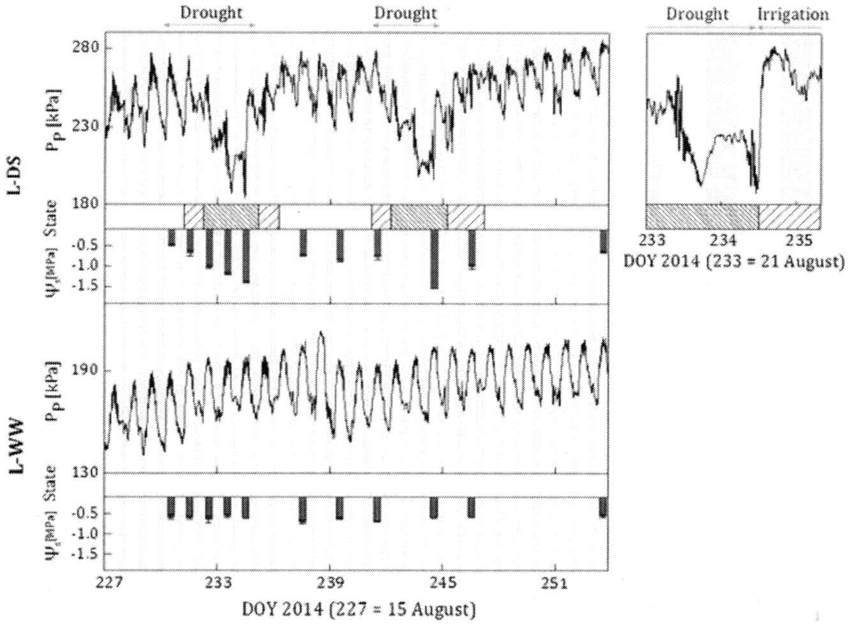

Figure 3. Stem water potential (Ψ_{stem}, columns) and patch pressure (P_p, *solid line*) evolution in one control (L-WW) and one non-irrigated (L-DS) tree during the two drought cycles. Different P_p*curve* shapes associated with different plant water status are also identified for each treatment (state I *without lines*; state II with low-density *diagonal lines*; and state III high-density *diagonal lines*). Ψ_{stem} values are means; *vertical lines* indicate the standard errors, $n = 4$. *Shaded background columns* indicate the nocturnal hours. The *additional graph* highlights the fast recovery of the plant water status when the drought period finishes. The P_p*curve* shows the evolution from the state III to state II in several hours

The assessment of the P_p curve shapes in both treatments during the experiment enabled the classification of each leaf turgor state within a range of Ψ_{stem}. Statistically significant differences in mean Ψ_{stem} were observed within trees from each leaf turgor state (Figure 4). State I of leaf turgor was observed in trees with a Ψ_{stem} higher than −0.80 MPa. The intermediate state of leaf turgor (state II) was observed in trees with a Ψ_{stem} compressed between −0.69 and −1.33 MPa. Finally, the state III of leaf turgor was observed in trees with a Ψ_{stem} ranging from −1.02 to −2.40 MPa.

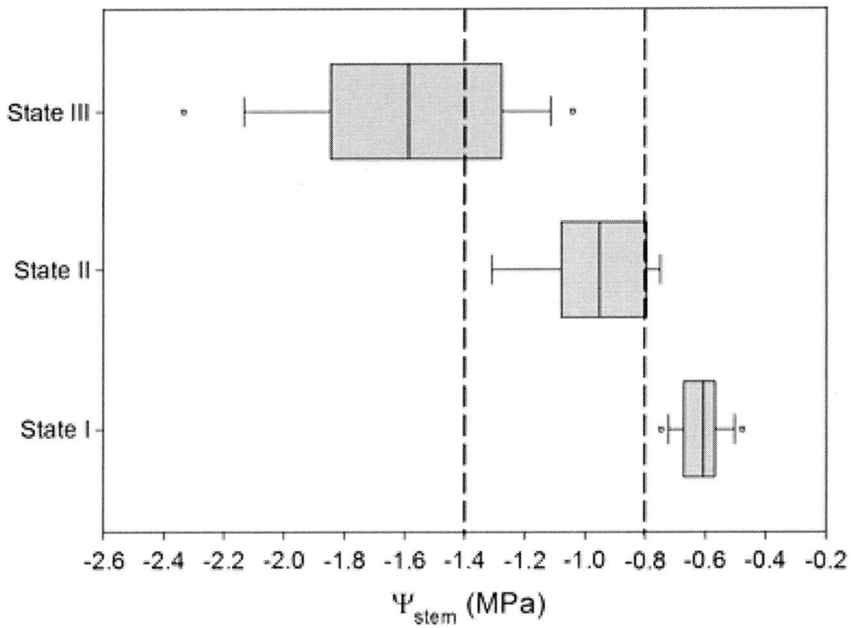

Figure 4. Range of stem water potential (Ψ_{stem}) values obtained during the two drought cycles within each leaf turgor. Ψ_{stem} values are the average of five trees. *Horizontal bars* in *boxes* indicate the maximum and minimum values of Ψ_{stem} for each state. *Dotted border lines* define Ψ_{stem} thresholds for plant water status: adequate (>−0.8 kPa), mild to critical stress (−0.8 to −1.4 kPa) and moderate stress (<−1.4 kPa)

Experiment 2: L-125-DS versus V-125-DS

Stem water potential and trunk diameter measurements

Water restrictions applied in both treatments offered different results in terms of mean and maximum Ψ_{stem} values. During the drought cycle, L-125-DS treatment registered an average Ψ_{stem} of −0.97 ± 0.38 MPa with a minimum value of −1.80 MPa. In contrast, V-125-DS treatment had a higher mean Ψ_{stem} value of −0.80 ± 0.16 MPa and reached a minimum value of −1.14 MPa. Significant differences were found in Ψ_{stem} between trees grafted onto L and V rootstocks from 2 days of the beginning of the water restrictions (Figure 5).

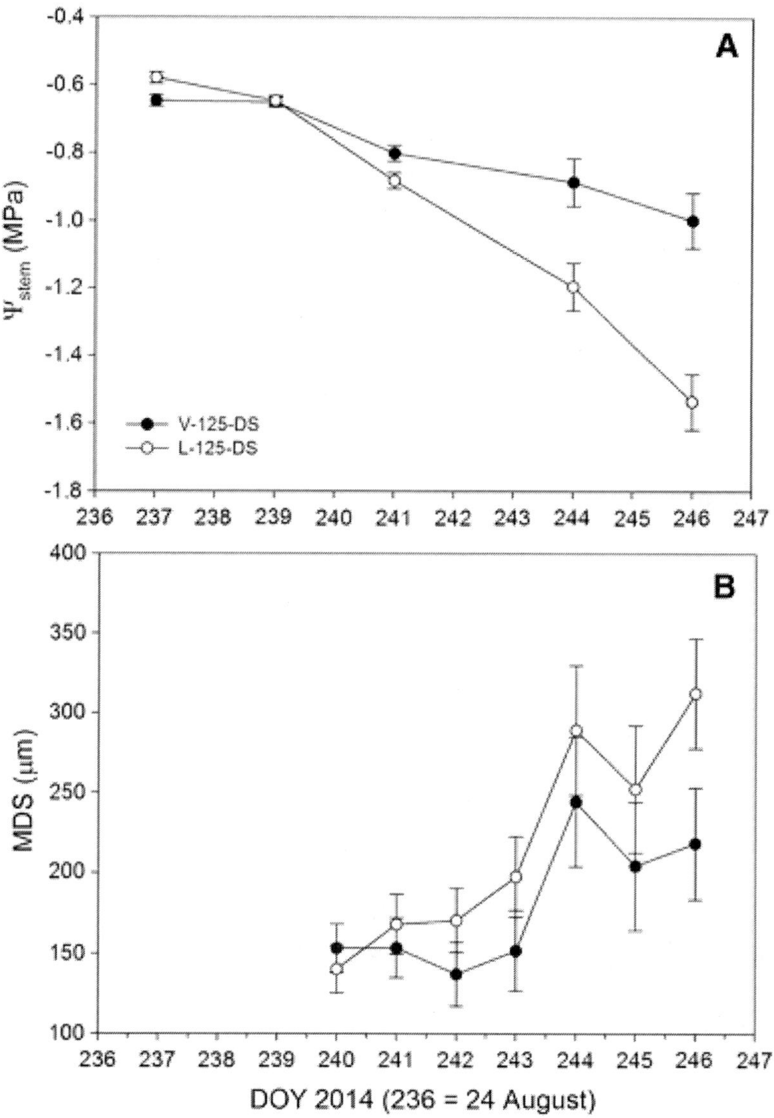

Figure 5. Stem water potential (Ψ_{stem}) and maximum daily trunk shrinkage (MDS) evolution in trees grafted onto lotus (L-125-DS) and virginiana (V-125-DS) during the drought cycle. *Each point* in figure (**A**, **B**) is the average of 10–20 leaves (5 trees per treatment) and 4 trees, respectively. *Vertical bars* represent the ± LSD intervals

Trunk diameter variations were monitored during the same period. MDS was lower in V (179.0 ± 48.0 μm) than in L (215.4 ± 67.8 μm) although no significant differences were observed between the rootstocks until the end of the drought cycle when Ψ_{stem} reached values around −1.6 MPa in L trees (Figure 5). The MXTD, on the other hand, was 42 % higher in V than in L rootstocks (Figure 6), and it is important to highlight that during the drought period, V maintained a constant growth rate slope in contrast to L where trunk growth rates decreased.

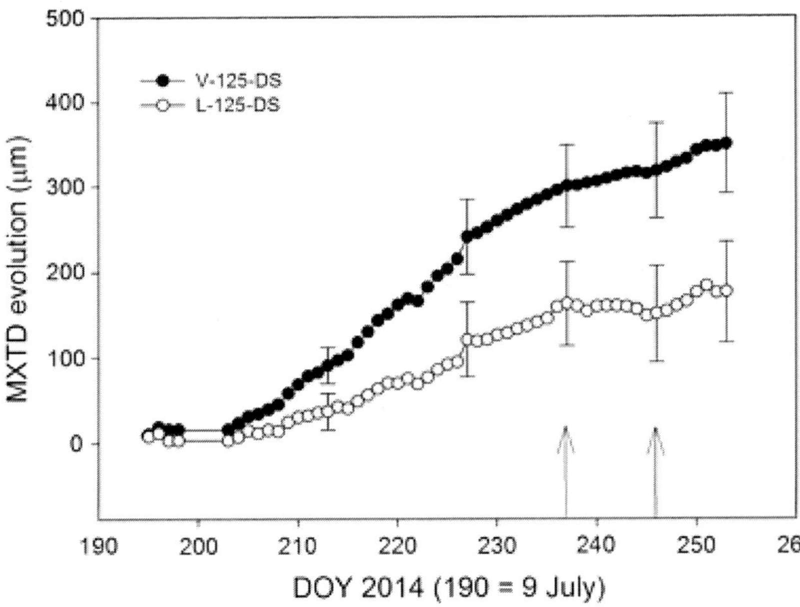

Figure 6. Maximum daily trunk diameter (MXTD) monitored during the crop season. *Each point* is the average of four diameter trees. *Vertical bars* represent the ± LSD intervals, and *arrows* indicate the starting of the drought cycles

Leaf turgor monitoring with the Yara ZIM-probes

The evolution of the P_p curve shape in the different leaf turgor states (I, II and III) was linked to the Ψ_{stem} measurements as did in experiment 1 (Figure 7). The P_p curve in V-125-DS trees was most of the time in state I, while L-125-DS trees showed a gradual evolution from state I to state II and III. On average, L-125-DS was in state I, II and III 30, 30 and 60 % of the time, respectively. However, V-125-DS was 80 % of the time in state I and just 20 % in state II and III. Each leaf turgor state within each rootstock corresponded to a range of Ψ_{stem} (Figure 8). L-125-DS reached lower values of Ψ_{stem} (−1.03 ± 0.18 and −1.41 ± 0.23 MPa) than V-125-DS (−0.91 ± 0.10 and −1.02 ± 0.29 MPa) for states II and III, respectively. However, state I in L-125-DS (−0.66 ± 0.10) was associated with higher Ψ_{stem} values than in V-125-DS (−0.74 ± 0.12).

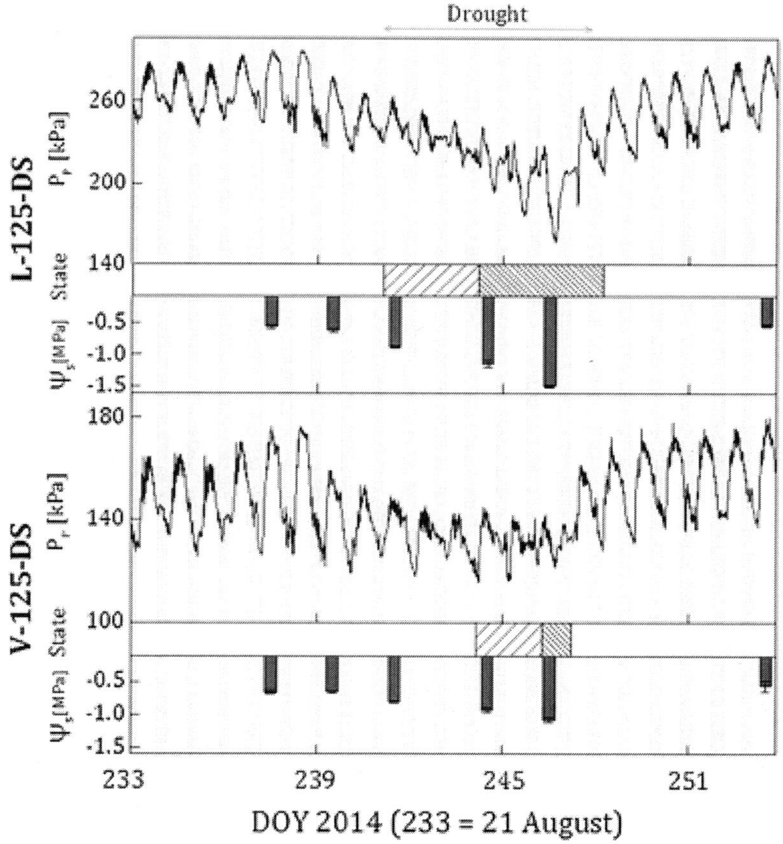

Figure 7. Stem water potential (Ψ_{stem}, *columns*) and patch pressure (P_p, *solid line*) evolution in a plant grafted onto lotus (L-125-DS) and virginiana (V-125-DS) during the drought cycles applied. Different P_p*curve* shapes associated with different plant water status are also identified for each treatment (state I *without lines*; state II with low-density *diagonal lines*; and state III high-density *diagonal lines*). Ψ_{stem} values are means; *vertical lines* indicate the standard errors, $n = 4$. The *shaded background columns* indicate the nocturnal hours

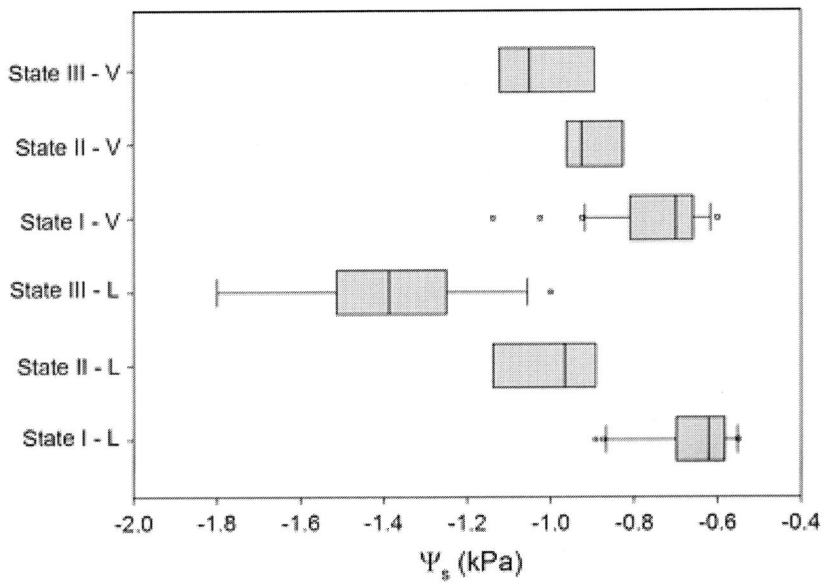

Figure 8. Stem water potential (Ψ_{stem}) values within each state of leaf turgor in trees grafted on lotus (L) and virginiana (V) trees. *Horizontal bars* indicate the maximum and minimum values of Ψ_{stem} for each state

DISCUSSION

The results presented in this work point out the Yara ZIM-probes as a reliable tool for continuously monitoring plant water status in Persimmon trees. Notwithstanding the $P_p$evolution observed just after clamping (April–May), in which daily and nightly P_p increased reaching the maximum pressure detected by the sensor in some of the probes (Figure 2), leaf turgor monitoring with the Yara ZIM-probes enabled the detection of mild to severe water stress in L-DS, L-125-DS and V-125-DS trees as well as their recovery when irrigation was resumed.

The particular P_p evolution observed just after clamping in Persimmon trees has not been reported for other fruit tree crops. This effect could be related to the structure of Persimmon leaves which is more complex than in other crops due to their density and thickness. Moreover, the foliar limb is slightly wavy and the main and secondary nervation stands out on the abaxial surface (Giordani et al. 2015), which most likely hamper the proper continuity between magnets. More studies would be needed regarding this matter to untangle this unusual behavior of P_p and be able to monitor leaf turgor during the whole growing period of Persimmon trees.

Different approaches can be followed to detect plant water stress when analyzing data from the Yara ZIM-probes (Zimmermann et al. 2013b). The assessment of the maximum daily P_p seems to be a useful method for clementine trees, in which a half or complete inversion of the P_p curve does not often occur even under severe (Ψ_{stem} of up to −1.9 MPa) water-stressed conditions (Ballester et al. 2016). Contrary to this and more in the line of what has been reported for other crops like almond, eucalyptus, avocado (Zimmermann et al. 2013b) and olive trees (Fernández et al. 2011; Ehrenberger et al. 2012; Padilla-Díaz et al. 2015), the change observed in the $P_p$curve profile was shown as a useful method to detect drought stress in Persimmon trees. The change in the P_p curve profile from a normal curve peaking at midday to a half and eventually complete inversion of the curve as a consequence of the water replacement for air into the parenchyma tissue of cells leaves [see Ehrenberger et al. (2012) for more details] made possible to differentiate between three ranges of Ψ_{stem} linked to well-watered, moderate and severe water-stressed trees. The P_p curve profiles were classified in three states of leaf turgor as described in Ehrenberger et al. (2012). Thus, each state was linked to a range of Ψ_{stem} and consequently to a plant water status (Table 1). State I, in which L-WW trees remained during the whole study and L-DS trees just less than 10 % of the time during the drought cycles, was linked to Ψ_{stem} values above −0.8 MPa, considered of well-watered Persimmon trees (Badal et al. 2010). The Ψ_{stem} thresholds for the intermediate state of leaf turgor (−0.69 and −1.33 MPa) were overlapped with the upper limit of state I (−0.79 MPa) and the lower limit of state III (−1.02 MPa) (Table 1). State II would include then trees with mild to moderate water stress (George et al. 1995), whereas trees within the state III, with Ψ_{stem} values ranging from −1.02 to −2.40 MPa, would indicate trees with moderate to severe water stress. Additionally, irrigation effect and water status recovery were also detected by the Yara ZIM-probes immediately. When irrigation was resumed after a drought cycle, the daily P_p curves evolved from the inverse shape to the half inverse or even the normal shape directly (see this behavior highlighted in Figure 3).

Table 1. Stem water potential (Ψ_{stem}) associated with each of the leaf turgor states observed in control (L-WW) and non-irrigated (L-DS) trees. Water status for the average, maximum and minimum Ψ_{stem} recorded within each state is also shown

State	Ψ_s (MPa)			Water status		
	Average	Maximum	Minimum	Average	Maximum	Minimum
I	−0.61 ± 0.08	−0.79	−0.47	Adequate	Adequate	Adequate
II	−0.98 ± 0.19	−1.33	−0.69	Mild stress	Moderate stress	Adequate
III	−1.58 ± 0.36	−2.40	−1.02	Moderate stress	Severe stress	Mild stress

The interpretation of the Ψ_{stem} values in relation to the degree of water stress is derived from previous experiments carried out in order to determine Persimmon trees responses to different irrigation regimes (Badal et al. 2010, 2013; Buesa et al. 2013)

These results show that leaf turgor monitoring with the Yara ZIM-probes could possibly be used in Persimmon trees for irrigation scheduling. Irrigation management has been proven as a useful tool to reduce fruit drop in Persimmon trees cv. 'Rojo Brillante' (Badal et al. 2013). Water restrictions applied during spring or summer have been reported as effective strategies to decrease fruit drop and significantly increase water use efficiency in this crop (Buesa et al. 2013). Nevertheless, fruit growth in this particular cv. ('Rojo Brillante') is highly sensitive to deficit irrigation and a proper management of the water stress reached by the trees is crucial to do not impair fruit size and reduce farm profitability (Buesa et al. 2013). The use of the Yara ZIM-probes in orchards under deficit irrigation strategies could provide continuous information of plant water status for an adequate management of water stress. Based on Buesa et al. (2013), fruit drop could be reduced by maintaining trees within the state II of leaf turgor during either spring or summer for a couple of weeks and then leading them to state I by increasing water allocations. The unusual P_p readings observed after clamping should not be a problem when using the probes in RDI orchards provided that these were installed in the trees with enough time to stabilize before the period of water restrictions, which in spring RDI strategies is recommended from late May to mid-July. Further research on how crop load and other factors apart from water status influence leaf turgor would be valuable in order to design deficit irrigation strategies based on Yara ZIM-probe readings.

Compared with the results obtained by Fernández et al. (2011) in a similar study conducted in Seville on olive trees, states I, II and III of leaf turgor in Persimmon trees were related to higher values of Ψ_{stem}. These results could be expected since olive is a well-adapted crop to water-limited environments (Connor 2005), while Persimmon has been reported as a crop not highly sensitive to vapor pressure deficit and with poor stomatal regulation (Badal et al. 2010; Ballester et al. 2013b).

In this study, the sensitivity of the Yara ZIM-probe to detect drought stress was tested in the experiment 2 comparing trees grafted on contrasting rootstocks to drought tolerance. V rootstock is known to produce more vigorous plants (larger root system) than L, which ensure a better performance of trees when are planted in heavy and dry soils (Badenes et al. 2015). Both rootstocks are used in the area of study although 90 % of the production stands on trees grafted onto L. Results obtained from the Ψ_{stem} and trunk diameter variation measurements showed that trees grafted onto V remained in a better plant water status than those grafted onto L during the drought cycle. These differences in water status were also reflected in the leaf turgor measurements since trees grafted onto V remained more days in state I (well-watered conditions, 8 out of 10 days) than those grafted onto L (3 out of 10 days). The use of the Yara ZIM-probes to monitor leaf turgor in combination with other plant physiological assessments may provide then useful information to assess the response of rootstocks to drought stress in order to identify those more adequate for a range of scenarios with different water-limiting conditions.

CONCLUSIONS

The results obtained from this study show that continuous leaf turgor monitoring with the Yara ZIM-probes enabled the detection of water stress in Persimmon trees. Three states of leaf turgor were identified depending on the shape of the P_p curve (normal, half inverted and total inverted) obtained from the probes. State I (Ψ_{stem} above −0.8 MPa) was observed during the whole experiment in the well-watered trees and when water restrictions were not applied in the drought-stressed treatments. States II and III of leaf turgor were observed just during the drought cycles when trees reached Ψ_{stem} values considered of mild to severe water stress.

These results suggest, on one hand, that the Yara ZIM-probe could be a possible tool to be used in this crop for scheduling irrigation. Further research, however, is needed to address different aspects such as the unusual P_p evolution observed just after clamping or how P_p is influenced by the seasonal variability in tissue water relations, or crop load before attempting to recommend its use for irrigation scheduling purposes. Its use along with other physiological measurements, on the other hand, may be used to assess the tolerance of rootstocks to drought stress with the aim to identify those more adequate for semiarid environments.

ACKNOWLEDGMENTS

The authors thank J. Castel, E. Badal and T. Yeves for their help with the installation and maintenance of the Yara ZIM-probe equipment. This experiment has been funded by the European Project WEAM4i (Water and Energy Advanced Management for Irrigation), grant agreement 619061 and FEDER-MINECO EASYRIEGO (IPT-2012-0950-310000). M. A. Martínez-Gimeno acknowledges the financial support received from the Spanish Ministry of Education, Culture and Sports (MECD) program Formación Profesoreado Univeritario (FPU).

REFERENCES

1. Allen RG, Pereira RS, Raes D, Smith M (1998) Crop evapotranspiration-guidelines for computing crop water requirements. Irrigation and drainage 56. FAO, Rome, p 56
2. Arnal L, Río MA (2003) Removing astringency by carbon dioxide and nitrogen-enriched atmospheres in Persimmon fruit cv. "Rojo Brillante". J Food Sci 68(4):1516–1518
3. Badal E, Buesa I, Guerra D, Bonet L, Ferrer P, Intrigliolo DS (2010) Maximum diurnal trunk shrinkage is a sensitive indicator of plant water, stress in *Diospyros kaki* (Persimmon) trees. Agric Water Manag 98(1):143–147
4. Badal E, El-Mageed TA, Buesa I, Guerra D, Bonet L, Intrigliolo DS (2013) Moderate plant water stress reduces fruit drop of "Rojo Brillante" persimmon (*Diospyros kaki*) in a Mediterranean climate. Agric Water Manag 119:154–160
5. Badenes ML, Naval MM, Martínez-Calvo J, Giordani E (2015) Material vegetal y mejora genética. In: Badenes ML, Intrigliolo DS, Salvador A, Vicent A (eds) El cultivo del caqui. Generalitat Valenciana, Valencia, pp 17–33
6. Ballester C, Castel J, Jimenez-Bello MA, Castel JR, Intrigliolo DS (2013a) Thermographic measurement of canopy temperature is a useful tool for predicting water deficit effects on fruit weight in citrus trees. Agric Water Manag 122:1–6
7. Ballester C, Jimenez-Bello MA, Castel JR, Intrigliolo DS (2013b) Usefulness of thermography for water stress detection in citrus and persimmon trees. Agric For Meteorol 168:120–129
8. Ballester C, Castiella M, Zimmermann U, Rüger S, Martínez-Gimeno MA, Intrigliolo DS (2016) Usefulness of the Yara ZIM plant technology for detecting plant water stress in Clementine and Persimmon trees. Acta Hortic (in press)
9. Buesa I, Badal E, Guerra D, Ballester C, Bonet L, Intrigliolo DS (2013) Regulated deficit irrigation in persimmon trees (*Diospyros kaki*) cv. 'Rojo Brillante'. Sci Hortic 159:134–142
10. Connor DJ (2005) Adaptation of olive (*Olea europaea* L.) to water-limited environments. Crop and pasture. Science 56(11):1181–1189
11. Ehrenberger W, Rüger S, Rodríguez-Domínguez CM, Díaz-Espejo A, Fernández JE, Moreno J, Zimmermann D, Sukhorukov VL, Zimmermann U (2012) Leaf patch clamp pressure probe measurements on olive leaves in a nearly turgorless state. Plant Biol 4(4):666–674
12. Fernández JE (2014) Plant-based sensing to monitor water stress: applicability to commercial orchards. Agric Water Manag 142:99–109
13. Fernández JE, Cuevas MV (2010) Irrigation scheduling from stem diameter variations: a review. Agric For Meteorol 150(2):135–151

14. Fernández JE, Rodríguez-Domínguez CM, Pérez-Martín A, Zimmermann U, Rüger S, Martín-Palomo MJ, Torres-Ruíz JM, Cuevas MV, Sann C, Ehrenberger W, Díaz-Espejo A (2011) Online-monitoring of tree water stress in a hedgerow olive orchard using the leaf patch clamp pressure probe. Agric Water Manag 100:25–35
15. George AP, Nissen RJ, de Kruiff HCB (1995) Effects of plant water stress on fruit set of non-astringent persimmon under controlled-environment conditions. Acta Hortic 409:117–121
16. Giordani E, Picardi E, Radice S (2015) Mofología y fisiología. In: Badenes ML, Intrigliolo DS, Salvador A, Vicent A (eds) El cultivo del caqui. Generalitat Valenciana, Valencia, pp 17–33
17. Intergovernmental Panel on Climate Change, IPCC (2014) Climate change 2014–impacts, adaptation and vulnerability: regional aspects. Cambridge University Press, Cambridge
18. Jones HG, Serraj R, Loveys BR, Xiong L, Wheaton A, Price AH (2009) Thermal infrared imaging of crop canopies for the remote diagnosis and quantification of plant responses to water stress in the field. Funct Plant Biol 36(11):978–989
19. Kant S, Burch D, Ehrenberger W, Bitter R, Rüger S, Mason J, Rodin J, Materne M, Zimmermann U, Spangenberg G (2014) A novel crop water analysis system: identification of water stress tolerant genotypes of canola (*Brassica napus* L.) using non-invasive magnetic turgor pressure probes. Plant Breed 133(5):602–608
20. Ortuño MF, Conejero W, Moreno F, Moriana A, Intrigliolo DS, Biel C, Mellisho CD, Pérez-Pastor A, Domingo R, Ruiz-Sánchez MC, Casadesus J, Bonany J (2010) Could trunk diameter sensors be used in woody crops for irrigation scheduling? A review of current knowledge and future perspectives. Agric Water Manag 97(1):1–11
21. Padilla-Díaz CM, Rodriguez-Dominguez CM, Hernandez-Santana V, Perez-Martin A, Fernández JE (2015) Scheduling regulated deficit irrigation in a hedgerow olive orchard from leaf turgor pressure related measurements. Agric Water Manag.
22. Perucho R (2015) El cultivo del caqui. antecedentes e importancia económica. In: Badenes ML, Intrigliolo DS, Salvador A, Vicent AEd (eds) El cultivo del caqui. Generalitat Valenciana, Valencia, pp 17–33
23. Rodriguez-Dominguez CM, Ehrenberger W, Sann C, Rüger S, Sukhorukov VL, Martín-Palomo MJ, Diaz-Espejo A, Cuevas MV, Torres-Ruiz JM, Perez-Martin A, Zimmermann U, Fernández JE (2012) Concomitant measurements of stem sap flow and leaf turgor pressure in olive trees using the leaf patch clamp pressure probe. Agric Water Manag 114:50–58
24. Rüger S, Netzer Y, Westhoff M, Zimmermann D, Reuss R, Ovadiya S, Gessner P, Zimmermman D, Schwartz A, Zimmermann U (2010) Remote monitoring of leaf turgor pressure of grapevines subjected to different irrigation treatments using the leaf patch clamp pressure probe. Aust J Grape Wine Res 16(3):405–412
25. Rüger S, Ehrenberger W, Zimmermann U, Ben-Gal A, Agam N, Kool D (2011) The leaf patch clamp pressure probe: a new tool for irrigation scheduling and deeper insight into olive drought stress physiology. Acta Hortic 888:223
26. Ruiz-Sanchez MC, Domingo R, Castel JR (2010) Review. Deficit irrigation in fruit trees and vines in Spain. Span J Agric Res 8(2):5–20
27. Turner NC (1981) Techniques and experimental approaches for the measurement of plant water status. Plant Soil 58:339–366
28. Westhoff M, Schneider H, Zimmermann D, Mimietz S, Stinzing A, Wegner LH, Kaiser W, Krohne G, Shirley S, Jakob P, Bamberg E, Bentrup F-W, Zimmermann U (2008) The mechanisms of refilling of xylem conduits and bleeding of tall birch during spring. Plant Biol 10:604–623
29. Westhoff M, Reuss R, Zimmermann D, Netzer Y, Gessner A, Geßner P, Zimmermann G, Wegner LH, Bamberg E, Schwartz A, Zimmermann U (2009) A non-invasive probe for online-monitoring of turgor pressure changes under field conditions. Plant Biol 11:701–712

30. Zimmermann D, Reuss R, Westhoff M, Gessner P, Bauer W, Bamberg E, Bentrup F-W, Zimmermann U (2008) A novel, non-invasive, online-monitoring, versatile and easy plant based probe for measuring leaf water status. J Exp Bot 59:3157–3167
31. Zimmermann U, Bitter R, Schüttler A, Ehrenberger W, Rüger S, Bramley H, Siddique K, Arend M, Bader MKF (2013a) Advanced plant-based, internet-sensor technology gives new insights into hydraulic plant functioning. Acta Hortic. In: IX international workshop on sap flow by ISHS, vol 991, pp 313–320
32. Zimmermann U, Bitter R, Marchiori PER, Rüger S, Ehrenberger W, Sukhorukov VL, Schüttler A, Ribeiro RF (2013b) A non-invasive plant-based probe for continuous monitoring of water stress in real time: a new tool for irrigation scheduling and deeper insight into drought and salinity stress physiology. Theor Exp Plant Physiol 25:3–12
33. Zimmermman U, Israeli Y, Zhou A, Schwartz A, Bamberg E, Zimmermann D (2009) Effects of environmental parameters and irrigation on the turgor pressure of banana plants measured using the non-invasive, online monitoring leaf patch clamp pressure probe. Plant Biol 12(3):424–436

CHAPTER 15

Effect of Deficit Irrigation on Water Productivity of Onion (Allium cepaL.)under Drip Irrigation

Enchalew B[1*], Gebre SL[2], Rabo M[2], Hindaye B[2], Kedir M[2], Musa Y[2] and Shafi A[2]

[1]*Melkassa Agricultural Research Center, Ethiopia*

[2]*Department of Natural Resources Management, Jimma University, Ethiopia*

ABSTRACT

Deficit irrigation (DI) improves water productivity and irrigation management practices resulting in water saving by maintaining soil moisture content below optimum level throughout growth season. Field study was carried out on clay loam soil at Melkassa Agricultural Research Center, Ethiopia with the objectives to estimate water productivity of onion and evaluate the effect of water deficit on onion yield and quality using drip irrigation. The experiment contained five DI treatments of 90%, 80%, 70%, 60%, and 50% Crop water use (ETc) and the control (100% ETc) laid out in RCBD design with three replications. Irrigation water was applied at allowable soil moisture depletion (p=0.25) of the total available soil moisture throughout the crops growth stage. Statistical analysis revealed that plant height was not affected by the level of DI while, leaf number, bulb diameter, marketable bulb yield and total bulb yield had shown a highly significant (P<0.01) differences among DI treatments. The highest bulb diameter was observed from a control treatment that was significantly different to all other treatments. The highest total bulb yield of 15,690 kg/ha was observed from a control treatment which was not significantly different with treatment receiving 90% ETc. Highest water productivity of onion bulb yield was observed from treatment receiving 70% ETc and better onion bulb diameter was observed from treatment receiving 100% ETc to 70% ETc. The yield response factor ranged between 0.8 and 1.7. Thus, DI practices should be avoided for Ky values that are less than unity. Considering yield response factors (Ky) is limiting factor, 80% ETc application was a marginal and beyond that yield losses are intolerable. Thus, the practice of DI application up to 20% saved 45 to 108 mm depth of water from the gross onion irrigation water requirement.

Keywords: Deficit irrigation; Drip irrigation; Melkassa; Onion; Water productivity

INTRODUCTION

Onion (Allium cepaL.) is the most important, widely grown vegetable crop throughout the world [1]. It is widely cultivated as a source of income by many **FARMERS** in many parts of the country. It is also one of the most important vegetable crops in Ethiopia. The crop is widely **CULTIVATED** as cash crop by small-scale and private large-scale farmers. The country has a great potential to produce onion throughout the year both for local consumption and export. It can be produced throughout the year provided dependable rain and/or irrigation water is available. The majority of onion production is found in the Central Rift Valley (CRV) of Ethiopia; however, rainfall is unreliable and insufficient to support onion production that makes irrigation an indispensable practice.

The crop is shallow rooted and sensitive to water stress. As result the crop is commonly given light and frequent irrigation to avoid water stress [2]. Maximum yield could be obtained with the achievement of the entire crop water requirements. The rift valley area is a semiarid with limited water resources and increasing demand for water combined with high evapo-transpiration rates limits the production and productivity of the crop. Hence, alternatives need to be explored for effective and efficient use of the existing water resources.

Under conditions of scarce water supply, application of deficit irrigation (DI) could provide greater economic returns than maximizing yields per unit of water. The DI has been considered worldwide as a way of maximizing water use efficiency (WUE) by eliminating irrigation that has little impact on yield [3-5]. With DI, the crop is exposed to a certain level of water stress either during a particular period or throughout the whole growing season [6].

A variety of **CROPS** have been found to benefit from DI strategy and many researchers pointed out that yield loss that may result from DI is offset by the benefits of reduced water use [7-10]. The response of Onion to water deficit has been reported by [11] and [12] that showed DI to increase the water use efficiency of onion.

Drip irrigation is one of the most efficient forms of irrigation technology that will allow to apply light and frequent irrigation. The experience from many countries showed that farmers who switch from surface irrigation to drip systems can cut their water use by 30% to 60% and crop yields often increase at the same time [12]. Therefore, the objectives of this study were to determine water productivity of onion and investigate the effect of DI levels on yield and quality onion bulb under drip irrigation.

DESCRIPTION OF THE STUDY AREA

A field experiment was conducted at **MELKASSA** Agricultural Research Center (MARC) of the Ethiopian Institute of Agricultural Research (EIAR) in CRV Ethiopia (8°24' N lat. 39°21' E long. 1550 m.a.s.l.). Central Rift Valley of Ethiopia is a semi-arid environment with mean monthly maximum and minimum temperature of 33°C and 10.8°C, respectively. It is characterized by uni modal low and erratic rainfall pattern with average annual rainfall of 767 mm [13]. The soil is a clay loam type with 35% sand, 28.5 and silt and 36.5% clay. The top 30 cm of the soil at the experimental site has a field capacity of 30.7%, **WILTING POINT** of 15.8% and bulk density of 1.1 g/cm3 while the total available water was about 49.4.

MATERIALS AND METHODS

EXPERIMENTAL LAYOUT AND DESIGN

The experiment was laid out in a Randomized Complete Block Design (RCBD) with three replicates. The treatments consisted of five soil moisture deficit levels, viz., 90% ETc (10% deficit); 80% ETc (20% deficit); 70% ETc (30% deficit); 60% ETc (40% deficit) and 50% ETc (50% deficit) and a control treatment of 100% ETc (no deficit) (**TABLE 1**). Drip irrigation system was used for applying the required quantity of irrigation water. Each irrigation treatment consisted of three lateral lines of 5 m length. Each lateral line contained emitters spaced at 30 cm interval.

Table 1: Soil moisture content before and after irrigation events and net irrigation requirements.

Treatment	Soil moisture content/(SMC (mm) Depth (cm) 0-30			30-45			45-60		
	Before irrig.	After irrig.	Net irrigation	Before irrig.	After irrig.	Net irrigation	Before irrig.	After irrig.	Net irrigation
100% ETc	89.3	102	12.4	129.1	146.1	17	167.3	189	21.6
90% ETc	89.3	101	11.2	129.1	144.4	15.3	167.3	187	19.4
80% ETc	89.3	99	9.9	129.1	142.7	13.6	167.3	185	17.3
70% ETc	89.3	98	8.7	129.1	141	11.9	167.3	182	15.1
60% ETc	89.3	97	7.4	129.1	139.3	10.2	167.3	180	12.9
50% ETc	89.3	96	6.2	129.1	137.6	8.5	167.3	178	10.8

Source: Own data, 2015

Onion variety Bombay red was raised on nursery bed and transplanted to field plots of 5 m × 2.7 m. Furrows spaced at 60 cm were used and transplanted on both side of a ridge at row and plant spacing of 20 and 10 cm, respectively. To ensure the plant establishment common irrigations was provided to all plots at two days interval before commencement of the differential irrigation. Irrigation water was applied at allowable soil moisture depletion (p=0.25) of the total available soil moisture throughout crops growth stage.

DATA COLLECTION AND ANALYSIS

Data collection comprised plant height, leaf number, bulb yield and yield components that include bulb diameter, marketable and unmarketable bulb yield. **WATER PRODUCTIVITY** and effect of water stress on crop performance were quantified from WUE and yield response factors (Ky), respectively. Estimation of WUE was carried out as a ratio of total bulb yield to the total water applied [14].

$$\text{WUE}\left(\text{kg mm}^{-1}\right) = \frac{\text{Total bulb yield}\left(\text{kg}\right)}{\text{Crop Water Use}\left(\text{mm}\right)}$$

Crop water use (ETc) was determined for each treatment for the growing period using the soil-water balance equation [15].

$$ET_c = I + R_e + \Delta S - D + G_e$$

Where: I is irrigation water (mm); Re is effective rainfall (mm) and ΔS is the change in soil water storage for the period (mm). D is drainage below the root zone (mm) and Ge is the **GROUNDWATER** contribution (mm). The contribution of D and Ge were assumed to be negligible. The ΔS were assumed the same at the beginning and at harvest and have no contribution to plant ET.

The **YIELD** response factor (Ky) was estimated from the relationship [16].

$$\left(1-\left(\frac{Y_a}{Y_m}\right)\right)=K_y\left(1-\left(\frac{ET_a}{ET_m}\right)\right)$$

Where,

Ya is actual harvested yield

Ym is maximum **HARVESTED** yield

Ky is yield response factor

ETa is actual evapo-transpiration

ETm is maximum evapo-transpiration

The data collected during the experimental period were subjected to statistical analysis using SAS computer program. Whenever treatment effects were significant, Least Significance Differences (LSD) test was used to assess the mean difference among treatments.

Table 2: Plant height and leaf number as influenced by DI.

Treatment	Plant height (cm)	Leaf number
100% ETc	41.2	7.33
90% ETc	46.3	8
80% Etc	41.7	7
70% Etc	43.8	6.67
60% Etc	42.9	6.33
50% Etc	38.2	6.67
Mean	42.3	7
LSD0.05	NS	0.66
0.01	NS	0.94
SE +	2.78	0.21
CV (%)	11.38	5.22
Source: Own data, 2015		

Table 3: Bulb diameter, bulb yield, marketable and unmarketable bulb yield as influenced by DI.

Treatment	Bulb diameter (cm)	Bulb yield (kg ha^{-1})	Marketable bulb yield (kg ha^{-1})	Unmarketable bulb yield (kg ha^{-1})	Percent unmarketable yield
100% ETc	6.01	15694.4	10902.8	4791.7	31
90% Etc	5.6	13958.3	9687.5	4270.8	31
80% Etc	5.4	12604.2	7923.6	4680.5	37
70% Etc	5.21	12395.8	8263.9	4131.9	33
60% Etc	4.95	11041.7	6847.2	4194.4	38
50% Etc	4.97	10034.7	6076.4	3958.3	39
Mean	5.4	12621.5	8283.6	4337.9	35
LSD0.05	0.15	1933	2171	NS	
0.01	0.22	2749	3087	NS	
CV (%)	1.6	8.42	14.4	11.85	
SE +	0.05	613.4	688.8	296.8	

RESULTS AND DISCUSSION

Statistical analysis has shown a highly significant (P<0.01) difference for leaf number, bulb diameter, total and marketable bulb yield under the different DI treatments. However, no significant difference was observed for **PLANT** height and unmarketable bulb yield. The data on **TABLES 2** and **3** provide plant height, leaf number, bulb diameter, total bulb yield, marketable and unmarketable bulb yield.

PLANT HEIGHT

The DI did not affect much onion plant height. The plant height ranged from 38 and 46 cm. The highest and lowest plant height was observed from treatment receiving 90% and 50% ETc, respectively. However, the control treatment and treatment receiving 80% ETc gave below the average plant height.

The increasing of plant height with adequate soil moisture application is related to water in maintaining the turgid pressure of the plant cells which is the main reason for the growth [17]. In the other side the shortening of plant height under less soil moisture stress may be associated due to the closure of stomata to conserve soil moisture evaporation, this leads to reduce uptake of CO_2 and nutrient. Therefore, photosynthesis and other biochemical reactions are hindred, eventually affecting plant growth [18]. This study outcome is in line with the research that has been done by [19], indicated that soil water supply is directly proportional with plant height growth.

LEAF NUMBER

The number of leave per plant ranged from 6 to 8. The highest leaf number was recorded from treatment receiving 90% ETc. This was significantly different to all other treatments at $p<0.05$ level and had no significant difference with the control at $P<0.01$ level. Treatment receiving 60% ETc was inferior to treatment receiving 100 to 80% ETc and had no significant difference with treatment receiving 70% and 50% ETc. From the result, it can be observed that up to 20% water deficit was tolerable to obtain at least seven leaf number per plant.

BULB DIAMETER

The DI water applications affected the size of onion bulb. The highest bulb diameter was recorded from the control treatment and this was significantly different to all other treatments. The least bulb diameter was recorded from treatment receiving 60% ETc and this was not significantly different to treatment receiving 50% ETc. Water deficit up to 20% gave bulb diameter above the mean value of 5.4 cm. This result is in agreement with that of a study conducted by [20], high amount of soil moisture application leads to large photosynthesis area (plant height and large number of leaves), results to large bulb diameter.

TOTAL BULB YIELD, MARKETABLE AND UNMARKETABLE BULB YIELD

The total bulb yield was highest for the control treatment and this was not significantly different to treatment receiving 90% ETc. The least bulb yield was recorded from treatment receiving 50% ETc and had no significant difference with treatments receiving 60% ETc at $p<0.05$ level and with treatment receiving 60% to 80% at $p<0.01$ level. From this finding, it can be observed that with 10% water deficit to result 11% yield reduction and above 10% water deficit to result total bulb yield below mean value. A study done by [20] also presented similar findings with this result.

Highest marketable bulb yield was recorded from the control treatment and had no significant difference with 90% ETc at $p<0.05$ level and with 70% to

90% ETc at $p<0.01$ level. The lowest marketable bulb yield was observed from treatment receiving 50% ETc and had no significant difference with 60% and 80% ETc at $p<0.05$ level and with 60% to 80% ETc at $p<0.01$ level. Here too the finding indicated that with 10% water deficit to result 11% marketable yield reduction and above 10% water deficit to result marketable bulb yield below mean value.

The highest and lowest unmarketable onion bulb yieldswere recorded from the control and 50% ETc treatments, respectively. However, the control and 90% ETc treatment gave the lowest percentage (31%) of unmarketable bulb yield and while 50% ETc treatment gave the highest percentage (39%) of unmarketable bulb yield. From the result, it can be observed that up to 30% water deficit was marginal to obtain the least unmarketable bulb yield. The results presented in this study is inclusive and similar with previous research done by [21], high soil moisture application attributes to vegetation growth and increases plant metabolic activities, which leads to marketable bulb yield increment.

WATER USE EFFICIENCY (WUE)

TABLE 4 shows the water-use efficiency, applied Water (AW), water saved for onion bulb yield. Treatment receiving 70% ETc resulted in higher WUE and saved 182.5 mm of water (**TABLE 4**). The control treatment gave practically similar WUE with 50% and 60% ETc treatments. The lowest WUE was from treatment receiving 90% and 80% ETc. However, 10% water deficit resulted 11% yield reduction and above 10% water deficit resulted bulb yields below mean value for the treatments. The difference in WUE among treatments is very small and considering the yield reduction is a limiting factor, 70% ETc application seems marginal. A study conducted by [22,23] also indicated that WUE is maximum at medium soil moisture level compare to high moisture level treatment.

Table 4: Applied water, water use efficiency, water saved and percent yield reduction under the control and DI practices.

Treatment	AW (mm)	Bulb yield (Kg ha^{-1})	WUE (kg mm^{-1})	Water saved (mm)	Yield reduction (%)
100%ETc	677.85	15694.4	23.2	0	0
90% ETc	632.74	13958.3	22.1	45.1	11.1
80% ETc	570.07	12604.2	22.1	107.8	19.7
70% ETc	495.4	12395.8	25	182.5	21
60% ETc	469.34	11041.7	23.5	208.5	29.6
50% ETc	429.18	10034.7	23.4	248.7	36.1

YIELD RESPONSE FACTOR (KY)

Observed yield response factors (Ky) for onion bulb production ranged between 0.8 and 1.7, the lowest and highest being for 70% and 90% ETc applications, respectively. The Ky observed was decreasing as irrigation water application decreasing. Treatments receiving 60% and 50% Etc water application showed almost similar yield response factor (**TABLE 5**).

Table 5: Crop water use and yield response factor under control and DI.

Treatments	ETc (mm)	Bulb yield (Kg/ha)	Yield response factor (ky)
100% Etc	677.85	15694.4	-
90% Etc	632.74	13958.3	1.7
80% Etc	570.07	12604.2	1.2
70% Etc	495.4	12395.8	0.8
60% ETc	469.34	11041.7	1
50% ETc	429.18	10034.7	1

The higher Ky values indicate that the crop will have a greater yield loss when the crop water requirements are not met. Generally, the result indicates the sensitivity of the crop to soil moisture deficit. Therefore, DI practices should be avoided for Ky values that are less than unity. This conclusion is in line with a statement given by [24], the decrease in yield is proportionally greater with increase in water deficit. Considering Ky is limiting factor, 80% ETc application was a marginal and beyond that yield losses are intolerable [25]. With DI application up 20%, hence, saved 45 to 108 mm depth of water from the gross IWR of 678 mm depth of water.

CONCLUSION

Water is scarce resource in Central Rift Valley of Ethiopia and is major limiting factor for crop production. The DI practice under drip irrigation is a suitable and most efficient practice for sustainable production in water scarce area. In this study, DI application of 90% ETc gave highest plant height and leaf number. The maximum bulb diameter, total bulb yield and marketable bulb yield were observed when 100% ETc irrigation water was applied. Drip irrigation with 10% water deficit resulted about 11% yield reduction and above 10% water deficit resulted bulb yields below mean value for the treatments. The highest water productivity of onion was observed at 70% ETc water deficit irrigation application.

The yield response factor of onion ranged between 0.8 and 1.7 and the higher Ky value indicated that onion will have a greater yield loss when the critical water requirements are not met. Thus, DI practices should be avoided for Ky values that are less than unity. Considering Ky is limiting factor, 80% ETc application was a marginal and beyond that yield losses are intolerable. With DI application up 20% saved 45 to 108 mm depth of water from the gross IWR (Irrigation Water Requirement) of 678 mm depth of water.

RECOMMENDATION

Water scarcity is the major limiting factor for increased production and productivity. Water is scarce resource in Central Rift Valley of Ethiopia and is major limiting factor for crop production. Onion is one of the major economically important **VEGETABLE**crops grown in this region. Therefore, DI practice under drip irrigation is a suitable and most efficient practice for sustainable production in water scarce area like Central Rift Valley of Ethiopia.

The maximum bulb diameter, total bulb yield and marketable bulb yield associated with application of 100% ETc irrigation water. However, to obtain the highest water productivity of onion, can be obtained when 70% ETc water deficit irrigation application. If yield response factor (Ky) is considered, 80% ETc application should be a marginal and beyond that yield losses are intolerable. With DI application up to 20%, 45 to 108 mm depth of **IRRIGATION** water can be saved from the gross irrigation water applied as in the case of no deficit. Deficit irrigation (DI) improves water productivity and irrigation management practices resulting in water saving by maintaining soil moisture content below optimum level throughout growth season.

ACKNOWLEDGEMENT

The authors grateful to Melkassa Agricultural Research Center (MARC, Ethiopia) Land and Water Resources Research Process section for providing financial and material support for conducting the research. Our gratitude also extends to Mr. Elias Meskelu, for his genuine advice in technical establishment of the research site.

REFERENCES

1. Brewster JL (1997) Onions and Garlic. In: Wien HC (ed.) The Physiology of Vegetable Crops.CAB International, UK. pp: 581-619.
2. Doornebos J, Kassam AH (1979) Yield response to water. FAO Irrigation and Drainage, Italy. p: 33.
3. English MJ (1990) Deficit irrigation. I: Analytical framework. J Am Soc Civil Eng116: 399-412.
4. English M, Raja SN (1996) Perspectives on deficit irrigation. Agric Water Manage 32: 1-14.
5. Kirda C, Moutonnet P, Hera C, Nielsen DR (1999) Crop yield response to deficitirrigation. Kluwer Academic Publishers, Dordrecht.
6. Kirda C (2000) Deficit irrigation scheduling based on plant growth stages showing water stress tolerance. Deficit Irrigation Practices.
7. Karam F, Breidy J, Stephan C, Rouphael Y (2003) Evapo-transpiration, yield and water use efficiency of drip irrigated corn in the bekaa Valley of Lebanon. Agric Water Manage 63: 125-137.
8. Karam F, Masaad R, Sfeir T, Mounzer O, Rouphael Y (2005) Evapo-transpirationand seed yield of field grown soybean under deficit irrigation conditions. Agric Water Manage 75: 226-244.
9. Karam F, Lahoud R, MasaadR, Daccache A, Mounzer O, et al. (2006) Water use and lint yield response of drip irrigated cotton to the length of irrigation season.Agric Water Manage 85: 287-295.
10. Kirnak H, Tas I, Kaya C, Higgs D (2002) Effects of deficit irrigation on growth, yield and fruit quality of eggplant under semi-arid conditions. Austr J Agric Res 53: 1367-1373.
11. Bekele S, Tilahun K (2007) Regulated deficit irrigation scheduling of onion in a semiarid region of Ethiopia. Agric Water Manage 89: 148-152.
12. Nigatu A, Hordofa T, Tesfaye K (2010) Effect of deficit irrigation at different growth stage of onion on bulb production in central rift valley of Ethiopia.
13. Sijali IV (2001) Drip irrigation: Options for smallholder farmers in eastern and southernAfrica. World Agroforestry Centre, Kenya.
14. Central Statistics Authority (2011) Atlas of agricultural statistics, Ethiopia.
15. Molden D, Oweis T, Steduto P, BindrabanP, Hanjra MA, et al. (2010) Improving agricultural water productivity: between optimism and caution. Agric Water Manage 97: 528-535.
16. Simonne EH, Dukes MD (2010) Principles and practices of irrigation management for vegetables UF. University of Florida. pp: 17-23.
17. Doorenbos J, Pruitt WO (1977) Crop water requirements. FAO Irrigation andDrainage Paper No. 24, Rome.
18. Vaux HJ, Pruit WO (1983) Crop water production function. In: DHillel (ed.)Advance irrigation. pp: 16-93.
19. El-Noemani AA, Aboamera AAA, Aboellil OM, Dewedar (2009) Growth, yield, quality and water use efficiency of pea plant as affected by evapotranspirationandsprinkler height. J Agric Res34: 1445-1466.
20. Al- Moshileh A (2007) Effects of planting date and irrigation water level on onion(Allium cepa L.)productionundercentralSaudiArabianconditions. J Basic ApplSci 8: 14-28.
21. Kumar S, Imtiyaz M, Kumar A (2007b)Effectofdifferentialsoil moisture andnutrient regimes on postharvest attributes of onion (Allium cepaL.).J HorticSci112: 121-129.
22. Neeraja G, Reddy KM, Reddy IP, Reddy YN (1999) Effect of irrigation andnitrogen on growth,yieldandyieldattributesofrabionion(AlliumcepaL.) inAndhra Pradesh. Vegetable Science 26: 64-68.

23. FabeiroC,MartindeSantaOlallaF,deJuanJA(2001)Yield andsizeofdeficitirrigatedpotatoes. Agric Water Manage 48: 255-266.
24. Sarkar S, Goswami SB, Mallick S, Nanda MK (2008) Different indices to characterize water use pattern of micro-sprinkler irrigated onion (Allium cepa L.). J Agric Water Manage 95: 625-632.
25. Kadayifci A, Tuylu GI, Ucar Y, Cakmak B (2005) Crop water use of onion (Allium cepaL.) in Turkey. Agric Water Manage 72: 59-68.

CHAPTER 16

The effects of applied water at various fractions of measured evapotranspiration on water relations and vegetative growth of Thompson Seedless grapevines

L. E. Williams Æ D. W. Grimes Æ C. J. Phene

[1.]*Department of Viticulture and EnologyUniversity of CaliforniaDavisUSA*

[2.]*Department of Land, Air and Water ResourcesUniversity of CaliforniaDavisUSA*

[3.]*Kearney Agricultural CenterParlierUSA*

[4.]*Water Management Research LaboratoryUSDA-ARSParlierUSA*

ABSTRACT

Vegetative growth and water relations of Thompson Seedless grapevines in response to applied water amounts at various fractions of measured grapevine ET_c were quantified. Treatments ranged from no applied water up to 1.4 times the water used by vines growing in a weighing lysimeter. All treatments were irrigated at the same frequency as the vines in the lysimeter (whenever they used 2 mm of water), albeit at their respective fraction. Soil water content and midday leaf water potential (Ψ_l) were measured routinely in four of the irrigation treatments across years. The amount of water depleted in the soil profile ranged from 190 mm for the 0.2 treatment in 1993 to no water depletion for the 1.4 treatment in 1992. The irrigation treatments significantly affected midday Ψ_l, total shoot length, leaf area per vine, pruning weights and trunk diameter; as applied water decreased so did vegetative growth. Pruning weights were a linear function of the seasonal, mean midday Ψ_l across growing seasons. The application of water amounts in excess of evapotranspiration negatively affected vegetative growth some of the years. A companion paper will demonstrate that over-irrigation can negatively affect reproductive growth of this grape cultivar due to excess vegetative growth.

INTRODUCTION

The competition among agricultural, societal and environmental interests for limited water supplies is increasing and this competition can be exacerbated in years where annual rainfall is below normal throughout the state of California and at other locations around the world (Laraus 2004; Morison et al. 2008). These circumstances require that agriculture becomes more efficient in the use of this natural resource (Morison et al. 2008). Increasing water use efficiency by agriculture can span various scales, going from individual farms to watersheds to regions and involve various disciplines including hydrology, engineering and soil and plant sciences (Hsiao et al. 2007). Future increases in water use efficiency at the farm level will require a more scientific basis for irrigation scheduling to include: (a) better estimates of crop water use, (b) yield response to applied water amounts or crop evapotranspiration and (c) a means to determine the timing of an irrigation event and/or assessing the level of plant stress (Fereres and Evans 2006).

Crop evapotranspiration (ET_c) can be estimated using reference evapotranspiration (ET_o) and reliable crop coefficients (Allen et al. 1998). Grapevine ET_c of Thompson Seedless has been determined with the use of a weighing lysimeter (Williams et al. 2003a) and seasonal crop coefficients developed for vines during the first 7 years of development (Williams and Ayars 2005a; Williams et al. 2003b). In addition, it was shown that grapevine ET_c and the crop coefficient are a linear function of the amount of shade cast on the ground at solar noon (Williams and Ayars 2005b). A similar type of relationship (K_c as a function of the amount of light intercepted by the canopy at midday) has been established for peach trees (*Prunus persica* (L.) Batsch) (Ayars et al. 2003). The relationship between the K_c and shaded area beneath grapevines has been used to schedule irrigation amounts at different locations in California (Williams and Baeza 2007). Therefore, more reliable estimates of grapevine water use in California can be obtained using the above information and should assist in using scientific means to schedule irrigations (Fereres and Evans 2006).

Plant and soil based methods to assess crop water status and their relationships with one another are needed to assist in irrigation scheduling (Jones 2004). It has been demonstrated that leaf (Ψ_l) and stem (Ψ_{stem}) water potentials of grapevines are highly correlated with one another (Intrigliolo et al. 2005; Stevens et al. 1995; Williams and Araujo 2002; Williams and Trout 2005). A fully irrigated baseline, using either Ψ_l or Ψ_{stem} as a function of vapor pressure deficit at the time of measurement, was recently developed for grapevines (Williams and Baeza 2007). Leaf and stem Ψs of grapevines are also highly correlated with soil water content (SWC) and soil matric potential and measures of leaf gas exchange (Williams and Araujo 2002; Williams and Trout 2005). Seasonal values of midday stomatal conductance and Ψ_l and crop water stress index (CWSI) have been shown to be highly correlated with Thompson Seedless yield (Grimes and Williams 1990). Therefore, there are numerous vine and soil based methods one can use to assist in developing a scientifically based vineyard irrigation management program in California.

A weighing lysimeter was installed at the Kearney Agricultural Center in 1987, planted with Thompson Seedless grapevines and seasonal ET_c measured and crop coefficients calculated (Williams and Ayars 2005a; Williams et al. 2003b). Subsequent to vine training (3 years after planting) irrigation treatments were established in the vineyard surrounding the lysimeter consisting of applied water amounts at various fractions (0, 0.2, 0.4, 0.6, 0.8, 1.0, 1.2 and 1.4) of measured lysimeter water use. This allowed us to determine the responses of vegetative and reproductive growth of Thompson Seedless vines to both under- and over-irrigation and then relate these responses to ET_c and the K_c. This paper describes the effects of these treatments on SWC, vine water status, grapevine phenology and vegetative growth, the former parameters with potential for use in a vineyard irrigation management program (Jones 2004). A companion paper describes the effects of these treatments on reproductive growth and water productivity of the same Thompson Seedless grapevines (Williams et al. 2009).

MATERIALS AND METHODS

The vineyard used in this study was planted on 9 April 1987 with cuttings of *Vitis vinifera* L (cv. 'Thompson Seedless' clone 2A) at the University of

California Kearney Agricultural Center located in the San Joaquin Valley of California (36°48′N, lat, 119°30′ W, long.). Vine and row spacings were 2.15 and 3.51 m, respectively (7.55 m^2 per vine). Row direction was 6° north of the east/west axis. The vineyard was approximately 1.4 ha (168 × 82 m). The soil was a Hanford fine sandy loam (coarse-loamy, mixed, nonacid, thermic Typic Xerorthent). Three trellis systems were used in the study: (1) A single wire placed atop a 2.13 m wooden stake driven 0.45 m into the soil at each vine, (2) A 0.6 m cross-arm placed atop the stake and wires attached at either end of the cross-arm to support the vine's fruiting canes and (3) A 1.2 m cross-arm placed atop the stake and wires attached at either end of the cross-arm to support the shoots and another two wires attached 0.3 m from the stake to support the fruiting canes. The vines were cane-pruned during the dormant portion of the growing season with each cane approximately 12–15 nodes in length. The numbers of canes per vine left after pruning were 4, 6, 8 and 8 in 1990, 1991, 1992 and 1993, respectively. The canes were then wrapped around the fruiting cane wires on each of the trellises. Cultural practices to control diseases and insect pests were performed by field station personnel as described previously (Williams et al. 2003b; Daane and Williams 2003).

The vineyard also contained a weighing lysimeter in which two grapevines were planted at the same time the rest of the vineyard was planted. The trellis system for the two, lysimeter vines consisted of a 0.6 m cross-arm, similar to that described above. The operation of the lysimeter and other technical details can be found in Williams et al. (2003a, b). Vines within the lysimeter and the surrounding vineyard were irrigated with 4 L h^{-1} in-line drip emitters, spaced every 0.30 m in the vine row. The drip tubing was attached to a wire suspended 0.4 m above the soil surface. The lysimeter was weighed hourly to determine ET of the two vines and when the decrease in weight exceeded an equivalent of 16 L (8 L $vine^{-1}$) threshold value the lysimeter was irrigated. The number of irrigations per day throughout each growing-season ranged from 0 to 7.

The irrigation pump for the rest of the vineyard was controlled by the lysimeter's datalogger (Campbell Scientific, 21X Micrologger). Whenever the lysimeter was irrigated the vineyard pump was activated and an irrigation event took place. The irrigation treatments were applied water amounts at various fractions (0.2, 0.4, 0.6, 0.8, 1.0, 1.2 and 1.4) of lysimeter water use. The irrigation treatments supplied with water amounts less than ET_c season long, as done in this study, have been termed 'sustained deficit irrigation' or 'SDI' (Fereres and Soriano 2007). A non-irrigated treatment was also included. The irrigation treatments were first imposed during the 1989 growing season. No supplemental irrigation water was applied during the dormant portion of the growing season regardless the amount of rain that fell prior to budbreak each year. Irrigation treatments within each of the 8 blocks of the experiment were set up in a line-source design, proceeding sequentially from the lowest to highest amount, the direction within each block randomly assigned. Two rows separated each block and the border row irrigated with the amount of water given the irrigation treatment assigned to the outside data row of the respective block. Each irrigation treatment plot consisted of 18 vines down a single row. Within each irrigation plot, the three trellises described above were installed using six vines per trellis sub-plot. The activation of solenoid valves at the head of each row for various times was used to provide the differing fractions of applied water. In-line water meters upstream from the solenoid valves in each row measured actual applied water amounts. The water meters were initially calibrated in 1989 and again before the 1992-growing season. Both times all meters were within 3% of the measured amounts.

The volumetric SWC in the 0.2, 0.6, 1.0 and 1.4 irrigation treatments was monitored using the neutron back-scattering technique with a neutron moisture probe (Model 503 DR Hydroprobe Moisture Gauge, Campbell Pacific Nuclear, Martinez, CA, USA). Nine access tubes were placed in one quarter of an individual vine's rooting volume and inserted to a depth of 3 m. Three access tubes were placed down the vine row (directly below the drip line), one close to the trunk, one midway between vines within the row and the third midway between the two previously mentioned tubes. Another three tubes were placed midway between rows, perpendicular to each of the three tubes placed within the row. The last three tubes were placed midway between the former two sets of tubes. Readings were taken at a depth of 0.23 and 0.45 m beneath the soil surface and then in increments of 0.3 m down to a depth of 2.90 m. Each access tube site was replicated three times, in three of the eight blocks, for each irrigation treatment mentioned previously. The neutron probe was calibrated according to Araujo et al. (1995) and SWC values expressed as percent by volume (θ_v). The SWC content at field capacity of this soil type was approximately 22.0% by volume while SWC at a soil moisture tension of −1.5 MPa was approximately 8.0% by volume. Therefore, total available water to a depth of 2.9 m for this soil at field capacity was approximately 400 mm.

The date of 50% budbreak was determined by marking a single vine in each irrigation treatment replicate using the 0.6 m cross-arm trellis ($n = 8$). Budbreak was assumed to have taken place once green tissue was observed through the bud scales. The buds on all canes were observed every couple of days until no further buds developed. Subsequently, the dates when 50% of the buds that actually grew were back calculated based upon the total number of buds that initiated growth. Dates of anthesis and veraison were estimated visually. The date of 50% anthesis occurred when 50% of the calyptras (fused petals of the grape flower) were estimated to have fallen. Veraison occurred when the berries began to soften.

Entire grapevines were harvested at various times throughout the course of the study for the 0.2, 0.6, 1.0 and 1.4 irrigation treatments. The vines used were border vines separating the different trellises within a given irrigation treatment plot. Shoots were removed from the vines, taken to the lab and separated into leaves, stems (main axis of the shoot) and clusters. Fresh weights of the different organs were measured and then sub-sampled. Area of the sub-sampled leaves was determined with an area meter (LiCor, model 3100 area meter) prior to drying. Leaf area of the entire vine was calculated using the leaf area per dry weight ratio for each vine.

Shoot length was measured on numerous occasions during the 1991, 1992 and 1993 growing seasons. Twenty shoots were marked early in the growing season for each irrigation treatment using the vines within the 0.6 m cross-arm trellis sub-plot (five shoots per plot in four of the blocks). If the shoots were cut due to hedging (a normal cultural practice to allow machinery continued access to travel between rows) shoot length included any growth arising from the uppermost lateral shoot. Trunk diameter was measured on each data vine within an irrigation/trellis subplot and the mean calculated. The mean of each trellis within an irrigation treatment plot was then used to calculate an irrigation treatment mean ($n = 24$). The measurements were taken just above the wire supporting the drip tubing on each trunk (0.4 m above the soil surface). Pruning weights were measured during the dormant portion of the growing season. The weight of the four middle vines of each trellis sub-plot within an irrigation treatment plot was used for data analysis ($n = 8$).

Water potential measurements were conducted according to the procedures of Williams and Araujo (2002). Pre-dawn Ψ (Ψ_{PD}) measurements began at ≈0330 h and were finished prior to sunrise using a pressure chamber (Model 1000, PMS Instrument Co., Corvallis, OR, USA). Midday measurements of leaf water potential (Ψ_l) generally were taken between 1230 and 1330 h Pacific Daylight Time. Leaf blades for Ψ_{PD} and Ψ_l determinations were covered with a plastic bag, quickly sealed, and petioles then cut within 1–2 s. The time between leaf excision and chamber pressurization was generally <10–15 s. Leaves, chosen for Ψ_{PD} and Ψ_l were fully expanded and mature. Pre-dawn Ψ_l was measured only once in 1991, just prior to harvest, and more frequently in 1992 and 1993. For midday and diurnal Ψ_l measurements, leaves exposed to direct solar radiation were used. A single leaf from a minimum of five individual vine replicates were measured and used for data analysis. Measurements of Ψ_l and Ψ_{PD} were made in three (same blocks that SWC was measured) of the eight irrigation blocks. Environmental and reference ET (ET_o) data were obtained from a California Irrigation Management Information System (CIMIS) weather station (# 39) located 2 km from the vineyard site.

Data were analyzed via regression analyses using linear, quadratic, and cubic terms. Regressions with the best fit are presented. The relationships among water status measurements (Ψ_l and/or Ψ_{PD}) were analyzed using the means of an individual irrigation treatment for a particular date or period of vine growth. Vegetative and reproductive growth were analyzed using analysis of variance for either irrigation or trellis treatments and means separated using Duncan's multiple range test. Differences were considered significant at $P \leq 0.05$.

RESULTS

Budbreak took place between March 11 and 19 across years for the vines irrigated at the 1.0 irrigation amount treatment (Table 1). Irrigation treatment significantly affected the date of budbreak with the earliest budbreak generally occurring for the vines that were deficit irrigated the previous year (Figure 1, unpublished data). Ten days separated the date of 50% budbreak for the 0 and 0.2 irrigated treatments compared to the vines irrigated at 1.0 ET_c and greater in 1992. The next greatest difference occurred in 1991 when 50% budbreak of the 0.2-irrigation treatment occurred 7 days before that of the 1.4-irrigation treatment. Similar comparisons for dates of anthesis and veraison were not attempted due to difficulty in quantifying those two phenological stages. In general it was observed that anthesis occurred a little earlier for the deficit irrigated vines than those receiving applied water amounts greater than ET_c.

Table 1. Phenological events of Thompson Seedless grapevines measured or estimated during the course of the growing season each year of the study

Year	Date of budbreak	Date of anthesis	Date of veraison	Date of harvest
1990	March 18 (77)	May 3 (123)	July 3 (184)	Sept. 24 (267)
1991	March 19 (78)	May 23 (143)	July 8 (189)	Sept. 4 and 24 (247 and 267)
1992	March 11 (71)	May 4 (125)	June 22 (174)	Sept. 3 (247)
1993	March 12 (71)	May 9 (129)	July 2 (183)	Sept. 21 (264)

Data represent calculated dates of 50% budbreak, approximate dates of anthesis and veraison for the 1.0 irrigation treatment and dates the fruit were harvested. Numbers following calendar dates in parentheses are days of year

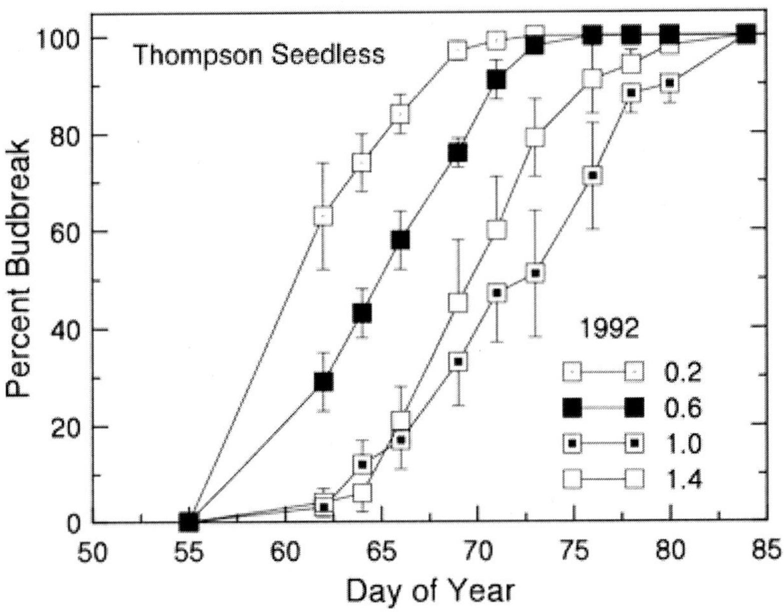

Figure 1. The percent budbreak of Thompson Seedless grapevines measured in 1992 as a function of four irrigation treatments. The values on a particular date represent the percentage of the total buds that had broken on that date compared to the total number of buds that actually grew. *Bars* represent one SE. Values in which the SE bars do not overlap were in general significantly different from one another

Irrigation of the vineyard did not commence until the last week of April to first week of May (Table 2). Midday Ψ_l ranged from −0.6 to −0.8 MPa for the 1.0 irrigation treatment, except in 1992 when it approached −1.0 MPa. Applied water amounts to the 1.0 irrigation treatment from the first irrigation of the season to harvest ranged from 72 to 96% of measured ET_c in 1990 and 1991, respectively. When averaged across years the 0.2, 0.4, 0.6, 0.8, 1.2 and 1.4 irrigation treatments received approximately 22, 40, 60, 81, 117 and 140% of the water applied to the 1.0 irrigation treatment, respectively. During years in which applied water to the 1.0 irrigation treatment was close to ET_c measured with the lysimeter, SWC remained level once irrigations began (Figure 2). The SWC of the 1.4 irrigation treatment increased while the SWC of the treatments being deficit irrigated decreased. When applied water amounts to the 1.0 treatment was 72–75% of ET_c (1990 and 1992), SWC generally decreased as the season progressed for all treatments except the 1.4 treatment in 1992. Water was depleted in the soil profile for all treatments across all years from budbreak until harvest except for the 1.4 irrigation treatment in 1991, where there was a slight increase in SWC between those two dates (Table 3). Vine evapotranspiration for the 1.0 irrigation treatment (applied water [Table 2] plus water depleted in the soil profile [Table 3]) between budbreak and harvest was 574, 673, 638 and 829 mm in 1990, 1991, 1992 and 1993, respectively. These values compared favorably with water use measured with the weighing lysimeter during the same time frame (580, 661, 634 and 732 mm in 1990, 1991, 1992 and 1993, respectively).

Table 2. Seasonal ET_c (budbreak to October 31), date of first irrigation, ET_c between the first irrigation of the season and harvest and applied water amounts between the first irrigation and harvest for 7 of the 8 irrigation treatments for each year of the study

Year	Seasonal ET_c (mm)	Date of 1st irr.	ET_c: 1st irr. To harvest (mm)[a]	Irrigation treatments (applied water amounts at various fractions of lysimeter water use) 0.2 (mm)[a]	0.4	0.6	0.8	1.0	1.2	1.4
1990	718	Apr 27	580	99	168	237	327	416	489	556
1991	865	May 8	661	134	260	383	520	632	694	884
1992	811	May 8	634	112	186	304	371	477	558	663
1993	857	May 3	732	149	269	432	561	698	864	1002

[a] Surface area per vine was 7.55 m^2

The eighth treatment received no applied water except in 1991 at which time it was irrigated at the 0.2 amount throughout the season. ET_c was determined with the weighing lysimeter. Irrigation to the vineyard was terminated on 30 August 1993, and resumed after harvest

The depletion of water in the soil profile was not necessarily related to applied water amounts. For example, 180 mm of water was depleted in the 0.6 irrigation treatment compared to 167 mm for the 0.2 treatment in 1990 while the depletion of water in the soil profile for the 1.0 irrigation treatment in 1992 was greater than those of all other treatments. Soil water was depleted to a depth of 2.9 m and out to the middle of the row for the period from May to September in 1991 for both the 0.2 and 0.6 irrigation treatments (data not given). The SWC for the 0.6 irrigation treatment at the highest depth measured within the row in September was close to what is was in May. Conversely, SWC increased at all depths (albeit a small increase for some of the depths) and distances from the vine row for the 1.4 irrigation treatment. There was a small decrease with depth and distance from the row for the 1.0 irrigation treatment during the same time frame with an increase in SWC at the higher depths where access tubes were placed within the row directly beneath the drip lines.

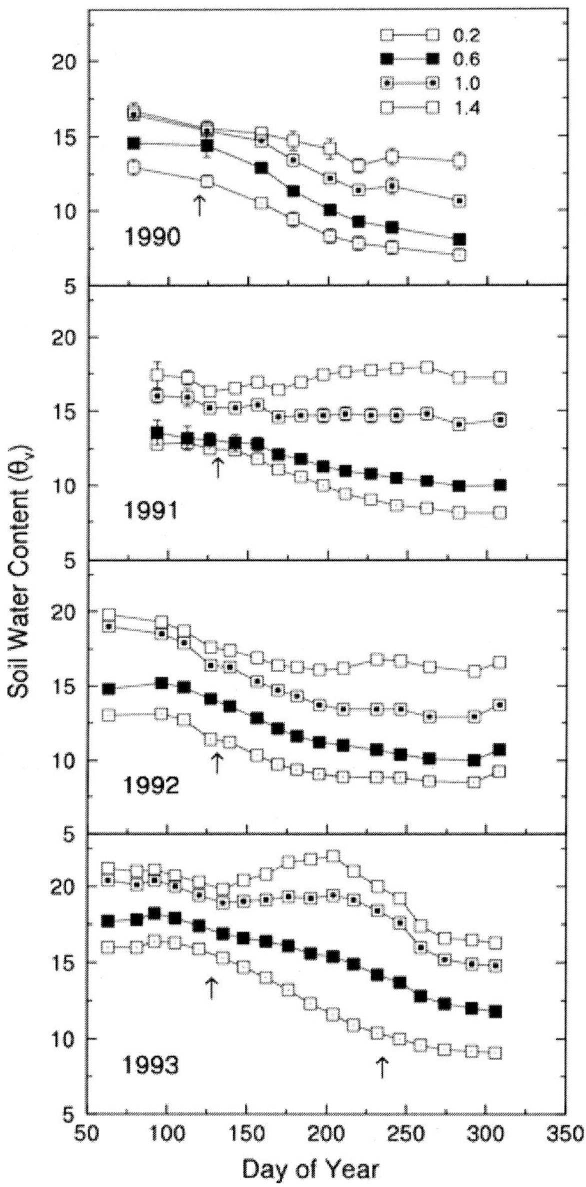

Figure 2. Soil water content measured in four of the irrigation treatments used in the study from 1990 to 1993. Values are the means of all access tubes per site (n = 3). *Bars*, when larger than the symbol, represent one SE. The *arrows* denote the approximate date when irrigation commenced each year. The second arrow in 1993 denotes irrigation cutoff before harvest

Table 3. The amount of water (mm)[a] depleted in the soil profile from budbreak until harvest for the 0.2, 0.6, 1.0 and 1.4 irrigation treatments across 4 years of the study

Year	Irrigation treatments			
	0.2	0.6	1.0	1.4
1990	167	180	158	93
1991	132	97	41	+15
1992	132	101	161	95
1993	190	150	131	117

[a] Surface area per vine was 7.55 m^2

Soil water content was measured using nine access tubes positioned in one quarter of a single vine's soil volume down to a depth of 3 m ($n = 3$)

There were significant differences in midday Ψ_l prior to the application of water in 1991 (Figure 3). Once irrigation commenced, midday Ψ_l would continue to decrease for the 0 and 0.2 irrigation treatments across years, while midday Ψ_l for the 0.6 treatment generally decreased in 1991 and 1992. Midday Ψ_l for the 1.0 and 1.4 irrigation treatments remained above −1.0 MPa across all years as did those of the 0.6 irrigation treatment in 1993. The mean values of midday Ψ_l for the specific irrigation treatments listed in Table 4, as a function of various phenological stages, were generally significantly different from one another. The mean values of all $\Psi_l$measurements in each irrigation treatment taken all season long were very similar to the anthesis to harvest values. The values from 1993, across all irrigation treatments, were higher than for similar intervals in 1991 and 1992. All values of midday Ψ_l, across irrigation treatments, within a particular interval (from one phenological stage to another) were linearly correlated (*r* values greater than 0.91 with correlations significant at the 0.001 level) with one another (data not given). In addition, the values of midday Ψ_l and Ψ_{PD} measured on the last date prior to harvest were also highly correlated with midday Ψ_l and Ψ_{PD} (using Ψ_{PD} data from 1992 and 1993) measured on other dates. Values of Ψ_{PD} for the 1.0 and 1.4 irrigation treatments prior to harvest across years ranged from −0.07 to −0.1 MPa while those of the non-irrigated treatment ranged from −0.5 to −0.8 MPa.

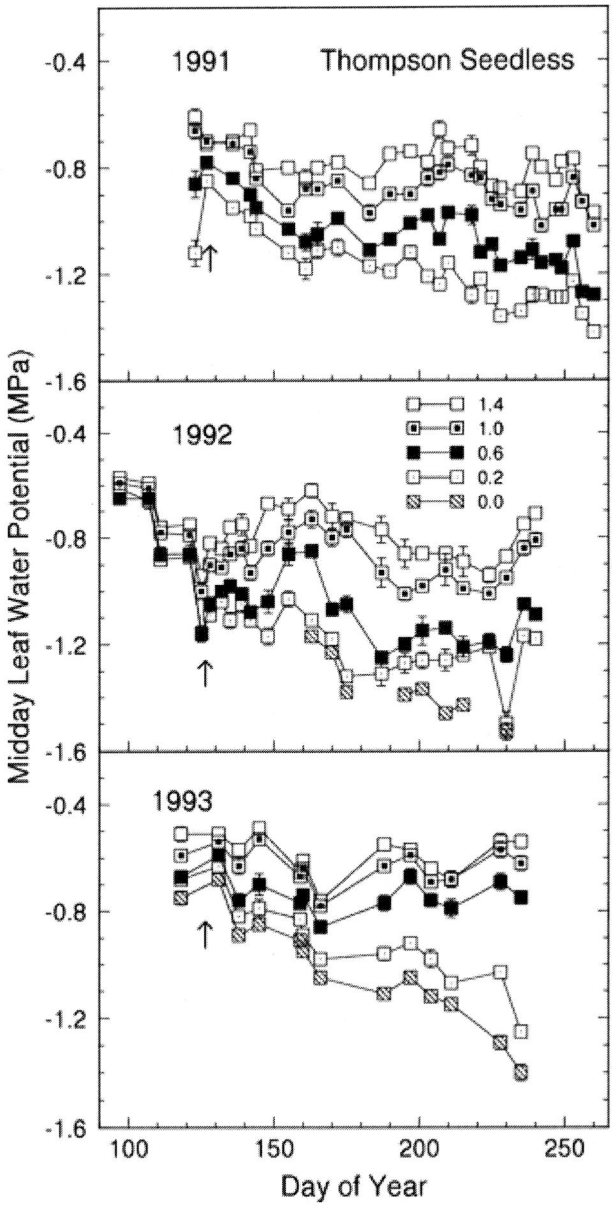

Figure 3. Midday Ψ_l measured throughout the 1991, 1992 and 1993 growing seasons as a function of irrigation treatment. Midday Ψ_l of the 0 applied water treatment was not routinely measured until 1992 and 1993. *Bars* represent one SE and are shown when larger than the symbol. *Arrows* denote when irrigation commenced

Table 4. The effect of year and irrigation treatment on mean midday leaf water potential (Ψ_l) measured from anthesis to veraison, veraison to harvest and anthesis to harvest on Thompson Seedless grapevines

Year[a]	Irrigation treatment	Anthesis to veraison Ψ_l (MPa)	Veraison to harvest Ψ_l (MPa)	Anthesis to harvest Ψ_l (MPa)
1991	0.2	−1.13 ± 0.02	−1.27 ± 0.02	−1.23 ± 0.02
	0.6	−1.04 ± 0.02	−1.11 ± 0.02	−1.09 ± 0.02
	1.0	−0.90 ± 0.02	−0.91 ± 0.01	−0.90 ± 0.01
	1.4	−0.81 ± 0.02	−0.82 ± 0.02	−0.81 ± 0.01
1992	0.0	−1.26 ± 0.06	−1.44 ± 0.03	−1.37 ± 0.04
	0.2	−1.10 ± 0.03	−1.27 ± 0.13	−1.17 ± 0.03
	0.6	−0.98 ± 0.03	−1.17 ± 0.02	−1.06 ± 0.03
	1.0	−0.85 ± 0.02	−0.95 ± 0.03	−0.89 ± 0.02
	1.4	−0.76 ± 0.03	−0.83 ± 0.02	−0.79 ± 0.02
1993	0.0	−0.89 ± 0.05	−1.19 ± 0.05	−1.04 ± 0.06
	0.2	−0.82 ± 0.05	−1.04 ± 0.05	−0.93 ± 0.05
	0.6	−0.74 ± 0.04	−0.74 ± 0.02	−0.74 ± 0.02
	1.0	−0.63 ± 0.04	−0.63 ± 0.02	−0.63 ± 0.02
	1.4	−0.60 ± 0.04	−0.54 ± 0.02	−0.59 ± 0.02

[a] The number of data points for the anthesis to veraison and veraison to harvest measurements were: n = 7 and 10 in 1991, 12 and 9 in 1992 and 6 and 6 in 1993, respectively. The number of data points for the anthesis to veraison and veraison to harvest for the 0.0 treatment in 1992 was 3 and 5, respectively

Values represent the means ± SE

There were significant ($P < 0.05$) differences in shoot length on the first measurement date in 1991 (Figure 4). Shoot growth of the 0 and 0.2 irrigation treatments had stopped by that date and midday Ψ_l of the 0.2 irrigation treatment was −1.12 MPa compared to values between −0.71 and −0.84 for the other three treatments. Clusters on the vines of the 0 and 0.2 irrigation treatments abscised during this period in 1991. Subsequently, shoot lengths of vines in the 0.2 and 0.6 irrigation treatments were significantly different from one another and from the 1.0 and 1.4 irrigation treatments. Shoot lengths were similar between the 0.2 and 0.6 irrigation treatments on all dates in 1992 as were lengths between the 1.0 and 1.4 irrigation treatments. Significant differences arose between the lower two irrigation treatments and the higher two treatments when midday Ψ_l values were −1.16 and −0.97 MPa, respectively. During the 1993 growing season, from DOY 125 onwards, shoot length of the 0.2 irrigation treatment was significantly less than shoot lengths of the other treatments. Midday Ψ_l on DOY 138 for the 0.2, 0.6 and 1.0 treatments were −0.82, −0.70 and −0.63 MPa, respectively.

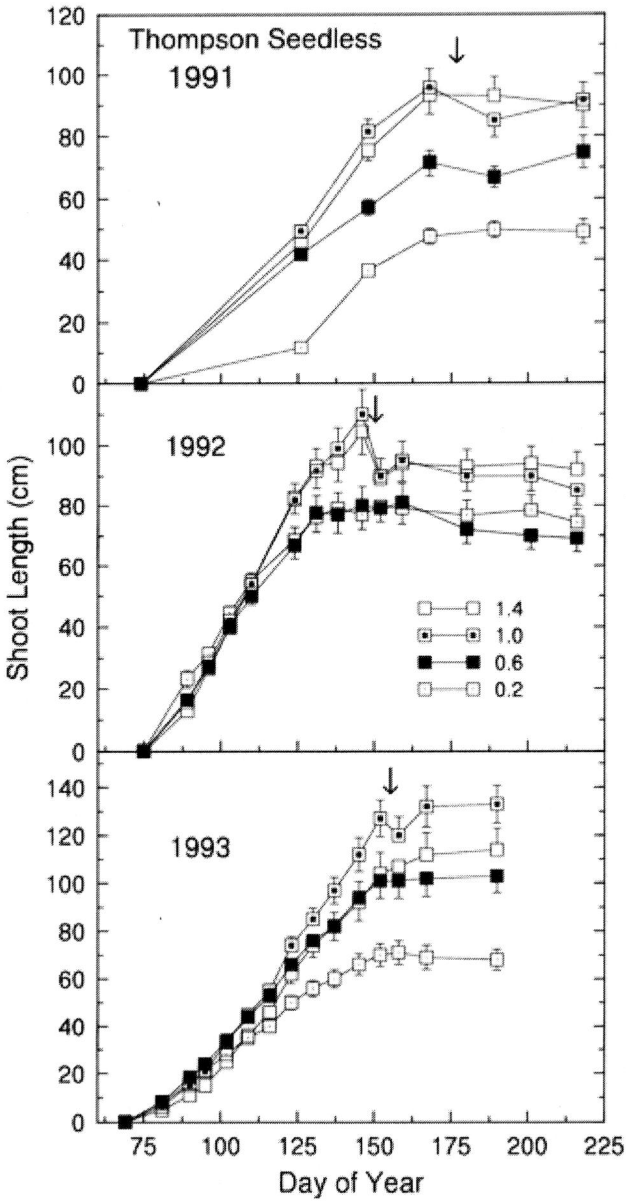

Figure 4. The seasonal progression of shoot length measured in four of the irrigation treatments from 1991 to 1993. The bars represent one SE and are shown when larger than the symbol. The *arrows* at the top of each graph represent the approximate dates the vines were first mechanically hedged

As applied water increased, leaf area per vine also increased (Table 5). Leaf area of vines irrigated at the 1.0 and 1.4 levels were 2.8 and 3.3 greater, respectively, than the leaf area of vines irrigated at the 0.2 applied water amount in August of 1991. Subsequent to the 5th June sample date in 1992, leaf area per vine decreased for both the 0.2 and 0.6 irrigation treatments. The decreases in leaf area per vine for the 0.2 and 0.6 treatments in 1992 and the 0.2 treatment in 1993 were due to leaf abscission. Any decrease in leaf area from one date to another in 1992 and 1993 for the 1.0 and 1.4 irrigation treatments were due to leaf area removed during mechanical shoot hedging.

Table 5. Leaf area (m^2 $vine^{-1}$) of Thompson Seedless grapevines measured as a function of year and date during the growing season and irrigation treatment

Year	Date (DOY)	Irrigation treatment			
		0.2	0.6	1.0	1.4
1991	Aug. 13 (225)	11.4 ± 1.4	22.8 ± 3.4	31.7 ± 2.1	38.0 ± 3.9
1992	May 4 (125)	11.9 ± 0.4	13.7 ± 0.7	18.1 ± 0.9	15.6 ± 1.5
	Jun. 5 (157)	14.6 ± 1.1	19.1 ± 0.8	27.0 ± 2.7	30.5 ± 2.9
	Jul. 8 (189)	11.3 ± 1.2	17.9 ± 2.6	24.5 ± 1.9	27.2 ± 2.1
	Aug. 12 (224)	8.4 ± 0.1	15.2 ± 0.4	23.7 ± 1.1	32.6 ± 1.3
	Sep. 7 (250)	8.2 ± 0.1	16.8 ± 0.6	23.3 ± 1.4	25.5 ± 1.4
1993	May 18 (128)	16.9 ± 0.7	17.2 ± 2.0	17.7 ± 1.3	18.9 ± 1.1
	Jun. 23 (143)	19.9 ± 1.6	26.8 ± 2.4	30.6 ± 2.6	36.6 ± 1.9
	Jul. 19 (200)	19.8 ± 3.9	26.5 ± 2.8	32.9 ± 2.1	33.0 ± 3.4
	Sep. 4 (247)	14.3 ± 2.3	25.5 ± 2.4	34.1 ± 3.0	37.0 ± 3.8

Values represent the means of three individual vines ± SE

Irrigation treatment had the greatest effect on pruning weights with little differences among trellis treatments with applied water amounts at less than full ET_c (Figure 5). At applied water amounts greater than full ET_c there were significant differences among the trellis treatments, especially in 1992 and 1993. In those 2 years, pruning weights of the single wire trellis was significantly different from those of the 0.6 and 1.2 m cross-arm trellises. The increase in pruning weights from lower to higher applied water amounts was a result of an increase in lateral shoot production (L.E. Williams, personal observation) and increased stem diameter (data not given). Pruning weights were a linear function of applied water amounts for vines trained to the 0.6 m crossarm with R^2 values in excess of 0.95 (data not shown). Pruning weights were also a linear function of the seasonal mean midday Ψ_l (Figure 6). While there were no significant differences in the slopes of this relationship from 1 year to the next, the 1993 values of pruning weights at any irrigation treatment were similar to those from 1991 and 1992 at greater values of midday Ψ_l. Pruning weights as a function of Ψ_{PD} measured prior to harvest, decreased sharply with only minimal decreases in Ψ_{PD}. Pruning weights were reduced by 70% at Ψ_{PD} values of −0.2 MPa compared to the irrigation treatments with Ψ_{PD} values at harvest of greater than −0.1 MPa under the conditions of this study.

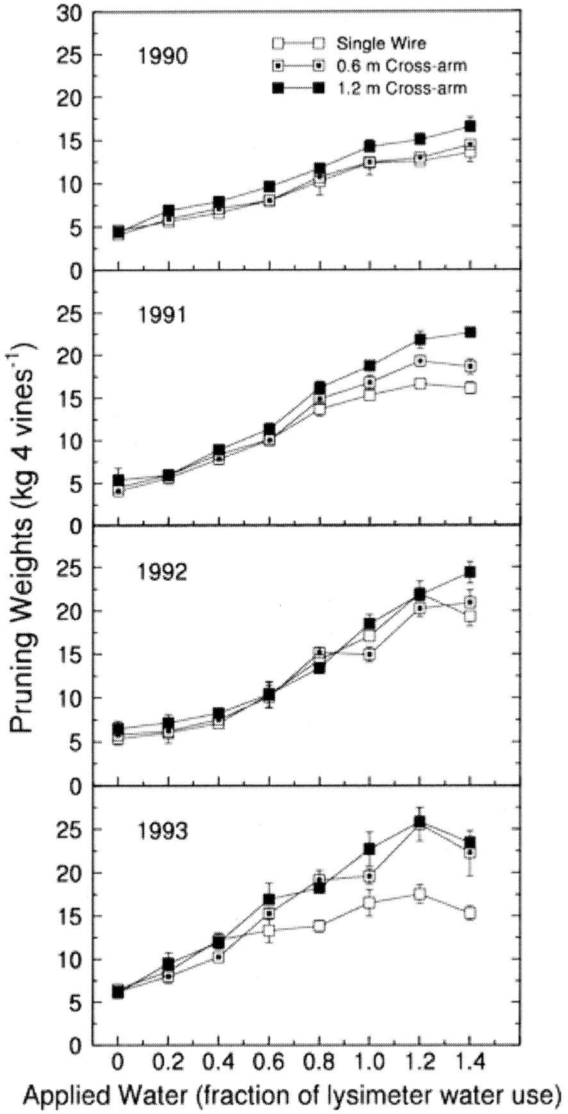

Figure 5. Pruning weights measured as a function of irrigation and trellis treatments across all years of the study. Other information is as given in previous figures

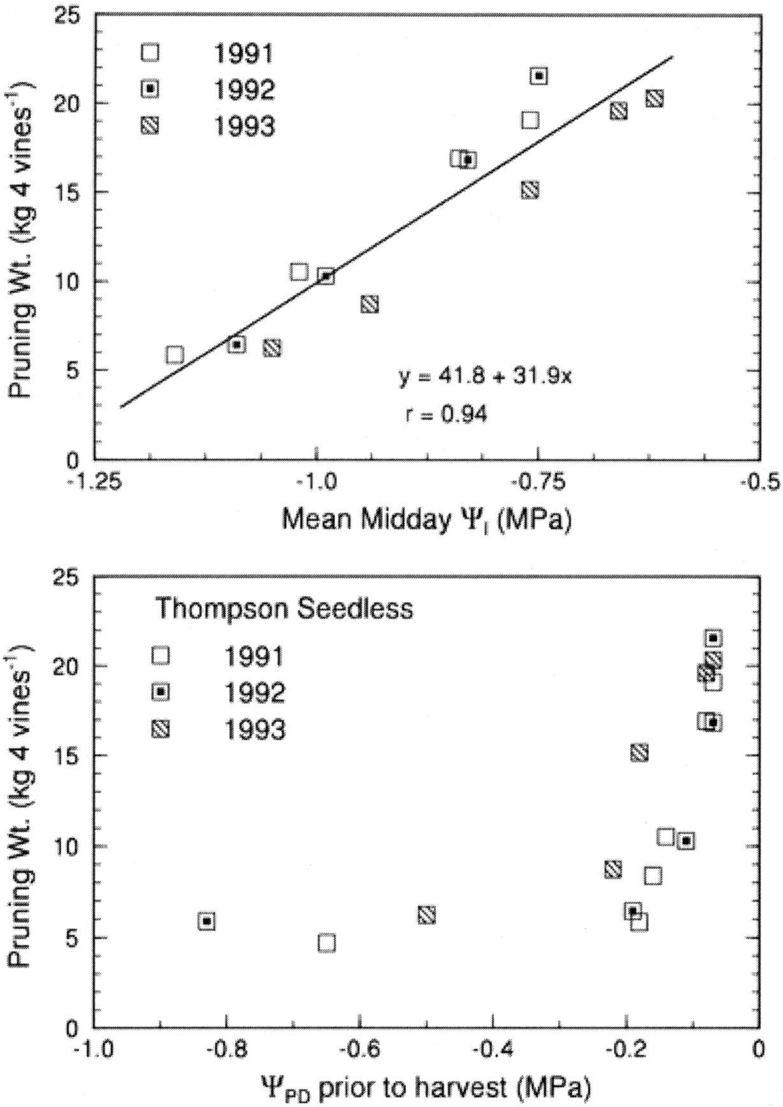

Figure 6. Pruning weights as a function of mean midday Ψ_l measured throughout the growing season and Ψ_{PD} measured prior to harvest for 3 years of the study. Pruning weights were taken from the 0.6 m trellis in each of the irrigation treatments

There were significant differences in trunk diameter among the irrigation treatments at the end of the 1991 season (Table 6). Similar results were also found 2 years later. Trunk diameter measurements were also taken at various times during the 1991 and 1993 growing seasons. In general, there were similar increases in trunk diameter across all treatments from after budbreak until mid-season (early July). There was no increase or sometimes a decrease in trunk diameters for the 0–0.6 irrigation treatments while the increase in diameter for the 0.8 treatment was less than the increase in trunk diameter for the 1.0–1.4 irrigation treatments from early July to the first of October. Trunk diameters of vines in all irrigation treatments decreased between October of 1991 and March 13th of 1992.

Table 6. Trunk diameter of Thompson Seedless grapevines as a function of irrigation treatment measured at the end of the 1991 and 1993 growing seasons and the change in diameter over the 2-year period

Irrigation treatment	Trunk diameter		Change in diameter (cm)
	30 September 1991 (cm)	8 October 1993 (cm)	
0	4.29 f	4.99 e	0.70
0.2	4.41 e	5.04 e	0.63
0.4	4.63 d	5.27 d	0.64
0.6	4.77 c	5.50 c	0.73
0.8	5.03 b	5.79 b	0.76
1.0	5.12 ab	6.01 a	0.89
1.2	5.15 ab	6.02 a	0.87
1.4	5.22 a	6.15 a	0.93

Means followed by a different letter within a column are significantly different at the $P < 0.05$ level. ($n = 24$)

DISCUSSION

The large number of irrigation treatments used in this study and the variability of climatic conditions across years allowed us to generate data providing a wide range of vine water statuses and enabled us to draw conclusions concerning how these treatments affected vegetative growth of Thompson Seedless grapevines and relate these changes to different measures of vine water status. The no applied water and applied water at 0.2 and 0.4 of ET_ctreatments were included for comparison purposes only and would probably never be used in a commercial Thompson Seedless vineyard in the San Joaquin Valley.

Irrigation treatments in this study were set up as a line-source in each experimental block of the vineyard without using border rows between treatments. This was possible as vines were irrigated at a high frequency (up to seven times daily at midseason across years), with a maximum of 16 L per vine per irrigation event for the 1.0 irrigation treatment (more or less for the other treatments) to minimize lateral water movement in the soil profile. Mean SWC decreased for the 0.2 and 0.6 irrigation treatments during the course of each growing season and their SWC values were significantly less than those of the 1.0 and 1.4 irrigation treatments. Deficit irrigating vines with applied water amounts at 0.2 and 0.6 of ET_c depleted water to a depth of 2.5 m and midway between rows from the first measurement of the season to harvest. It was also demonstrated that SWC remained constant throughout the course of the season if the vines were irrigated at 100% of ET_c while it increased if they were irrigated at 140% of ET_c. It is interesting to point out that SWC for the vines that had been deficit irrigated the previous year was always lower than the SWC for vines irrigated at ET_c or greater. This even occurred when rainfall was in excess of 350 mm during the dormant portion of the growing season from 1992 to 1993 (Williams et al. 2003b). The above information should be helpful in future studies where SWC is monitored and conclusions drawn concerning whether applied water amounts were close to estimated ET_c.

The daily and mean values of midday Ψ_l for the 1.0 and 1.4 irrigation treatments reported herein are generally greater than those previously reported for vines that were assumed to be well-watered (Correia et al. 1995), were the well-watered treatment vines under furrow irrigation (Smart 1974) or irrigated at 100% of estimated ET_c (Salón et al. 2005). They are similar to midday Ψ_l values reported by Grimes and Williams (1990) and Girona et al. (2006) for treatments irrigated at 100% of estimated ET_c and van Zyl (1987) for vines in his wet treatment (10% allowable soil water depletion). The lowest value of midday Ψ_l measured in this study (~−1.5 MPa) was that for the no applied water and the 0.2 irrigation treatments close to harvest in 1992. It would appear that a Ψ_{PD} value >−0.1 MPa indicates that the vines were not stressed for water under the conditions of this study. This value is greater than the Ψ_{PD} value (≥−0.2 MPa) often assumed to indicate the absence of or only minimal water stress (Deloire et al. 2004). The lowest Ψ_{PD} value measured for the 0.2 irrigation treatment in this study was −0.22 MPa, only because irrigation was cut off prior to harvest, otherwise it was generally >−0.2 MPa.

Deficit irrigation of grapevines in this study generally advanced the date of budbreak compared to vines irrigated at ET_c or greater. Year did affect the absolute number of days separating the treatments. A review of the literature did not find comparable data published elsewhere. Thompson Seedless grapevines are cane pruned unlike many of the other cultivars of *V. vinifera* that are spur pruned. It is unknown whether this may explain the results reported here. In addition, a difference of a week to 10 days in the date of budbreak would not result in comparable difference in dates of anthesis and/or veraison as found in this study and elsewhere (Williams 1987).

As expected, the increase in and final shoot length was significantly affected by irrigation treatment. Shoot length increased almost linearly for vines that were irrigated at ET_c or greater up until the time the vines were hedged. There were several anomalies regarding shoot length across years. Shoots on vines in the 0.2 treatment in 1991 initiated growth but then shoot growth stopped for a while. At this time, irrigation had not commenced and midday Ψ_l was close to −1.2 MPa for that treatment. The increase in shoot length of the 1.4 irrigation treatment lagged behind that of shoots in the 1.0 irrigation treatment subsequent to DOY 125 in 1993 and remained such throughout the remainder of the year. Lastly, prior to DOY 125 in 1993 shoot lengths of the 0.2 irrigation treatment were significantly less than those of the 1.0 irrigation treatment despite the fact that midday Ψ_l at that time was approximately −0.7 MPa. The above indicates that over irrigation of Thompson Seedless grapevines can negatively affect shoot growth. In addition, the reduction in shoot growth for the 0.2, deficit irrigated vines can occur at Ψ_l values greater than often assumed to stop or slow shoot elongation rate. It has been demonstrated that there is a linear reduction in the growth rate of leaves and shoot apices when tissue Ψ decreases from −0.4 to −1.0 MPa and that growth of leaves, internodes and tendrils can be reduced 100, 60 and 50% at a tissue Ψ of −1.0 MPa with a complete inhibition of internode growth at a tissue Ψ of −1.2 MPa (Schultz and Matthews 1988). Alternatively, it has been shown that non-hydraulic signals, perhaps originating in the roots (abscisic acid, ABA), may also be responsible for the inhibition of shoot growth (Dry and Loveys 1999; Dry et al. 2000). Such may be the cause in reducing shoot length for the 0.2 irrigation treatment in 1993 in this study.

The canopies of the vines used in this study were quite large even for vines in the 0.2 irrigation treatment. Leaf areas for the vines in the 1.0 and 1.4 irrigation treatments were close to or often exceeded 30 m^2 $vine^{-1}$ and the leaf areas for the vines in 0.6 irrigation treatment were in excess of

25 m^2 $vine^{-1}$ during the 1993 growing season. The great amount of leaf area for the deficit, irrigated vines may have been due to the high frequency with which the vines were irrigated. The reduction in leaf area for vines in the 0.2 irrigation treatment from mid-season onwards was due to leaf abscission, despite the availability of irrigation water daily and throughout the growing season. Canopy size or leaf area per unit row length has important implications for next year's bud fruitfulness in grapevines, especially Thompson Seedless. It has been demonstrated that shading compound buds of Thompson Seedless (syn. Sultana) reduced cluster primordia formation (May 1965). Indeed, the numbers of clusters per vine were reduced as applied water increased, especially those vines irrigated at ET_c or greater (Williams et al. 2009).

Pruning weights (a measure of vegetative growth) were a linear function of applied water amounts and seasonal, mean midday Ψ_l in this study for vines trained to the 0.6 m cross-arm trellis. These results indicate that vegetative growth was highly sensitive to vine water status in this study despite results given previously concerning a possible lack of a relationship between hydraulic signals and shoot length. This may be due to the fact that radial shoot growth is less affected by water deficits than apical shoot growth (Matthews et al. 1987; Williams and Matthews 1990).

Trunk growth is affected by water deficits in grapevines (Intrigliolo and Castel 2007; Myburg 1996; Sellés et al. 2004; van Zyl 1984). Trunk diameter decreased significantly for the sustained deficit irrigated treatments compared to those irrigated at ET_c or greater in this study. Trunk diameter did not increase further with the application of water at amounts greater than ET_c. The increase in trunk diameter averaged 4.5 mm per year for the 1.0, 1.2 and 1.4 irrigation treatments while the increase averaged 3.3 mm per year for the no applied water, 0.2 and 0.4 irrigation treatments. It is interesting to note that maximum daily water use during the 1993 growing season was almost 50 L d^{-1} with maximum hourly water use greater than 7.1 L (Williams et al. 2003b). If the trunk diameter of the vines in the lysimeter were similar to those reported in Table 6 for October 1993 (~6.0 cm), then the cross-sectional area of the vines in the lysimeter would be 28.4 cm^2. The velocity of sap through the trunk using the previously given data, assuming the entire cross-sectional area was active in the transport of water, would be 250 cm per hour.

CONCLUSIONS

This study is the first the authors are aware in which irrigation treatments consisted of applied water amounts at various fractions of measured ET_c by grapevines grown in a weighing lysimeter. In addition, the various treatments were irrigated when the vines within the lysimeter were irrigated (whenever they used 2 mm of water), albeit at their respective fraction. The treatments provided a wide array of SWC and vine water statuses. Midday Ψ_l of vines irrigated at ET_c or greater was rarely lower than −0.9 MPa throughout the growing season while Ψ_{PD} was always greater than −0.1 MPa. Sustained deficit irrigation significantly reduced leaf area per vine and pruning weights compared to those irrigated at ET_c or greater. Pruning weights were a linear function of both applied water amounts and seasonal, mean midday Ψ_l. Shoot growth was adversely affected by the application of water in amounts greater than ET_c, in some years and as a function of trellis system. The results indicate that either measurements of soil or plant water status could be used to determine when to begin irrigating each year. Subsequently, water amounts at a specific fraction of estimated vineyard ET_c could be applied to obtain a desirable canopy and conserve water.

ACKNOWLEDGMENTS

We thank Peter Biscay, Weigang Yang, Nona Ebisuda and Paul Wiley for their technical assistance and R. M. Mead and D. A. Clark for their assistance with the weighing lysimeter. We also thank Dr. A. Patakas for his editorial comments. The research was funded in part by grants from the American Vineyard Foundation and the California Raisin Advisory Board to L. E. Williams. Mention of trade names or proprietary products is for the convenience of the reader only and does not constitute endorsement or preferential treatment by the University of California or USDA/ARS.

OPEN ACCESS

REFERENCES

1. Allen RA, Pereira LS, Raes D, Smith M (1998) Crop evapotranspiration: guidelines for computing crop water requirements. (FAO irrigation and drainage paper 56) FAO, Rome

2. Araujo F, Williams LE, Grimes DW, Matthews MA (1995) A comparative study of young 'Thompson Seedless' grapevines under drip and furrow irrigation. I. Root and soil water distributions. Sci Hortic 60:235–249
3. Ayars JE, Johnson RS, Phene CJ, Trout TJ, Clark DA, Mead RM (2003) Water use by drip irrigated late season peaches. Irrig Sci 22:187–194
4. Correia MJ, Pereira JS, Chaves MM, Rodrigues ML, Pacheco CA (1995) ABA xylem concentrations determine maximum daily leaf conductance of field-grown *Vitis vinifera* L. plants. Plant Cell Environ 18:511–521
5. Daane KM, Williams LE (2003) Manipulating vineyard irrigation amounts to reduce insect pest damage. Ecol Appl 13:1650–1666
6. Deloire A, Carbonneau A, Wang Z, Ojeda H (2004) La vigne et l'eau. J Int Sci Vigne Vin 38:1–13
7. Dry PR, Loveys BR (1999) Grapevine shoot growth and stomatal conductance are reduced when part of the root system is dried. Vitis 38:151–156
8. Dry PR, Loveys BR, Düring H (2000) Partial drying of the rootzone of grape. I. Transient changes in shoot growth and gas exchange. Vitis 39:3–7
9. Fereres E, Evans RG (2006) Irrigation of fruit trees and vines: an introduction. Irrig Sci 24:55–57
10. Fereres E, Soriano MA (2007) Deficit irrigation for reducing agricultural water use. J Exp Bot 58:147–159
11. Girona J, Mata M, Del Campo J, Arbonés A, Bartra E, Marsal J (2006) The use of midday leaf water potential for scheduling deficit irrigation in vineyards. Irrig Sci 24:115–127
12. Grimes DW, Williams LE (1990) Irrigation effects on plant water relations and productivity of 'Thompson Seedless' grapevines. Crop Sci 30:255–260
13. Hsiao TC, Steduto P, Fereres E (2007) A systematic and quantitative approach to improve water use efficiency in agriculture. Irrig Sci 25:209–231
14. Intrigliolo DS, Castel JR (2007) Evaluation of grapevine water status from trunk diameter variations. Irrig Sci 26:49–59
15. Intrigliolo DS, Pérez D, Castel JR (2005) Water relations of field grown drip irrigated 'Tempranillo' grapevine. Acta Hortic 689:317–323
16. Jones HG (2004) Irrigation scheduling: advantages and pitfalls of plant-based methods. J Exp Bot 55:2427–2436
17. Laraus JL (2004) The problems of sustainable water use in the Mediterranean and research requirements for agriculture. Ann appl Biol 144:259–272
18. Matthews MA, Anderson MM, Schultz HR (1987) Phenologic and growth responses to early and late season water deficits in Cabernet Franc. Vitis 26:26–30
19. May P (1965) Reducing inflorescence formation by shading individual Sultana buds. Aust J Biol Sci 18:463–473
20. Morison JIL, Baker NR, Mullineaux MP, Davies WJ (2008) Improving water use in crop production. Philos Trans R Soc B 363:639–658
21. Myburg PA (1996) Response of *Vitis vinifera* L. cv. Barlinka/Ramsey to soil water depletion levels with particular references to trunk growth parameters. S Afr J Enol Vitic 17:3–1
22. Salón JL, Chirivella C, Castel JR (2005) Response of cv. Bobal to timing of deficit irrigation in Requena, Spain: Water relations, yield and wine quality. Am J Enol Vitic 56:1–
23. Schultz HR, Matthews MA (1988) Vegetative growth distribution during water deficits in *Vitis vinifera* L. Aust J Plant Physiol 15:641–656
24. Sellés G, Ferreira R, Muñoz I, Silva H (2004) Physiological indicators of plant water status as criteria for irrigation scheduling in table grapes cv. Crimson Seedless, irrigated dry drip. Acta Hortic 664:599–60
25. Smart RE (1974) Aspects of water relations of the grapevine (*Vitis vinifera*). Am J Enol Vitic 25:84–9

26. Stevens RM, Harvey G, Aspinall D (1995) Grapevine growth of shoots and fruit linearly correlate with water stress indices based on root-weighted soil matric potential. Aust J Grape Wine Res 1:58–66
27. van Zyl JL (1984) Response of Colombar grapevines to irrigation as regards quality aspects and growth. S Afr J Enol Vitic 5:19–28
28. van Zyl JL (1987) Diurnal variation in grapevine water stress as a function of changing soil water status and meteorological conditions. S Afr J Enol Vitic 8:45–52
29. Williams LE (1987) The effect of cyanamide on budbreak and vine development of Thompson Seedless grapevines in the San Joaquin Valley of California. Vitis 26:107–11
30. Williams LE, Araujo F (2002) Correlations among predawn leaf, midday leaf, and midday stem water potential and their correlations with other measures of soil and plant water status in *Vitis vinifera* L. J Am Soc Hortic Sci 127:448–45
31. Williams LE, Ayars JE (2005a) Water use of Thompson Seedless grapevines as affected by the application of gibberellic acid (GA3) and trunk girdling–practices to increase berry size. Agric For Meteor 129:85–94
32. Williams LE, Ayars JE (2005b) Grapevine water use and the crop coefficient are linear functions of the shaded area measured beneath the canopy. Agric For Meteor 132:201–211
33. Williams LE, Baeza P (2007) Relationships among ambient temperature and vapor pressure deficit and leaf and stem water potentials of fully irrigated, field-grown grapevines. Am J Enol Vitic 58:173–181
34. Williams LE, Matthews MA (1990) Grapevines. In: Stewart BA, Nielsen DR (eds) Agronomy monograph #30 irrigation of agricultural crops. ASA-CSSA-SSSA Publishers, Madison, Wisconsin, pp 1019–1055
35. Williams LE, Trout TJ (2005) Relationships among vine and soil based measures of water status in a Thompson Seedless vineyard in response to high frequency drip irrigation. Am J Enol Vitic 56:357–366
36. Williams LE, Dokoozlian NK, Wample RL (1994) Grape. In: Shaffer B, Anderson PC (eds) Handbook of environmental physiology of fruit crops. Volume 1. Temperate crops. CRC Press, Orlando, Florida, pp 83–133
37. Williams LE, Phene CJ, Grimes DW, Trout TJ (2003a) Water use of young Thompson Seedless grapevines in California. Irrig Sci 22:1–9
38. Williams LE, Phene CJ, Grimes DW, Trout TJ (2003b) Water use of mature Thompson Seedless grapevines in California. Irrig Sci 22:11–18
39. Williams LE, Grimes DW, Phene CJ (2009) The effects of applied water amounts at various fractions of measured evapotranspiration on reproductive growth and water productivity of Thompson Seedless grapevines. Irrig Sci.

CHAPTER 17

Wheat Yield Response to Water Deficit under Central Pivot Irrigation System Using Remote Sensing Techniques

M. A. El-Shirbeny*, A. M. Ali, A. Rashash, M. A. Badr

National Authority for Remote Sensing and Space Sciences (NARSS), Cairo, Egypt

ABSTRACT

Scarcity of rainfall and limited irrigation water resources is the main challenge for agricultural expanding policies and strategies. At the same time, there is a high concern to increase the area of wheat cultivation in order to meet the increasing local consumption. The big challenge is to incerese wheat production using same or less amount of irrigation water. In this trend, the study was carried out to analyze the sensitivity of wheat yield to water deficit using remotely sensed data in El-Salhia agricultural project which located in the eastern part of Nile delta. Normalized Difference Vegetation Index (NDVI) and Land Surface Temperature (LST) were extracted from Landsat 7. Water Deficit Index (WDI) used both LST minus air temperature (Tair) and vegetation index to estimate the relative water status. Yield response factor (ky) was derived from relationship between relative yield decrease and relative evapotranspiration deficit. The relative Evapotranspiration deficit was replaced by WDI. Linear regression was found between predicted wheat yield and actual wheat yield with 0.2^{-6}, 0.025, 0.252 and 0.76 as correlation coefficient on 30th of Dec. 2012, 15th of Jan. 2013, 16th of Feb. 2013 and 20th of Mar. 2013 respectively. The main objective of this study is using a combination between FAO 33 paper approach and remote sensing techniques to estimate wheat yield response to water.

Keywords: Normalized Difference Vegetation Index (NDVI), Land Surface Temperature (LST), Water Deficit Index (WDI), Yield Response Factor (ky), Arid Region and Egypt

1. INTRODUCTION

About 21% of the world's food depends on the wheat crop, which grows on 200 million hectares of farmland worldwide. Most of developing countries including Egypt are wheat importers. About 81% of wheat in the developing world is produced and utilized within the same country, if not the same community [1]. With a rapidly growing world population, the pressure on limited fresh water resources increases. Egypt consumes 14 million ton of wheat yearly, on the other hand, it produces about 8 million ton yearly. Although, wheat is low water consumption crop compared with other crops in Egypt, it is sensitive to water stress in some phonological stages.

Agriculture is the largest water consuming sector. It faces competing demands from other sectors, such as the industrial and the domestic sectors. With an increasing population and less water available for agricultural production, the food security for future generations is at stake. The great challenge of the agricultural sector is to produce more food from less water, which can be achieved by increasing Crop Water Productivity (CWP) [2]. Agronomical research and improving land and water management practices succeeded to increase CWP significantly during the years. Limited water is the principal factor responsible for reduced cereal yields globally and especially in Mediterranean environment [3]. Crop productivity is determined by the total amount of precipitation and also by its distribution during the growing season [4] [5].

Yield is defined as the marketable part of the total above ground biomass production; for wheat, maize and rice total grain yield is considered, and for cotton the total lint yield and total seed yield. Unfortunately, very few sources give the moisture content at which the yield is measured, which inevitably means an error that exists in the final results [2]. Grain yield depends on number of plants per unit land area, spikes per plant, spikelets per spike, grains per spikelet, and single grain weight (SGW). These yield determining components are interrelated and increase a crop's capacity to compensate for losses in any of the components caused by temporary, unfavorable conditions [6]. The response of yield to water supply is quantified by [7] through the yield response factor (ky) which relates relative yield decrease (1-Ya/Ym) to relative evapotranspiration deficit (1-ETa/PET). Water deficit of a given magnitude was expressed in the ratio actual evapotranspiration (ETa) and potential evapotranspiration (PET) may either occur continuously over the total growing period of the crop or it may occur during any of the individual growth periods (i.e. establishment (0), vegetative (1), flowering (2), yield formation (3), or ripening (4) period). Determination of PET that is substantially different from ETa has been made on the basis of a so-called two-step approach. ETo is first estimated and semi-empirical coefficient (crop coefficient, Kc) is then applied to take into account all other crop and environmental factors [8]-[11]. Similarities between Kc curve and a satellite-derived vegetation index showed potential for modeling a Kc as a function of the vegetation index [11]. Therefore, the possibility of directly estimating Kc from satellite data was investigated [9] [12]-[15].

Crop development and growth are subject to drought stress at different stages of the growth cycle, which results in differences in composition of yield components. This interrelationship was also evident when the role of grain number and weight was compared in yield determination in winter and spring wheat (Triticum aestivum L.) [16]. Generally, water deficit during the rapid spike-growth phase from booting to anthesis reduces floret set due to decreased shoot water status and increased accumulation of abscisic acid [17]. Number of grain bearing tillers and grain set may be also reduced [18] [19]. WDI is a function of actual evapotranspiration (ETa) to potential evapotranspiration (PET) ratio [20]:

$$WDI = 1 ? ETa/PET \qquad (1)$$

where: ET (mm/day) is the product of an uptake coefficient (α, mm/day) and available water (θ ? θWP) when ET is less than PET (mm/day) [21]: If ET < PET, ET = $\alpha(\theta ? \theta WP)$ and If ET $\geq$ PET, ET = PET. PET occurs when the availability of soil water does not limit transpiration. It is estimated using the FAO 56 Penman-Mon- teith model [8]. [22] developed the WDI that used both surface minus air temperature and a vegetation index to estimate the relative water status of a field. The crop begins to experience some level of stress when the WDI falls to the right of a line formed between points 1 and 4 [23].

Productivity response to water stress is different for each crop and this response is expected to vary with the climate. Therefore, the critical values of WDI should be determined for a particular crop in different climates and soils to use it in yield prediction and irrigation scheduling. Many satellite data were used to calculate WDI. [24] used NOAA/AVHRR to calculate WDI in eastern part of Nile delta-Egypt. The main objective of this work is studying wheat yield response to water deficit under central pivot irrigation system using remote sensing techniques.

2. MATERIALS AND METHODS

2.1. Study Area Location

El-Salhia project is located at the eastern part from Nile Delta. It is bounded by 30°22'35" and 30°31'19" latitudes and 31°55'24" and 32°02'38" longitudes as shown in (Figure 1). The whole area of the project is about 13,800 ha. Two irrigation systems are used in the project; the central pivots and the drip irrigation. The project has about 100 pivots irrigation units. Each pivot unit irrigates an area of about 63.6 ha when pivot length is 450 meter. The farms that are cultivated by orchards are irrigated by drip irrigation system. Climate in the study area is Dry Arid according to Köppen Climate Classification System, where precipitation is less than 50% of potential evapotranspiration. Annual average temperature is over 18°C. The average rainfall is approximately 20 mm/year. The maximum values of rainfalls are registered in January with average of 6.9 mm. The average maximum temperatures reach (34.6°C) in June and January represents the coldest month (19.0°C). The minimum temperatures range between 8.0°C in January to 21.5°C in August.

2.2. Remote Sensing Data

Remote sensing provides spatial coverage by measurement of reflected and emitted electromagnetic radiations, across a wide range of wavebands, from the earth's surface and surrounding atmosphere. Landsat ETM+ imageries, (path 176/row 039) around 10 a.m. local time with 30 meter ground resolution, on 30th of Dec. 2012, 15th of Jan. 2013, 16th of Feb. 2013 and 20th of Mar. 2013 were used in the current study to estimate LST, NDVI and WDI.

2.3. Land Surface Temperature (LST)

For landsat ETM+ data, the recorded digital numbers (DN) were converted to radiance units (Rad) using the calibration coefficients specific for each band.

$$\text{Radiance} = \text{Gain}^* \text{ DN} + \text{offset} \tag{2}$$

Surface emissivity (E_o) was estimated from the NDVI using the empirical equation developed from raw data on NDVI and thermal emissivity [25].

$$E_o = 0.9932 + 0.0194 \ln \text{NDVI} \tag{3}$$

The radiant temperature (T_o) can be calculated from band 6 radiance (Rad6) using calibration constants K1 = 666.09 and K2 = 1282.71 [26].

$$T_o = K2/\ln((K1/Rad6) + 1) \tag{4}$$

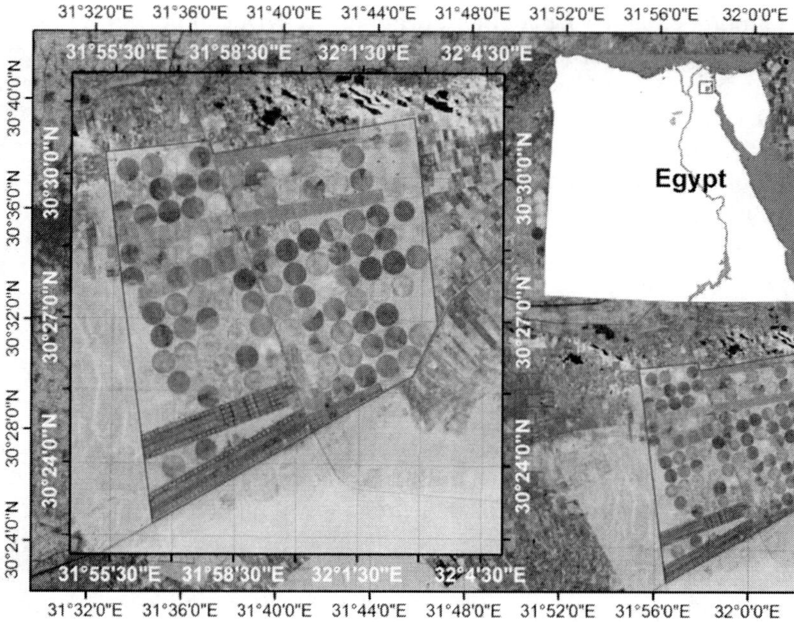

Figure 1.Location map of the study area.

The resulting temperature (Kelvin) is satellite radiant temperature of the viewed Earth atmosphere system, which is correlated with, but not the same as, the surface (kinetic) temperature. The atmospheric effects and surface thermal emissivity have to be considered in order to obtain the accurate estimate of surface temperature from satellite thermal data [27]. The surface temperature is calculated from the top of atmosphere radiant temperature (T_o) and estimated surface emissivity (E_o) as:

$$T = T_o/E_o \tag{5}$$

2.4. Water Deficit Index (WDI)

[22] developed the WDI (Equation (6)) that uses both surface minus air temperature and a vegetation index to estimate the relative water status of a field (Figure 2).

$$WDI = (dT - dTL_{13})/(dTL_{24} - dTL_{13}) \quad (6)$$

where: dT is the measure of surface subtracting air temperature at a particular percent cover, dTL_{13} is the surface minus air temperature determined by the line from points 1 to 3 for the percent cover of interest ("wet" line), and dTL_{24} is the temperature difference on the line formed between points 2 and 4 ("dry" line). Graphically, WDI can be viewed as the ratio of the distances AB to AC in the previous figure. As the WDI considers evaporation from a soil surface as well as the crop, it can be interpreted as a measure of the amount of ETa occurring relative to PET (Equation (1)). While WDI could be used to estimate ET, it does not provide a direct measure of crop water stress. As an index, it is vary according to soil-water evaporation as well as crop transpiration. The crop begins to experience levels of stress when the WDI falls to the right of a line formed between points 1 and 4 [23].

2.5. Yield Response to Water

The major importance in production planning is the yield response to water deficit. The response of yield to water supply is quantified through the yield response factor (ky) which relates relative yield decrease to relative evapotranspiration deficit. Water deficit of a given magnitude, is expressed in the ratio ETa and PET, may either occur continuously over the total growing period of the crop or it may occur during any one of the individual growth periods. The yield response to water deficit in different individual growth periods has a major importance in the scheduling of limited supply in order to obtain highest yield. Generally, crops are more sensitive to water deficit during emergence, flowering and early yield formation than early (vegetative, after establishment) and late growth periods (ripening) [7]. For Wheat, spring wheat is more sensitive than winter wheat and flowering period is more sensitive than yield formation and yield formation is more sensitive than vegetative period. In order to quantify the effect of water stress, it is necessary to derive the relationship between relative yield decrease and relative evapotranspiration deficit given by the empirically derived yield response factor (ky) [7].

$$(1 - Ya/Ym) = ky\,(1 - ETa/PET) \quad (7)$$

where: Ya is actual harvested yield; Ym is maximum harvested yield; Ky is yield response factor; ETa is actual evapotranspiration; PET is potential evapotranspiration. Relative Evapotranspiration deficit could be replaced by WDI. From equation (1 and 7) remote sensing can take a place in FAO 33 equation as follows.

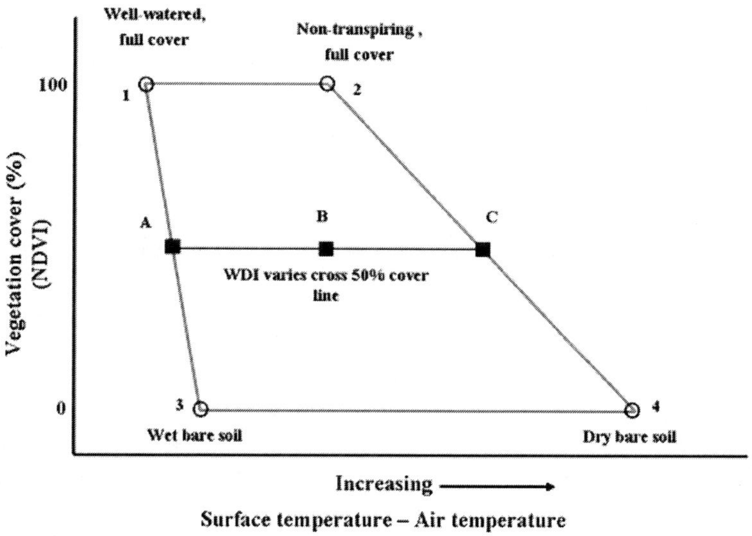

Figure 2. Water Deficit Index (WDI) trapezoid.

$$(1 - Y_a/Y_m) = Ky\ (WDI) \quad (8)$$

Figure 3 is a flowchart that shows the whole process to derive yield response from remote sensing data.

3. RESULTS AND DISCUSSION

3.1. Water Deficit Index (WDI)

WDI is a function of ETa to PET ratio [28]. WDI of 0 indicates no water stress, and a value of 1 represents maximum water stress (Figure 4). Water stress causes stomatal closure and interruption in energy dissipation which results in rise of leaf temperature. The leaf or canopy temperature is used as an indicator of plant water stress [29].

WDI has been developed for the reference crop as a generic index for quantifying crop water stress for various crops. It explores how reliable of water stress estimations would be for various crops. WDI represents the suffering of crop from water shortage or/and thermal stress. WDI in the study area varied from stage to another and from year to year. It is affected by applied irrigation system, soil type and climatic conditions.

The minimum values of WDI for wheat in study area were 0.08, 0.03, 0.03, and 0.02 and maximum values were 0.51, 0.27, 0.2, and 0.14 on 30th of Dec. 2012, 15th of Jan. 2013, 16th of Feb. 2013 and 20th of Mar. 2013 respectively. The values of WDI were high in the first stage because the canopy was not 100% coverage and the temperature of soil was higher than temperature of canopy.

3.2. Yield Prediction

Actual wheat yield can only be determined by accurately measuring the area and determining the weight of grain harvested. Environmental stress always reduces Ym. In (Figure 5) the predicted Wheat yield was plotted against actual yield, the relations were varied during different phenological stages where R^2 were 0.2^{-6}, 0.03, 0.3 and 0.76. The yield response factor (ky) for most crops is derived through the assumption that the relationship between relative yield (Ya/Ym) and relative evapotranspiration (ETa/ETm) is linear and is valid for water deficits of up to about 50 percent or (1 – ETa/ETm = 0.5). The values of ky are based on an analysis of experimental field data covering a wide range of growing conditions. The experimental results used represent high-producing crop varieties, well-adapted to the growing environment and grown under a high level of crop management [7]. In the current work, yield response factor (ky) values of FAO 33 paper for wheat were used to predict the wheat yield through remotely sensed data.

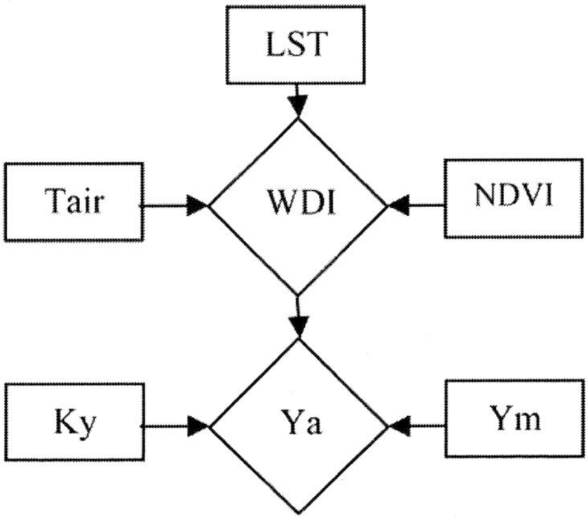

Figure 3. Illustrates the combination of FAO 33 paper approach and remote sensing techniques to estimate yield response to water.

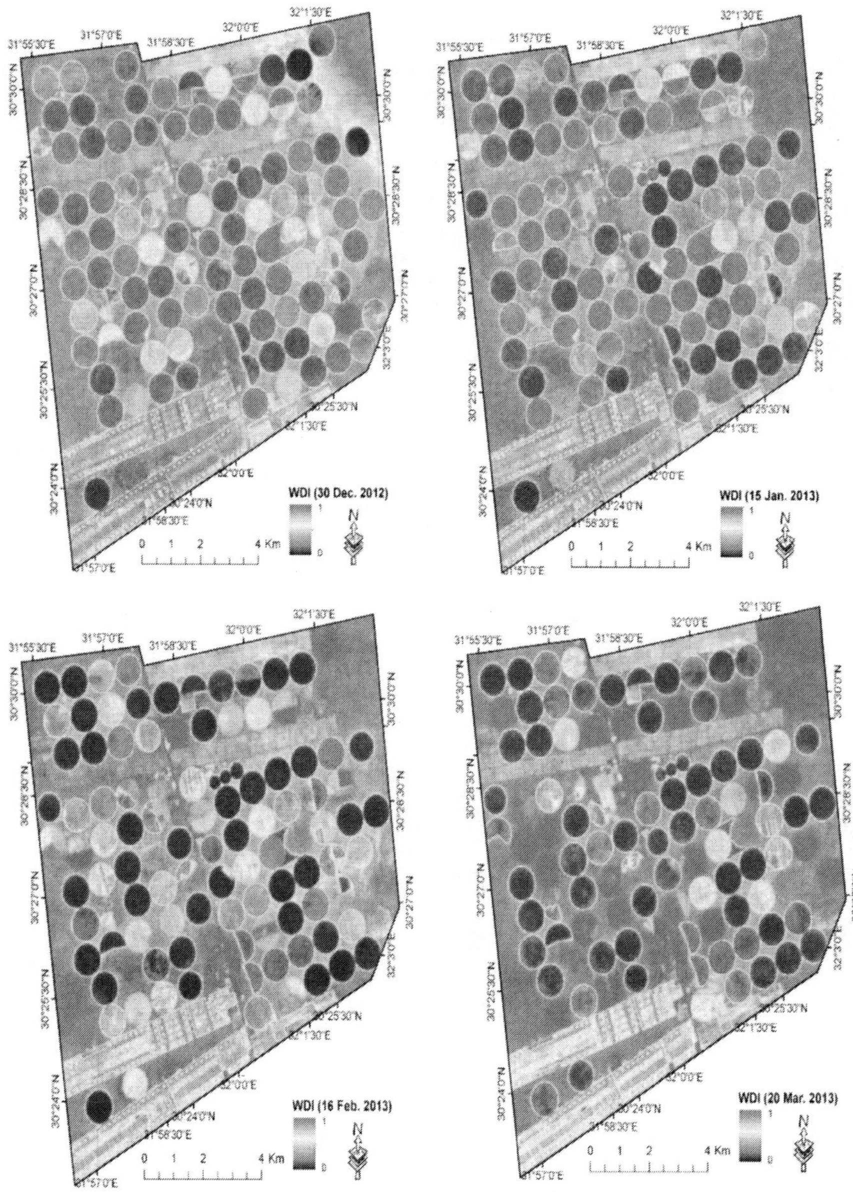

Figure 4. WDI calculated from remotely sensed data for study area.

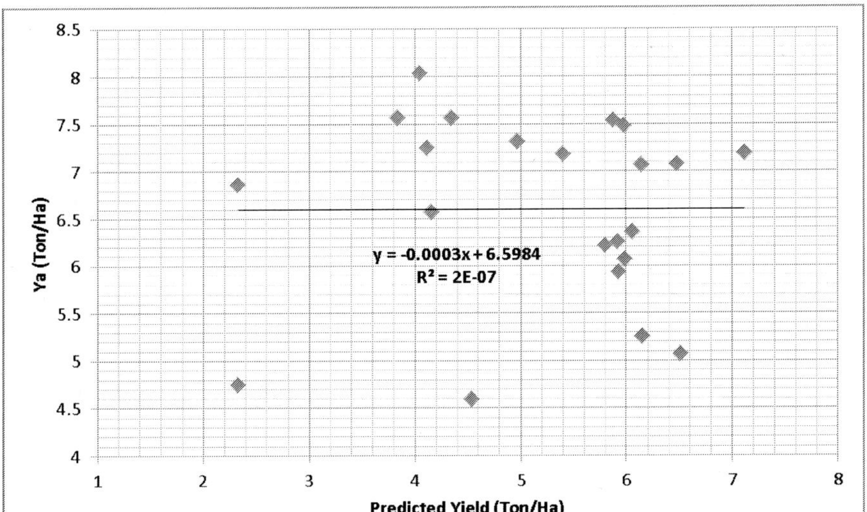

(30 Dec. 2012)

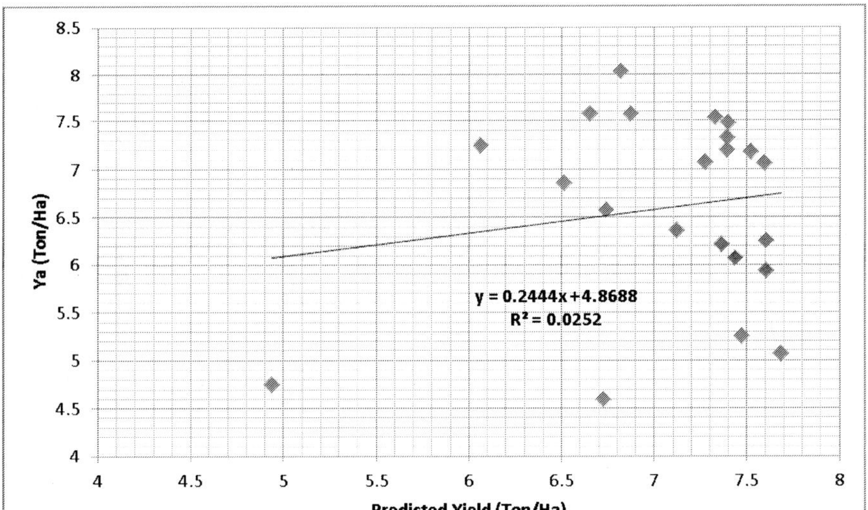

(15 Jan. 2013)

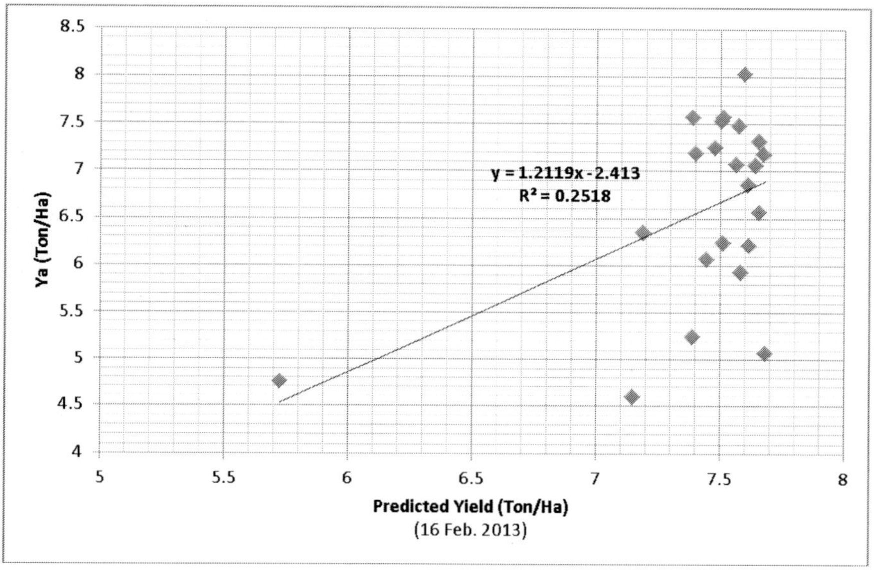

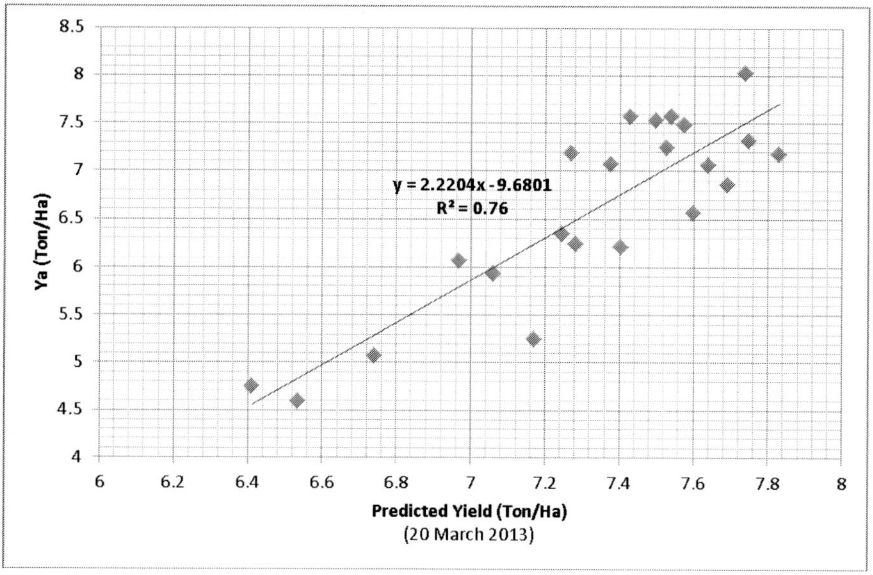

Figure 5. shows the relation between predicted wheat yield (calculated according to Equation (8)) and actual wheat yield.

4. CONCLUSION

The values of WDI were higher in establishment and vegetative stages than flowering and yield formation because the soil was not fully covered with canopy and the temperature of soil was higher than temperature of canopy. This factor reduces the accuracy of this method in partial canopy coverage case. The predicted wheat yield was plotted against actual yield, the relations were varied during different phenological stages where $R^2 = 0.2^{-6}$, 0.03, 0.3 and 0.76. There was no correlation in establishment and vegetative stages but it improved in flowering stage and it was good in yield formation stage. The combination of FAO 33 paper approach and remote sensing techniques is a good idea to estimate yield response to water but it needs to improve.

CITE THIS PAPER

M. A. El-Shirbeny, A. M. Ali, A. Rashash, M. A. Badr (2015) Wheat Yield Response to Water Deficit under Central Pivot Irrigation System Using Remote Sensing Techniques. *World Journal of Engineering and Technology*,**03**,65-72.

REFERENCES

1. http://www.fao.org
2. Zwart, S.J. and Bastiaanssen, W.G.M. (2004) Review of Measured Crop Water Productivity Values for Irrigated Wheat, Rice, Cotton and Maize. Agricultural Water Management, 69, 115-133.
3. Acevedo, E.H., Silva, P.C., Silva, H.R. and Solar, B.R. (1999) Wheat Production in Mediterranean Environments. In: Satorre, E.H. and Slafer, G.A., Eds., Wheat Ecology and Physi-ology of Yield Determination, Food Products Press, New York, 295-323.
4. Loss, S.P. and Siddique, K.H.M. (1994) Morphological and Physiological Traits Associated with Wheat Yield Increases in Mediterranean Environments. Ad-vances in Agronomy, 52, 229-276.
5. Slafer, G.A. (2003) Genetic Basis of Yield as Viewed from a Crop Physiologist's Perspective. Annals of Applied Biology, 142, 117-128.
6. Royo, C., Villegas, D., Rharrabti, Y., Blanco, R., Martos, V. and Garcia del Moral, L.F. (2006) Grain Growth and Yield Formation of Durum Wheat Grown at Contrasting Lati-tudes and Water Regimes in a Mediterranean Environment. Cereal Research Communications, 34, 1021-1028.
7. Doorenbos, J. and Kassam, O.W. (1979) Yield Response to Water. FAO Irrigation And Drainage, Paper No. 33, FAO, Rome.
8. Allen, R.G., Perrier, L.S., Raes, D. and Smith, M. (1998) Crop Evapotranspiration: Guidelines for Computing Crop Requirements,. FAO Irrigation and Drainage, Paper No. 56, Rome.

9. Magliulo, V., d'Andria, R. and Rana, G. (2003) Use of the Modified Atmometer to Estimate Reference Evapotranspiration in Mediterranean Environments. Agricultural Water Management, 63, 1-14.
10. da Silva, V.de.P.R., Borges, C.J.R, Farias, C.H.A., Singh, V.P., Albuquerque, W.G. and da Silva, B.B. (2012) Water Requirements and Single and Dual Crop Coefficients of Sugarcane Grown in a Tropical Region, Brazil. Agricultural Sciences, 3, 274-286.
11. Kamble, B., Kilic, A. and Hubbard, K. (2013) Estimating Crop Coeffi-cients Using Remote Sensing-Based Vegetation Index. Remote Sensing, 5, 1588-1602.
12. El-Shirbeny, M.A., Ali, A. and Saleh, N. (2014) Crop Water Requirements in Egypt Using Remote Sensing Techniques. Journal of Agricultural Chemistry and Environment, 3, 57-65.
13. El-Shirbeny, M.A., Ali, A.M., Badr, M.A. and Bauomy, E.M. (2014) Assessment of Wheat Crop Coefficient Using Remote Sensing Techniques. World Research Journal of Agricultural Sciences, 1, 12-17.
14. El-Shirbeny, M.A., Saleh, N.H. and Ali, A.M. (2014) Estimation of Potential Crop Evapotrans-piration Using Remote Sensing Techniques. Proceedings of the 10th International Conference of AARSE, 460-468.
15. El-Shirbeny, M.A., Alsersy, M.A.M., Saleh, N.H. and Abu-Taleb, K.A. (2015) Changes in Irrigation Water Consumption in the Nile Delta of Egypt Assessed by Remote Sensing. Arabian Journal of Geosciences (in Press).
16. Peltonen-Sainio, P., Kangas, A., Salo, Y. and Jauhiainen, L. (2007) Grain Number Dominates Grain Weight in Temperate Cereal Yield Determination: Evidence Based on 30 Years of Multi-Location Trials. Field Crops Research, 100, 179-188.
17. Westgate, M.E., Passioura, J.B. and Munns, R. (1996) Water Status and ABA Content of Floral Organs in Drought- Stressed Wheat. Australian Journal of Plant Physiology, 23, 763-772.
18. Foulkes, M.J., Sylvester-Bradley, R., Weightman, R. and Snape, J.W. (2007) Identifying Physiological Traits Associated with Improved Drought Resistance in Winter Wheat. Field Crops Research, 103, 11-24.
19. Rajala, A., Hakala, K., Makela, P., Muurinen, S. and Peltonen-Sainio, P. (2009) Spring Wheat Response to Timing of Water Deficit through Sink and Grain Filling Capacity. Field Crops Research, 114, 263-271.
20. Hiler, E.A. and Clark, R.N. (1971) Stress Day Index to Characterize Effects of Water Stress on Crop Yields. Transactions of Hydrology (210-VI-NEH).
21. Dardanelli, J.L., Ritchie, J.T., Calmon, M., Andrianiand, J.M. and Collino, D.J. (2004) An Empirical Model for Root Water Uptake. Field Crops Re-search, 87, 59-71.
22. Moran, M.S., Clarke, T.R., Inoue, Y. and Vidal, A. (1994) Estimating Crop Water Deficit Using the Relation between Surface Air Temperature and Spectral Vegetation Index. Remote Sensing of Environment, 49, 246-263.
23. Clarke, T.R. (1997) An Empirical Approach for Detecting Crop Water Stress Using Multi-Spectral Airborne Sensors. Horticulture Technology, 7, 9-16.
24. El-Shirbeny, M.A., Aboelghar, M.A., Arafat, S.M. and El-Gindy, A.G.M. (2014) Assessment of the Mutual Impact between Climate and Ve-getation Cover Using NOAA-AVHRR and Landsat Data in Egypt. Arabian Journal of Geosciences, 7, 1287-1296.
25. Valor, E. and Caselles, V. (1996) Mapping Land Surface Emis-sivity from NDVI: Application to European, African and South American Areas. Remote Sensing of Environment, 57, 167-184.
26. Goetez, S.J., Halthore, R.N., Hall, F.G. and Mark-ham, B.L. (1995) Surface Temperature Retrieval in Temperate Grassland with Multi-Resolution Sensors. Journal of Geophysical Research, 100, 397-410.
27. Norman, J.M., Divakarla, M. and Goel, N.S. (1995) Algorithms for Extract-ing Information from Remote Thermal-IR Observations of the Earth's Surface. Remote Sensing of Environment, 51, 157-168.
28. Hiler, E.A. and Clark, R.N. (1971) Stress Day Index to Cha-racterize Effects of Water Stress on Crop Yields. Transactions of Hydrology (210-VI-NEH).

29. Jackson, R.D., Idso, S.B., Reginato, R.J. and Pinter, P.J. (1981) Canopy Temperature as a Crop Water Stress Indicator, Water Resource Research, 17, 1133-1138.

CHAPTER 18

Effect of Reference Conveyance Parameter Usage on Real Time Canal Performance: The Case of Fentale Irrigation Scheme in Ethiopia

Mebruk Mohammed[1,2], Adam Tefera[3]

[1]*Addis Ababa Institute of Technology, Addis Ababa University, Addis Ababa, Ethiopia*

[2]*Federal Policy Study and Research Center, Addis Ababa, Ethiopia*

[3]*Madawalabu University, Bale Robe, Ethiopia*

ABSTRACT

In designing a canal system, a major problem is to decide what conveyance parameter to apply in the calculations. Since basic knowledge on this subject is lacking, it is usually taken from literatures. Most of the irrigation projects in Ethiopia are found to work below their expectation. One of the main reasons is the conveyance parameters variation from the expected (design) value which ultimately affects the envisioned conveyance efficiency. To evaluate this variation, Fentale irrigation scheme was used as case study. The conveyance efficiency used at the design stage was 80%, which was within Food and Agricultural Organization of the united nations recommendations; while the field survey value was 17%. Such huge variation was due to the fact that the assigned conveyance parameter values (roughness coefficient, hydraulic radius and bed slope) no longer represent the current situation of the scheme. Such variation has resulted in increase in the depth and top width of the water surface which further resulted in 13% and 3% increase in wetted perimeter and top width of the canal, respectively. Thus this study suggests that conveyance parameters shall be derived from history of existing irrigation schemes in a country, rather than adopting it from standard literatures. As such construction quality, maintenance activities and technological transfer activities in a country shall be seen in deciding the conveyance parameters. The study also suggests that the ever increasing water shortage in an irrigation project could be managed by proper maintenance of the entire irrigation system.

Keywords: Conveyance Efficiency, Conveyance Parameters, Fentale Irrigation Scheme, Irrigation Canal Design

1. INTRODUCTION

Different irrigation schemes did not satisfy the net irrigation demand even though the available water is more than that, due to high losses from storage, conveyance, and application to irrigation plots. Identifying the various components with their level of losses and knowing what improvements can be made are essential in making most effective use of vital water resource. As available water resources become scarcer, more emphasis is given to efficient use of irrigation water for maximum economic return and water resources sustainability. This requires appropriate methods of measuring and evaluating how effectively water extracted from a water resource is used to produce crop yield [1] , which intern shall start from the planning and design stage of an irrigation project.

In planning and designing a canal system, a major problem is to decide what conveyance parameters to apply in the calculations. Since basic knowledge on this subject is lacking, it is common practice that this efficiency is usually assumed. Obviously, the parameters thus obtained are unlikely to suit the conditions of the canal in its future state. Because conveyance parameters future state is usually the "guess" factor in the design of canal system, engineers are facing the problem of uncertainty in their calculations. Due to this uncertainty, canals and their appurtenant hydraulic structures are being given a greater or lower capacity than would be necessary. Apart from harmful side effects, this way of doing things leads to investments that may be considerably higher than would otherwise be necessary [2] .

Stakeholders in Fentale Irrigation scheme complain that water stress is increasing from year to year as information obtained from staff members of supervisory bureaus. The same idea was supported by other studies ([3] [4] [5]). Water conveyance loss consists mainly of operation losses, evaporation and seepage into the soil from the sloping surfaces and bed of the canal. The most important of these is seepage. The seepage loss in the irrigation canals accounts for the major portion of water conveyance loss (98.37%) while approximately 0.3 per cent of the total stream is lost due to evaporation [6] . Evaporation loss in irrigation networks is generally not taken into consideration [7] [8] .

Seepage losses could be enhanced due to inappropriate conveyance parameter estimation which leads to smaller irrigation efficiency. Among the causes of irrigation inefficiency, conveyance efficiency is one of them. Main causes for low canal conveyance efficiency would be hydrologic, hydraulic and/or technological transfer. This research limits the hydrologic and hydraulic reasons as it would be hydrologic in the sense that the evaporation and seepage losses are under estimated. It would be hydraulic, if the designed hydraulic parameter values of the conveyance system might not reflect the real values. If the trained farmers in the water use association failed to operate the irrigation system, the effect of technological transfer in having low conveyance efficiency is an obvious consequence.

Appropriate value for conveyance parameters in planning and design will lead to improve equity in water distribution and minimize the gap between crop water requirement and actual water use. Therefore, the assessment of the potential cause of canal conveyance inefficiency is vital for sustainable functionality of irrigation schemes. Two main principal factors affecting conveyance efficiency (hydraulic and hydrologic factors) will be studied in this paper by taking a main canal as a case study. The envisioned (designed) and actual conveyance characters of the main canal in Fentale Irrigation scheme found in Awash Basin of Ethiopia will be assessed for possible hydraulic and hydrologic reasons of inefficiency.

2. MATERIALS AND METHOD

2.1. Location of the Study Area

In order to evaluate canal conveyance parameters, Fentale irrigation scheme in Awash River Basin was selected. Awash River basin, geographically located in the world's main rift system, covers a total area of 110,000 km^2 and is by far the most utilized river basin of Ethiopia which lends itself for usual water shortage complaints [9] . Fentale Irrigation Scheme is found at a distance of 200 km the capital of Ethiopia, Addis Ababa. Generally, the project is in low land agro-ecological zone characterized by low annual precipitation (~500 mm) and high temperature (10°C to 38.3°C) and. It falls in between 8°50'N and 9°04'N latitude and 39°41'E and 39°52'E longitude, with altitude ranging from 970 to 1020 m above mean sea level. The headwork is at an elevation of 1050 m amsl [9] .

The planned irrigation area is about 27,000 hectares (ha). Irrigation water for abstraction is raised by weir at the headwork of the project and conveyed through main canal of 49.3 km length and two Branch Canals (38 and 31 km length) with pertinent structures including Electro-mechanical equipment for Kawa command area [9] [10] . The whole canal system uses gravity driven flow except Kawa command area, which is irrigated by pump. Except the major conveyance canal, all the project distribution systems were subsurface pipes.

Fentale was designed and constructed to take maximum of 18 m^3/sec of water from the river by Diversion Weir with conveyance efficiency of 83% (see Table 1). The diverted irrigation water is conveyed through the 49.3 km Main Canal. At 34.650 km a Branch Canal (BC-I) bifurcates towards the Haro-arbo area and the other Branch Canal-II (BC-II) going towards Galcha (North Block) after 49.3 km. The designed discharge of branch Canal-I and Branch canal-II were 6.72 m^3/sec and 6.0 m^3/sec respectively [9] (Figure 1).

2.2. Data Collection

The first activity done was the reconnaissance survey in order to get the general insight of irrigation projects in the basin. During which water use associations, professional staffs and some farmers were consulted about water use practices, highlight of the status of the projects and inefficiencies encountered by different projects. Such consultation and reconnaissance survey results were used to identify the data type, location and time to collect these data. Frequent field observation at Fentale project was made to judge places where canal discharge measurement must be taken. After that, main canal discharge at carefully selected points of the project was taken by using the Current Meter in collaboration with the Hydrology Department of the Ministry of Water Resources and Energy of Ethiopia. With this Current meter, canal discharge was measured at different depths across the width of the canal.

Table 1. Computed and designed Ec at different locations along the main canal.

Location at	Slope		Manning's n		Discharge		Ec	
	Existing	Design	Existing	Design	Existing	Design	Existing	Design
HR	0.001	0.0012	0.049	0.02	4.351	18	-	-
17.52 km	0.0012	0.001	0.073	0.03	3.355	17	0.77	0.94
26.9 km	0.001	0.001	0.048	0.02	1.239	15	0.37	0.88
42.7 km	0.0008	0.001	0.026	0.017	0.536 + 0.2**	7.2 + 6.72*	0.59	0.93
Canal conveyance efficiency							**0.17**	**0.77**

*The discharge carried by branch canal (BC-I) located at 34th km. **The discharge observed in the branch canal I (BC-I) during the study time.

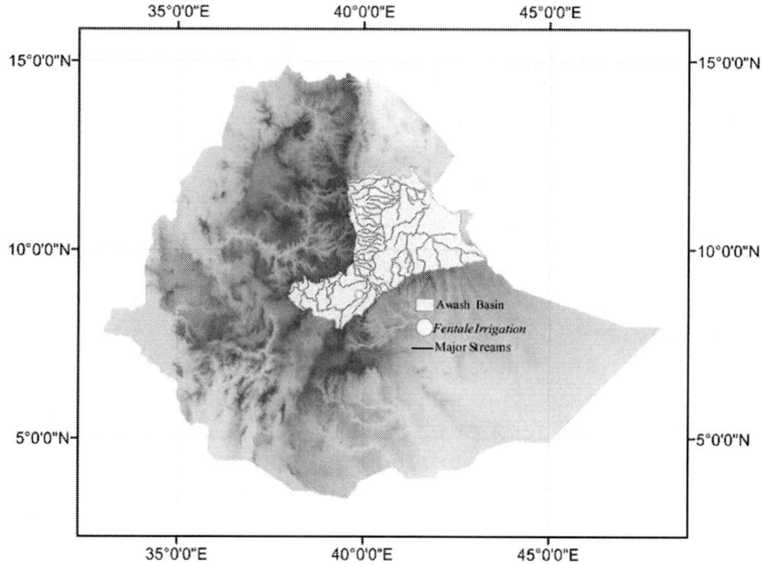

Figure 1. Location of the study area.

The discharge measured by this equipment was obtained by dividing the number of clicks by the duration in which the number of clicks was counted. To bring about high precision in measurement, the duration in which the number of clicks counted was made to be greater than forty seconds, which is in accordance with [11] . The points of measurement taken were at straight reaches of the canal. Besides the canal slope and cross-sectional survey was also conducted along and across the location where discharge measurement was taken. The discharge is measured through careful measurement of velocity, slope, wetted area and perimeter of the canal.

While measuring the velocity, the canal section was first divided in to 0.5 m - 1 m interval cross-sectional lengths. Then on the basis of the existing depth of flow at each interval, velocities were measured at one or two depth locations. Such velocity data were collected at four locations (see Table 2) along the main canal of Fentale irrigation scheme.

Slope of the canal were measured through surveying the canal bed profile for 200 m upstream and 150 m downstream of the canal section at which the velocity is measured. Wetted area and perimeter were also obtained through careful survey of the canal cross section.

Table 2. Observed data location naming and canal section description.

Main canal reach	Designed Q (m^3/sec)	Canal shape*	Canal-section taken at	Canal bed elevation (m amsl)
From the head regulator (HR) to about 17.52 km	18	Trapezoidal 1.5 H × 1 V and 1.0 H × 1.0 V (short distance for stretch covered by sound rock)	HR	1050.0
From 17.52 km to 26.9 km	17		17.52 km	1032.48
From 26.9 km to 42.7 km	15		26.9 km	1023.1
From 42.7 km to 49.3 km	7.2	Trapezoidal 1.5 H × 1 V except rectangular** for stretches covered by sound rocks	42.7 km	1007.3

*Canal stretches of permeable soil formation; geo-membrane lining with a protection of boulders covered with geo-mesh was adopted. **The particularity of this part of the canal is that, it was constructed to serve as storage by widening the canal section and providing head regulating weir at the end.

2.3. Data Analysis

Irrigation water is normally conveyed from a water source to the farm or field through natural drainage ways, constructed earthen or lined canals, or pipelines. Many conveyance systems have transmission losses, meaning that water delivered to the farm or field is usually less than the water diverted from the source. Water losses in the conveyance system include canal seepage, canal spills, evaporation losses from canals, and leaks in pipelines. These losses have to be taken into account when calculating the gross irrigation requirements of an irrigation project. This can be done through the use of an efficiency factor called conveyance efficiency, which has to be estimated at the planning stage. The higher the conveyance efficiency, the lesser the water loss, thus the larger the area that can be irrigated from a given finite water source, and the less the leaching of nutrients and damage to the soil and the more environmentally friendly the irrigation system will become. Canal conveyance efficiency, for the purpose of this study being defined as the ratio of the volume of water delivered at the downstream end to the volume of water released at the upstream end of the canal section, can be derived by the use of the following equation [2] :

$$Ec = \frac{V_d + V_2}{V_c + V_1} \times 100 \quad (1)$$

where

Ec = conveyance efficiency;

V_c = volume of water at upstream canal end (m^3);

V_1 = inflow, if any, from other sources within the section (m^3);

V_d = volume of water delivered to the farmer (at the downstream end of the section) (m^3);

V_2 = non-irrigation deliveries from the conveyance system (m^3).

Conveyance efficiency could also be analyzed on the basis of measured discharge data, which is the volume of water per unit time. In this study discharge was measured at different locations over the length of the canal in order to calculate the Ec_i at different sections. The overall conveyance efficiency of the canal is then the product of Ec_i calculated at different sections over the length of the canal.

Many factors affect the conveyance efficiency of a canal, among which, hydraulic and hydrologic causes are presented in this paper. To evaluate the effect of change in the conveyance efficiency the study and design documents of Fentale irrigation scheme was thoroughly assessed. The study and design documents of the project have used manning equation for canal design. This is acceptable, as the canals could be treated as uniform flow where in manning's equation is popular [12] [13] [14] [15] . In using the manning's equation for conveyance efficiency evaluation, conveyance parameters like manning's roughness coefficient, canal slope and canal hydraulic radius were scrutinized. Thus observed values of manning's roughness coefficient, wetted perimeter and bed slope shall be compared with the designed values. Observed values of these conveyance characters were obtained through careful measurement of velocity, canal slope, wetted area and wetted perimeters of the canal at different locations and use these in manning equation for manning's roughness coefficient estimation as follows [16] .

$$Q = \frac{1}{n} S^{1/2} A R^{2/3} \Rightarrow n = \frac{S^{1/2} A R^{2/3}}{Q} \quad (2)$$

where,

Q is discharge of the canal (m^3/s) ? Measured in a field;

R = (A/P) is Hydraulic radius (m) ? Computed from field observed values;

A = Wetted cross-sectional area (m^2) ? Surveyed in filed;

P = Wetted perimeter (m) ? Surveyed in field;

S = bed slope (-) ? Surveyed in field;

n = Manning's roughness coefficient (-) ? Computed from field observed values.

In evaluating the hydrologic effects, the resulting canal dimension change effect on the possible hydrologic water loss reason were analyzed. Thus the possible evaporation and seepage loss along the canal length is evaluated and compared with the expected (designed) one.

3. RESULTS AND DISCUSSIONS

The first reconnaissance study to the site has consulted the potential stakeholders of Fenatale irrigation scheme (District Agricultural office, Irrigation Water users Association, and Oromia irrigation development authority). From the consultation, in addition to the canal cross section near by the head regulator, it was decided to observe canal discharges at three different locations along the main canal. The same consultation is used to decide the time and location on which non-irrigation delivery along these canals is minimal (V_2 in Equation (1) is zero). Besides the main canal do not have any other inflow along the reach selected (V_1 in Equation (1) is zero). Accordingly the data were collected during such defined time and locations. The locations of the measurements are shown in Table 2.

Fentale was designed and constructed to take maximum of 18 m^3/sec of water from the river by Diversion Weir. The main canal off taking from the diversion weir and serving the command area was designed and constructed for 24-hour continuous flow [9] . Nevertheless, during site observation (January, 2013), the discharge released from the weir was measured to be 4.513 m^3/s and the total command area developed was estimated to be 8000 ha.

The four locations naming, design capacity and observed discharges during field study were shown in Table 2 and the observed canal section profile with the corresponding velocity is depicted in Figures 2-5 below in which the scale bars show the velocity in m/s unit. For better depiction of the flow velocity, Figure 4 and Figure 5 are drawn by making the scale in the depth direction exaggerated.

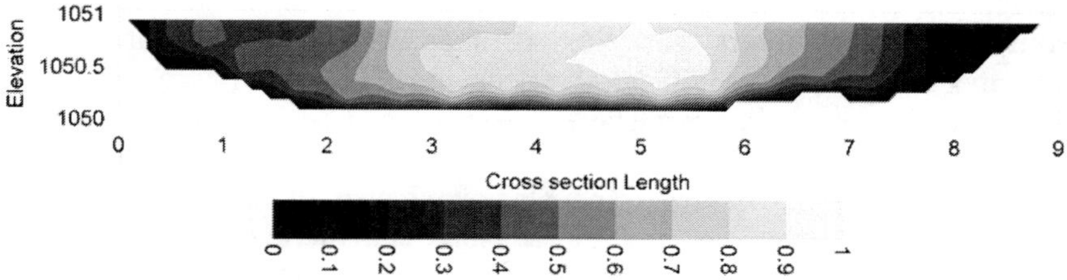

Figure 2. Observed canal section and velocity distribution at head regulator.

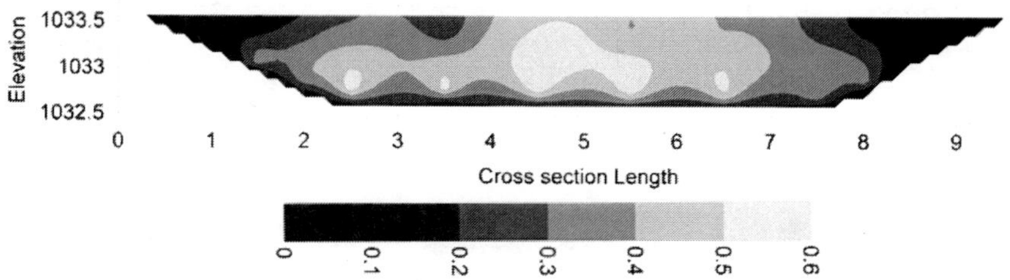

Figure 3. Observed canal section and velocity distribution at 17.52 km.

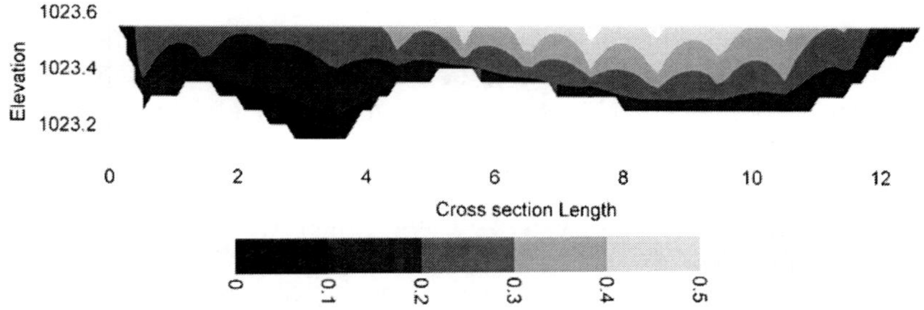

Figure 4. Observed canal section and velocity distribution at 26.9 km.

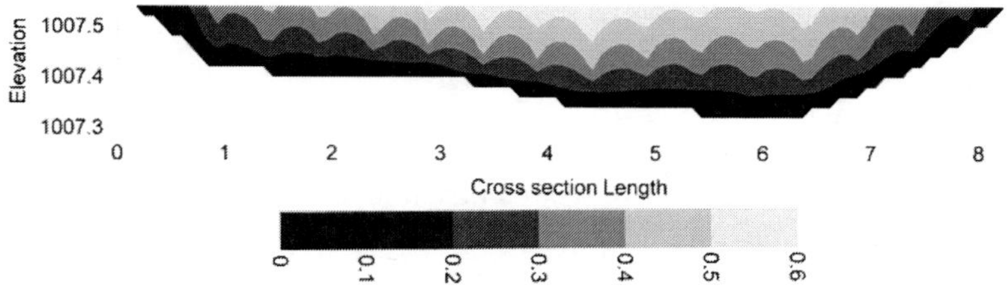

Figure 5. Observed canal section and velocity distribution at 42.7 km.

Conveyance Parameters

Hydraulic parameters variation from what has been estimated during design has a great effect on the conveyance efficiency (Ec) of canals. The conveyance parameters variation is seen by carefully analyzing the manning's roughness coefficient, wetted perimeter of the canal and top width of the canal water surface. The values of the existing manning's roughness coefficient for different sections along the main canal were between 0.026 - 0.073 (see Table 3); however, at the same locations the designed values range from 0.017 - 0.03. The increase in the value of roughness coefficient is mainly due to the canal section material change which ultimately decreases the discharge carrying capacity of the canal.

Table 3. Conveyance parameters at the four sections along the main canal.

Location at	S Bed slope*	A (m^2)*	P (m)*	R (m) (A/P)	Q (m^3/s)*	Manning's n**
HR	0.0012	7.409	9.682	0.636	4.351	0.049
17.525 km	0.001	8.706	10.428	0.683	3.355	0.073
26.9 km	0.001	4.118	13.201	0.29	1.239	0.048
42.7 km	0.001	1.445	8.431	0.16	0.536	0.026

*Measured at field level. **Computed by making use of Equation (2).

The conveyance efficiency of the main canal is also compared in Table 1 for the existing and designed condition. Conveyance efficiency is affected by several factors among which are size irrigated area, size of rotational unit, number and types of crops grown, type conveyance system and the technical and managerial facilities for water control [14] . Conveyance efficiency could be as high as 0.9, if continuous supply with no substantial change in flow occur. According to Food and Agricultural Organization of the united nations (FAO) [14] the lower limit for conveyance efficiency is 0.65, which presumed to exist if rotational supply in large schemes (>10,000 ha) and small schemes (<1000 ha) with respective problematic communication and less effective management.

The design Ec value for Fentale Irrigation project is within the recommended range, as far as the FAO recommendations are accepted. However, the existing conveyance efficiency could not add up to the expectations. The low efficiency indicates possible assumption errors in the design or construction of infrastructure. This variation complicates operation and maintenance of the irrigation system. The difference between the planned and actual efficiency is partly due to a too optimistic estimate of the post-con- struction efficiency the main canal.

It could be seen here that the values of Ec are by far different from one another. The decrease in efficiency is larger in the main canal reach after the 17.52 km length. This is mainly due to high silt deposition and wide coverage of the canal section with grass and weed as observed during the field survey.

To evaluate the possible water loss through seepage and evaporation due to the altered conveyance parameters, the wetted perimeter and top width variation of the canal were also computed for the design discharges by taking the bed width and side slopes of the design parameters. Accordingly Table 4 shows the results, from the table it is possible to deduce that the design dimensions will no longer carry the discharges, which exacerbate the seepage and the evaporation losses. The same table show that the design water depth has increased by 0.66 m, 0.43 m, 0.47 m, and 0.41 m at the head regulator at 17.52 km, at 26.9 km and at 42.7 km, respectively. This increase in water level will bring about overflow of canal water which is again another cause for conveyance Efficiency variation.

Table 4. Top width (TW) and wetted parameter (P) increase on the main canal.

Location	Design Q	Existing				Designed			% increase	
		Slope	y (m)	TW (m)	P (m)	y (m)	TW (m)	P (m)	TW	P
HR	18	0.0012	1.79	10.2	15.5	1.13	9.7	13.1	5	18
17.52 km	17	0.001	1.65	11.0	15.9	1.22	10.7	14.3	3	11
26.9 km	15	0.001	1.33	13.9	17.8	0.86	13.6	16.1	2	10
42.7 km	7.2	0.001	1.1	9.13	12.4	0.69	8.86	10.9	3	14
		Average increase in percent							3	13

4. CONCLUSIONS

The expected (designed) and the actual value of conveyance parameters obtained for the main canal of Fentale irrigation scheme show great difference; 77% during design against the 17% existing conveyance efficiency obtained during the study time. Such huge variation in conveyance efficiency resulted from variation in the hydraulic parameters like manning's roughness coefficient and canal slope; the former being the main contributor to the decrease in conveyance efficiency. Such variation will result in 42 - 66 cm water level rise, if the canal is expected to carry the design discharge. This rise in the water level has resulted in 3% increase in the top width of the canal water surface and 13% increase in wetted perimeter across the canal section. This in turn results in loses like evaporation from the top water surface and seepage through the increased wetted perimeter. Besides the increase in water depth could bring about over bank flow along the main canal section.

Such differences are basically either be improper conveyance parameter estimation during design and planning phase or improper operation and maintenance of the canal. The study clearly shows that, irrespective of the cause, there is a lot of water which is lost in the main canal. This is also true for other canal systems (secondary and lower order canals) which were not seen in this study, which makes the total loss a much larger threat to the overall benefit of the project. Thus this study suggests that conveyance parameters shall be derived from experience of existing irrigation systems, rather than adopting it from standard literatures. As such construction quality and maintenance activities shall be seen in deciding the conveyance efficiency estimation. The study also suggests that the ever increasing water shortage in an irrigation project could be managed by proper maintenance of the entire irrigation system.

ACKNOWLEDGEMENTS

The authors are grateful to Addis Ababa Institute of technology, Federal Policy study and research center for their strong support of the research publication. We would also like to acknowledge Ministry of water irrigation and Electricity for giving us the current meter for the discharge measurement. Last but not least the authors are grateful to the anonymous reviewers for their valuable comments that improve this publication.

CITE THIS PAPER

Mohammed, M. and Tefera, A. (2017) Effect of Reference Conveyance Parameter Usage on Real Time Canal Performance: The Case of Fentale Irrigation Scheme in Ethiopia. Computational Water, Energy, and Environmental Engineering, 6, 79-88.

REFERENCES

1. McCornic, P.G., Kamara, A.B. and Tadesse, G. (2003) Integrated Water and Land Research and Capacity Building Priorities for Ethiopia.

2. Bos, M.G. and Nugteren, J. (1990) On Irrigation Efficiencies. ILRI Publication.
3. Abdu, B. and Bekele, S. (2007) Analysis of Irrigation Systems Using Comparative Performance Indicators: A Case Study of Two Large-Scale Irrigation Systems in the Upper Awash Basin. International Water Management Institute, Addis Ababa.
4. Yesuf, K. (2004) Assessment of Small-Scale Irrigation Using Comparative Performance Indicators on Two Selected Schemes in Upper Awash River Valley. Unpublished MSc Thesis, Alemaya University.
5. OWWDSE (2009) Fentale Irrigation Based Integrated Project: Socioeconomic Study. Unpublished, Finfinne.
6. Badenhorst, J.W., De Lange, M., Mokwena, M.E. and Rutherford, R.J. (2002) Water Conservation and Water Demand Management in Agriculture: Development of Water Management Plans by Irrigation Water Suppliers in South Africa. ICID Eighteenth Congress Best Paper for 6th Hassan Ismail Award Paper, Montreal.
7. Xie, M., Kuffner, U. and Le Moigne, G. (1993) Using Water Efficiently: Technological Options. World Bank Technical Paper Number 205, Washington DC.
8. Australian National Committee on Irrigation and Drainage (ANCID) (2000) Open Channel Seepage and Control—Literature Review of Channel Seepage Identification and Measurement. Vol. 1.1, Victoria, Australia.
9. OWWDSE (2009) Fentale Irrigation Based Integrated Project: Engineering Report. Unpublished, Fifinne.
10. OWWDSE and Synergics Hydro (India) (2009) Fentale Irrigation Scheme Operation and Management Study Report. Unpublished, Finfinne.
11. Turnipseed, D.P. and Sauer, V.B. (2010) Discharge Measurements at Gaging Stations. USGS, Virginia.
12. Michael, A.M. (2009) Irrigation Theory and Practice. Vikas Publication House, New Delhi.
13. Bos, M.G., Wolters, W., Drovandi, A. and Morabito, J.A. (1991) The Viejo Retamo Secondary Canal-Performance Evaluation (Case Study). Irrigation and Draniage Systems, Mendiza.
14. Food and Agricultural Organization of the United Nations (FAO) (1992) Irrigation Scheme Operation and Maintenance: Irrigation Water Management. Manual No. 10, Rome.
15. FAO (1989) Guidelines for Designing and Evaluating Surface Irrigation Systems. Irrigation and Drainage Paper No. 45.
16. VenTe Chow (1959) Open-Channel Hydraulics. McGraw-Hill, New York.

CHAPTER 19

Determination of Evapotranspiration and Water Use Efficiency in Crop Production

Kazeem O. Rauff[1*], Saheed A. Shittu[2]

[1]*Federal University, Kashere, Nigeria*

[2]*Federal College of Animal Health & Production Technology, Ibadan, Nigeria*

ABSTRACT

The main concern in cultivating crops has always been water availability. To increase food production, water plays a major role after securing a large portion of land area. Knowledge of the factors influencing crop water use efficiency and hope to increase the efficiency has continued to be an objective in many modern studies. The gap between irrigation supply and demand is increasing from year to year as a result of an increase in population growing rate and people moving from place to place. Searching for new water may be a too difficult and very expensive process, so the shortest and easiest way is to maximize the water use efficiency throughout optimizing water use efficiency and the first step on that is determination of the actual crop water requirements.

Keywords: Evapotranspiration, Micrometeorological, Bowen Ratio, Microclimate, Turbulent Transport

1. INTRODUCTION

The crop evapotranspiration is defined like the water transferred to atmosphere by plant transpiration and surface evaporation. The evapotranspiration term was proposed by [1] to conceptualize the process of plant transpiration and surface evaporation which occurs simultaneously and no easy way to distinguish them. The evapotranspiration process occurs naturally only if there is inflow of energy in the system, from the sun, atmosphere, or both, and is controlled by the rate of energy in the form of water vapor that spreads from the surface of the Earth [2] . This transfer takes place physically in the forms of molecular and turbulent diffusion.

According to [3] , evaporation is the process whereby liquid water is converted to water vapor (vaporization) and removed from the evaporating surface (vapor removal). Water evaporates from a variety of surfaces, such as lakes, rivers, pavements, soils and wet vegetation. Transpiration, in turn, is the transfer of water from plants through their aerial parts. Water transfer from plant to atmosphere occurs mainly through the stomata through which they pass more than 90% water transpired.

The evapotranspiration can be derived from a range of measurement systems including lysimeters, eddy covariance, Bowen ratio, water balance (gravimetric, neutron meter, other soil water sensing), sap flow, scintillometer and even satellite-based remote sensing and direct modeling [4] . The micrometeorological method of Bowen Ratio Energy Balance (BREB) has been widely applied due to its relative simplicity, practicality, robustness and accuracy. Recently, the Eddy Covariance method has been increasingly applied mainly after the sensors have reduced costs.

The science of micrometeorology, which deals with atmospheric processes near the earth's surface, is providing increased knowledge of the mechanisms controlling our environment. Micrometeorological methods have only recently been applied to agronomic problems. Such application has increased our understanding of the response of crop plants to their physical environment, thereby permitting us to develop improved agronomic practices aimed at achieving more efficient use of water with increased and more economical production of crops [5] .

The main aim of this discussion is to summarize some of the key factors in the application of the meteorological methods to agronomic problems. The application of the methods to the problems of increasing water-use efficiency and crop production will be emphasized rather than the details of the methods. This will include a review of some of the basic physical factors controlling the microclimate at any point on the earth's surface and the application of the meteorological approach in studying this microclimate [5] .

The meteorological method has a twofold application to agronomic problems. First, it serves as a useful research tool, permitting the study of basic physical processes within the microclimate. Second, it has the potential for direct application to farm management practices, such as measurement of the evaporative loss of water and determination of when to irrigate. The expenditure of considerable time, effort and money in the development of these methods is fully justified because only by gaining an increased understanding of the natural physical environment of soils and plants and their response to this environment can we hope to understand and predict plant responses to microclimatic conditions. The resulting information will permit us to develop methods of altering the natural environment to increase crop production and water-use efficiency [5] .

2. BOWEN RATIO ENERGY BALANCE (BREB)

The Bowen-ratio method is one of the most common methods used to determine the fluxes of sensible and latent heat. The method is based on Bowen-ratio similarity and the energy balance equation. The energy balance equation provides a sound theoretical basis for estimating evaporation from natural surfaces. For a natural surface the complete equation can be written as

$$Q_s^* = Q_H + Q_E + Q_G + Q_V + Q_W + Q_P \tag{1}$$

where:

Q_H = turbulent transfer of sensible heat flux which is responsible for heating the atmosphere from the surface up to some 100 m during the day, except for days with strong convection,

Q_E = latent heat energy loss by evapotranspiration,

Q_G = vertical heat flux downward into the ground,

Q_V = horizontal heat flux, or advection, out of the region,

Q_W =net heat loss due to vertical water storage exchange,

Q_P = net energy used in such natural processes as photosynthesis.

The vertical water storage exchange term refers to the transfer through evaporation and precipitation at temperature different than the mean surface temperature. The relative magnitudes of terms in equation above are highly dependent on the space and time scales under consideration, the characteristics and homogeneity of the surface, and atmospheric conditions.

The crop evapotranspiration is estimated by use of the BREB technique from latent heat flux density which is derived from energy balance equation (Equation (2)). For many practical purposes the last three terms are negligible and the equation reduces to [6] :

$$-Q_s^* = Q_H + Q_E + Q_G \tag{2}$$

where: $-Q_s^*$ is the net radiation, Q_H is the sensible heat flux which is responsible for heating the atmosphere from the surface up to some 100 m during the day, except for days with strong convection, Q_E is the latent heat energy loss by evapotranspiration and Q_G is the vertical heat flux downward into the ground.

On the other hand [4] lists some potential advantages of BREB method, which are:

- non-destructive, direct sampling of the turbulent boundary layer;
- no aerodynamic data are required;
- simple measurement of temperature and vapor pressure at two heights;
- can measure ET over both potential and non-potential surfaces;
- gradient-based fluxes are averaged over a medium sized area (200 - 100,000 m^2); automated.

Additionally it is relative simplicity of method made it widely applied [7] - [11] .

The Bowen ratio method can be used to access the apportionment of the remaining available energy between the sensible heat term Q_H and the latent heat term Q_E. The Bowen ratio can be determined from the gradients of temperature and vapor pressure above the evaporating surface, using either the equation quoted by [6] :

$$B = \frac{Q_H}{Q_E} = \frac{0.62P}{1000} \frac{T_2 - T_1}{e_2 - e_1} \tag{3}$$

or that by [12] :

$$B=\left[\left(\frac{S}{\gamma}+1\right)\left(\frac{T_{w2}-T_{w1}}{T_2-T_1}\right)-1\right]^{-1}=\left[\left(\frac{S}{\gamma}+1\right)\left(\frac{\Delta T_w}{\Delta T}\right)-1\right]^{-1} \quad (4)$$

where:

B = Bowen ratio,

P = atmospheric pressure (mb),

T_2, T_1 = potential temperature (°C) at heights z_1 and z_2,

T_{w2}, T_{w1} = wet bulb temperatures at z_1 and z_2,

S = $\frac{\partial e}{\partial T_W}\left[\mathrm{mb}/(^{\circ}\mathrm{C})^{-1}\right]$ at the mean wet bulb temperature $\bar{T}_W$,

γ = psychrometric constant,

$$B=\frac{c_p}{\lambda}\frac{\Delta T}{\Delta q}=\frac{c_p}{\lambda}\frac{p}{0.622}\frac{\Delta T}{\Delta e}=\gamma\frac{\Delta T}{\Delta e} \quad (5)$$

where the psychrometric constant is $\gamma=0.667\ \mathrm{K}\cdot\mathrm{hPa}^{-1}$, the pressure p = 1000 hPa, the specific heat for constant pressure $c_p=1004.834\left[\mathrm{J}\cdot\mathrm{K}^{-1}\cdot\mathrm{kg}^{-1}\right]$, the evaporation heat of water $\lambda=2500827-2360(T-273.1)\left[\mathrm{J}\cdot\mathrm{kg}^{-1}\right]$,

the air density $\rho=\frac{p[\mathrm{hPa}]\cdot 100}{R_L\cdot T_v}\left[\mathrm{kg}\cdot\mathrm{m}^{-3}\right]$, and $t=20^{\circ}\mathrm{C}$,

From both Equations (1) and (2), the sensible and latent heat fluxes can be determined:

$$\frac{-Q_s^*}{Q_E}=\frac{Q_H}{Q_E}+\frac{Q_E}{Q_E}+\frac{Q_G}{Q_E} \quad (6)$$

$$Q_E=\frac{\left(-Q_s^*-Q_G\right)}{B+1} \quad (7)$$

$$\frac{-Q_s^*}{Q_E} = B + 1 + \frac{Q_G}{Q_E} \tag{8}$$

$$\frac{-Q_s^*}{Q_E} - \frac{Q_G}{Q_E} = B + 1 \tag{9}$$

$$\frac{1}{Q_E}\left(-Q_s^* - Q_G\right) = B + 1 \tag{10}$$

$$Q_H = \left(-Q_s^* - Q_G\right)\frac{B}{B+1} \tag{11}$$

The latent heat flux can also be expressed as $Q_E = \lambda E$, where $\lambda \approx 2.45 \times 10^6\ \mathrm{J \cdot kg^{-1}}$ is the latent heat of evaporation/condensation and E is the rate of evaporation/condensation. The accuracy of the Bowen ratio has been discussed by [13] (Tanner, 1960) among others. It is apparent that Equations (7) and (11) do not include the wind velocity and do not prescribe a certain difference between the measurement heights.

The crucial disadvantage of the Bowen-ratio method is that because of the apparent unclosed energy balance the residual is either added to the net radiation or distributed according to the Bowen ratio to the sensible and latent heat flux. In general, the fluxes determined with the Bowen-ratio method are larger than those determined with the eddy-covariance method.

The sensible heat flux can be directly determined with a sonic anemometer and a correction of the buoyancy flux [14] (Liu and Foken, 2001). The Bowen ratio can be calculated with Equation (3), and the latent heat flux can be calculated with:

$$Q_E = \frac{Q_H}{B} \tag{12}$$

3. METEOROLOGICAL METHODS AND THE PLANT MICROCLIMATE

3.1. The Plant Microclimate

The physical processes of the atmosphere, called meteorological processes, establish the existing climate or microclimate. Although it does not act alone, climate, in turn, determines the regime of soils and plants at a given location. Wind, precipitation, sunshine, temperature, humidity, and soil moisture are the primary factors involved. The profitable production of crops and efficient use of water require a microclimate suitable for plant growth. What makes the microclimate suitable? The meteorological methods are helping us answer this question. Recent studies of the nature of the microclimate have yielded a wealth of information on the interaction of the physical processes of the environment and the response of plants to these processes. A brief review of the results from these studies will be presented here [15] -[18] . The meteorological processes are dynamic, each reacting to the changes in the others. The dynamic balance involves a continual exchange of mass and energy. Of particular importance to us are the exchanges of molecular heat energy, radiant energy, momentum, water vapor, carbon dioxide, and oxygen between the plant or soil surface and the atmosphere. We refer to this exchange as the vertical flux of the respective quantities. The meteorological method currently offers one of the best approaches to the study of these fluxes. It also permits an evaluation of the rates of the associated plant physiological processes such as photosynthesis and transpiration under natural field conditions [5] .

The energy exchange processes are basic to the rest of the dynamic system. The sun provides the source of radiant energy called solar radiation. Of the total solar energy reaching the outer edge of the earth's atmosphere, about, 65% is available at the earth's surface for processes requiring an energy input. The other 35% is reflected by the atmosphere and the earth's surface. The solar energy absorbed by the earth causes it to be warmed. The heated surface emits thermal radiation back into space, resulting in a loss of energy and a cooling of the surface. We refer to the portion of the radiant energy absorbed at the surface and remaining after the emitted thermal radiation portion is subtracted as the net radiation component. This is the important component so far as the energy-utilizing processes at the earth's surface are concerned because it is partitioned among the various physical and physiological processes existing here [5] .

Net radiation can be measured with reliable instruments. In the absence of direct measurements, net radiation can be obtained using the radiation balance equation if other components are known. The radiation in the atmosphere is divided into shortwave (solar) radiation and long-wave (heat) radiation [19] . Radiation balance at the surface of the Earth is composed of four spectral radiant fluxes; shortwave down welling irradiation from the Sun $(K\downarrow)$ in the wavelength range 0.15 - 4 mm, the amount of energy that is reflected $(K\uparrow)$, the incoming long-wave in the wavelength range from (4 - 100 mm) radiation from the atmosphere $(I\downarrow)$; and the amount of long-wave radiation emitted from the surfaces $(I\uparrow)$ [19] . The net radiation at the ground surface is given by:

$$Q_S^* = K\downarrow + K\uparrow + I\uparrow + I\downarrow \tag{13}$$

From Equation (13) above we see that the net radiation is the sum of the shortwave down-welling radiation mainly from the sun (global radiation), the long-wave down welling infrared (heat) radiation emitted by clouds, aerosols, and gases, the shortwave up-welling reflected (solar) radiation, and the long wave up-welling infrared (heat) radiation. The shortwave radiation can be divided into the diffuse radiation from the sky and direct solar radiation. The down-welling radiation may be greater if clouds appear in a previously clear sky and the ground surface has cooled. It is only in the case of fog, that up-welling and down-welling radiations are equal. The up-welling long wave radiation is greater than the down-welling long wave radiation, because of the earth's surface is warmer than clouds and aerosols. For clear sky without clouds and dry air, the radiation temperature is approximately ?55°C [19] .

Expressing this in terms of an energy balance, we can say that the net radiation is equal to the total incoming solar radiation minus that which is reflected and that which is lost by radiation from the surface. We can further say that the net radiation is partitioned among the components: 1) sensible heat, utilized in the heating of the air; 2) latent heat, utilized in the evaporation of water; 3) stored heat in the soil and plant materials; and 4) the photochemical energy, utilized in photosynthesis.

As an example of information gained from studies of the microclimate on the interaction of the radiation energy exchanges and the other processes existing within the microclimate, let us consider what happens within a growing crop on a clear summer day. Beginning at sunrise, the soil and plant surfaces are rapidly heated by the absorption of radiant solar energy. Because the air is nearly transparent to shortwave radiation, it does not absorb as much of the radiant energy as do the surfaces. Therefore, the surfaces are much warmer than the air and there is a resulting flow of heat from the surfaces to the air, referred to as sensible heat flux. Because of this, a strong temperature gradient develops very near the surface and the rising of the warmer air parcels causes the air to become unstable. This greatly increases the mixing motion or turbulence of the airstream, which serves to cool the surface and warm the air at even faster rates.

In addition, the solar energy absorbed provides energy for the evaporation of water from the surface. The increased turbulence of the unstable air increases the rate of transfer of water vapor away from the surface, so that there is a rapid drying of soil surfaces and wilting of plants if moisture is limiting.

During photosynthesis, photochemical energy is used by the crop leaves to synthesize organic matter using carbon dioxide from the air. This results in a net transfer of carbon dioxide from the air to the leaves of the plant. Turbulent air motion enhances this transfer rate [5] .

The net result is that the surface gains energy from the sun then loses energy in heating the air, in the evaporation of water and in the flow of heat to the deeper soil layers. This is accompanied by a flow of water vapor away from and a flow of carbon dioxide to, the surface [5] .

After sunset, the situation quickly reverses. The surface radiates heat away rapidly and cools to a temperature below that of the air. There is then a flow of heat from the deeper soil layers and from the air to the cooling surface. A strong temperature gradient develops between the air and the soil or plant surface opposite in direction to that existing during the day-time. This condition, called temperature inversion, results in the air mass becoming more stable, with decreased turbulent mixing. Evaporation rapidly decreases, and photosynthesis ceases so there is then an uptake of oxygen and a release of carbon dioxide associated with respiration [5] .

The plant and soil surfaces thus act as sources and sinks for the various energy and molecular entities throughout the day. There is a flux of energy and molecular quantities either away from or toward the surface at almost all times, the direction depending upon the daytime conditions. The microclimatic conditions determine the flux rates, but the plant can modify these by stomata closure, shading the ground, and increasing the surface roughness.

Our desire is to be able to quantitatively measure these flux rates on a continuous basis. The meteorological approach serves as a useful tool in enabling us to do this. A complete study of the microclimate involves the measurement of net radiation, total incoming incident radiation, wind speed, air temperature, air humidity, soil temperature, the flow of heat into the soil, the soil moisture content, and the various properties of the surface or plant community itself [5] .

3.2. The Meteorological Approach

The meteorological approach to determining the vertical flux of energy and mass from a crop surface requires the measurement of the respective gradients, with height, of the energy and molecular entities and the determination of a transfer coefficient. The plant and soil surfaces acting as sources and sinks for the various entities establish the concentration or energy differences, which we call gradients, along which the respective entities are transferred.

The transfer rate, or flux, of the particular entity is related to its gradient by a transfer coefficient. This coefficient, which is similar to a diffusion coefficient, characterizes more than any other single parameter the effectiveness of the exchange process [5] .

Of particular interest to the micrometeorologist is the flux of carbon dioxide and water vapor associated with photosynthesis and evapotranspiration. The evaporation rate can be calculated under natural field conditions without altering the crop or soil system from measurements of the water vapor gradient and the transfer coefficient. Short period results so necessary in the evaluation of the important aspects of the highly transient system are possible. The meteorological approach also lends itself to mobility. It does, however, require careful application, considerable instrumentation, and involved analysis procedures. Recent advances in instrumentation, electronic technology, and computer developments have enabled use of this approach in spite of its complexity [5] . However, an example of one of the meteorological methods is presented to show how they can be applied to problems of crop production and water-use efficiency [5] .

The energy balance approach is a suitable example because of its basic nature and its prominence in studies of evapotranspiration. This approach involves determination of the partitioning of energy among the various physical processes contributing to the microclimate, and can be used to measure daily evaporation with little more effort than that required to measure weekly evaporation by soil moisture sampling methods. It has been applied to the determination of evapotranspiration from crops only within the past few years [13] .

4. THE ENERGY BALANCE METHOD

A brief theory of the energy balance is presented here to illustrate what is involved in using such a method. In the total balance, the earth's surface receives more radiation energy than is lost, i.e. the net radiation at the ground surface is positive $-Q_s^*$. The surplus of supplied energy will be transported back to the atmosphere due to two turbulent energy fluxes, the sensible heat flux Q_H and the latent heat flux Q_E, A small portion of the net radiant energy is utilized in the photochemical processes of photosynthesis. A larger portion goes into the heating of the soil, plants, and air. A fairly large portion provides evaporation. Furthermore, energy is transported into the soil due to the ground heat flux Q_G and will be stored by plants, buildings, etc. ΔQ_S. The sensible heat flux is responsible for heating the atmosphere from the surface up to some 100 m during the day, except for days with strong convection. A simplified energy balance equation for a crop community, neglecting the relatively small photosynthetic energy component and assuming horizontal uniformity, can be written as Equatian (2) above.

The terms for sensible heat and latent heat can also be represented by diffusion type equations:

$$Q_H = -\rho C_p K_H \frac{\partial T}{\partial z} \tag{14}$$

$$Q_E = -\rho \lambda K_E \frac{\partial q}{\partial z} \tag{15}$$

where ρ : the air density $(\mathrm{g \cdot cm^{-3}})$; C_p: the specific heat capacity for constant pressure $(\mathrm{calg^{-1}})$; λ : evaporation heat of water; K_H heat transfer coefficient $(\mathrm{cm^2 \cdot s^{-1}})$; K_E water vapor transfer coefficient; z height above ground (cm); and T temperature (°C).

The ground heat Q_G, is based mainly on molecular heat transfer, and is proportional to the temperature gradient times the thermal molecular conductivity a_G

$$Q_G = a_G \frac{\partial T}{\partial z}. \tag{16}$$

This molecular heat transfer is so weak that during the day only the upper decimeters are heated. When considering the annual cycle of ground temperature, maximum temperature is at the surface during the summer, but 10 - 15 m below the surface during winter [20] . Table 1 shows different values for different thermal molecular conductivity a_G of Equation (16).

Table 1. Thermal molecular conductivity a_G, volumetric heat capacity C_G, and molecular thermal diffusivity v_T for different soil and ground properties [21] .

Ground surface	$a_G\ (W \cdot m^{-1} \cdot K^{-1})$	$C_G\ (10^6\ W \cdot s \cdot m^{-3} \cdot K^{-1})$	$v_T\ (10^{-6}\ m^2 \cdot s^{-1})$
rocks (granite)	2.73	2.13	1.28
moist sand (40%)	2.51	2.76	0.91
dry sand	0.30	1.24	0.24
sandy clay (15%)	0.92	2.42	0.38
swamp (90% water)	0.89	3.89	0.23
old snow	0.34	0.84	0.40
new snow	0.02	0.21	0.10

From Equation (16) the determination of Q_G is not practicable because the temperature profile must be extrapolated to the surface to determine the partial derivative there. This can be uncertain because of the high temperature gradients near the surface and the difficulties in the determining thermal heat conductivity. According to [22] using sensitivity analysis, the soil heat flux plate should be buried rather deeply (10 - 20 cm) with some temperature measurements (water protected thermometers) made above it to calculate the heat storage. A similar accuracy can be archived if a temperature profile is used to calculate both the soil heat flux according to Equation (16) at a certain depth and the heat storage between this depth and the surface [19] .

The heat exchange in the air due to turbulence is much more effective. This is because turbulent exchange occurs over scales of motions ranging from millimeters to kilometers. Turbulent elements can be thought of as air parcels with largely uniform thermodynamic characteristics. Small-scale turbulence elements join to form larger ones and so on. The largest eddies are atmospheric pressure systems. The heated turbulent elements transport their energy by their random motion. The larger turbulent elements receive their energy from the mean motion, and deliver the energy by a cascade process to smaller elements. Atmospheric turbulence is a specialty of the atmospheric motion consisting in the fact that air volume (much larger than molecules: turbulent elements, turbulent eddy) achieve irregular and stochastic motions around a mean state. They are of different order with characteristic extensions and lifetimes ranging from centimeters and seconds to thousands of kilometers and days [19] .

The characteristic distribution of turbulent elements (turbulent eddies) takes place according to their size and is represented by the turbulence spectrum: The turbulence spectrum is a plot of the energy distribution of turbulent elements (turbulent eddies) according to their wavelength or frequency. Depending on the frequency, the distribution is classified as macro, meso or micro turbulence (Foken, 2008). Turbulence refers to the apparently chaotic nature of many flows, which is manifested in the form of irregular, almost random fluctuations in velocity, temperature, and scalar concentrations around their means values in time and space [23] . The calculation of the heat fluxes (sensible and latent) caused by turbulent elements is analogous to Equation (16) using the vertical gradients of temperature T and specific humidity q respectively. The sensible heat flux, Q_H, describes the turbulent transport of heat from and to the earth's surface.

Turbulent transfer is a bulk transfer, so the transfer coefficients K_H and K_E are nearly equal because they are not as much a function of molecular properties as are molecular diffusion coefficients. Substitution of Equations (14) and (15) for the sensible and latent heat terms of Equation (2) and solving for the latent heat term gives:

$$Q_E = \frac{-\left(Q_s^* + Q_G\right)}{1 + \frac{C_p}{\lambda}\left(\frac{\partial T/\partial z}{\partial q/\partial z}\right)} \tag{17}$$

Therefore, if we measure the net radiation, soil heat storage, and the air temperature and humidity gradients all of which we are presently able to do, then we can calculate the latent heat flux and water vapor flux which gives the evaporation rate.

5. THE APPLICATION OF METEOROLOGICAL METHODS

The energy balance method has much potential in the study of rate of evaporative loss of water from cropped surfaces because of its fundamental nature and because it can be employed in various other forms. It lends itself to estimating evapotranspiration by approximation procedures when complete data are not available. Such approaches have been suggested and successfully used by [24] .

The meteorological methods serve as tools for increasing crop production and water-use efficiency by providing a means of obtaining fundamental information about the controlling physical processes involved in the microclimate.

The economic use of water, for example, requires knowledge of the water evaporated by various vegetated or treated surfaces. It requires an understanding of the effects of stage of growth, plant characteristics, and various management practices on water use and crop production. An understanding of plant response to the microclimate in the evaporation of water is necessary for predicting the amount of water that will be evaporated under given conditions and for controlling evaporation [5] .

Furthermore, the meteorological method has a potential for more direct application to these problems. It was suggested several decades ago that the careful determination of evaportranspiration by such a method could be used to determine the time to irrigate [25] (Van Bavel, 1956). This has been tested on an experimental basis and found to be a usable technique. With more information about crop factors and their relationship to water use, the utility of this approach will be increased [5] .

As a further development along this line, the scheduling of irrigation of several crops on farms in this area is now being cooperatively carried out on a trial basis. The scheduling is based on meteorological parameters and previously obtained information about the response of crops to certain conditions [5] .

This procedure offers an alternative to present methods of determining when to irrigate. The method most commonly used is that of judging irrigation needs by the plant or soil appearance, which often results either in irrigation after the plant has already suffered from lack of water or in the waste of water by irrigating before it is required. Another method is that of using soil moisture measuring instrumentation or sampling procedures to monitor the water content of the soil. These methods can be used satisfactorily on a research basis or on large- scale operations where individuals highly skilled in the interpretation of these instruments are available. However, it is not likely that these instruments will ever be used extensively on the small farming scale. It is highly probable that within the near future the meteorological approach will find widespread application and that with further developments it may be possible to forecast irrigation requirements from crop data, meteorological data, and predicted weather patterns.

Meteorological methods may also be used to study the effects of irrigation on the microclimate and the plant response to these effects. Irrigation decreases air and soil temperatures and increases the humidity of the atmosphere near the surface. It, therefore, exerts a profound influence upon the disposition of energy at the surface and upon the climate near the ground. Such side effects of irrigation may have great importance in the management of suitable microclimates [5] .

Another interesting application is that of studying the modification of the microclimate by such management practices as planting of windbreaks. In another application, the meteorological approach has been used to gain valuable information about the distribution of photosynthesis within a crop throughout the day [26] [27] . Information gained from these studies could lead to the development of an ideal model of a plant so far as leaf arrangement, leaf characteristics and other plant morphological characteristics are concerned. This might help plant breeders in the development of improved plants [5] .

6. CONCLUSIONS

The probable advent of increasing water scarcity in this century will see less increase in irrigated land available for food production than in the past. Novel irrigation technologies need to be tested under local environments and particular agricultural production systems of developing countries like Nigeria. While irrigation can benefit yields and enhance in water-limited environments, the potential for full irrigation is decreasing, with increased competition from the domestic and industrial sectors. Thus the main challenge confronting both rain fed and irrigated agriculture is to improve sustainable water use for agriculture. This can be achieved through 1) an increase in crop water productivity (an increased in marketable crop yield per unit of water taken up by crop), 2) a decrease in water outflows from the plant rooting zone other than that required by plants and 3) an increase in soil water storage within the plant rooting zone through better soil and water management practices at farm and catchment scales.

The micrometeological method of studying the microclimate and all its important physical processes is serving as a useful tool for gaining information about these processes. With this information, we will be better able to predict the response of plants as we alter the natural environmental conditions in an attempt to provide an ideal environment for plant growth. Also, the micrometeorological method may in the near future serve as the basic tool by which we manage our crops and decide when to irrigate our agricultural lands (Wright and Lemon, 1967).

CITE THIS PAPER

Kazeem O. Rauff, Saheed A. Shittu (2015) Determination of Evapotranspiration and Water Use Efficiency in Crop Production. *Agricultural Sciences*,**06**,1058-1067.
doi: 10.4236/as.2015.69101

REFERENCES

1. Thornthwaite, C.W. (1948) An Approach toward a Rational Classification of Climate. Geographical Review, 38, 55-94.
2. Tucci, C.E.M. and Beltrame, L.F.S. (2009) Evaporaç&atildeo e evapotrans piraç&atildeo, In: Tucci, C.E.M., Ed., Hidrologia: Ciência e Aplicaç&atildeo, UFRGS Editora, Porto Alegre, 253-287.
3. Allen, G.R., Pereira, S.L., Raes, D. and Smith, M. (1998) Crop Evapotranspiration: Guidelines for Computing Crop Water Requirements. Food and Agrocultural Organization of the United Nations, FAO-56, Rome, 300 p.
4. Allen, R.G., Pereira, L.S., Howell, T.A. and Jensen, M.E. (2011) Evapotranspiration Information Reporting: I. Factors Governing Measurement Accuracy. Agricultural Water Management, 98, 899-920.
5. Wright, J.L. and Lemon, E.R. (1967) Micrometeorological Methods as Tools for Increasing Crop Production and Water-Use Efficiency. Proceedings of 8th Annual Fertilizer Conference of the Pacific Northwest, Twin Falls, 11-13 July 1967. [Citation Time(s):17]
6. Munn, R.E. and Davidson, B. (1957) Exploring the Atmosphere's. Pergamon Press, New York, 376 p.
7. Bezerra, J.R.C., de Azevedo, P.V., da Silva, B.B. and Dias, J.M. (2010) Evapotranspiraç&atildeo e coeficiente de cultivo do algodoeiro BRS-200 Marrom, Irrigado. Revista Brasileira de Engenharia Agrícola e Ambiental, 14, 625-632.
8. Gavilán, P. and Berengena, J. (2007) Accuracy of the Bowen Ratio-Energy Balance Method for Measuring Latent Heat Flux in a Semiarid Advective Environment. Irrigation Science, 25, 127-140.
9. Hou, L.G., Xiao, H.L., Si, J.H., Zhou, M.X. and Yang, Y.G. (2010) Evapotranspiration and Crop Coefficient of Populus euphratica Olivi Forest during the Growin Season in the Extreme arid Region Northwest China. Agricultural Water Management, 97, 351-356.
10. Steduto, P. and Hsiao, T.C. (1998) Maize Canopies under Two Soil Water Regimes: IV. Validity of Bowen Ratio-Energy Balance Technique for Measuring Water Vapor and Carbon Dioxide Fluxes at 5-min Intervals. Agricultural and Forest Meteorology, 89, 215-228.
11. Todd, R.W., Evett, S.R. and Howell, T.A. (2000) The Bowen Ratio-Energy Balance Method for Estimating Latent Heat Flux of Irrigated Alfalfa Evaluated in a Semi-Arid, Advective Environment. Agricultural and Forest Meteorology, 103, 335-348.
12. Sergeant, D.H. and Tanner, C.B. (1967) A Simple Psychrometric Apparatus for Bowen Ration Determinations. Journal of Applied Meteorology, 6, 414-418.
13. Tanner, C.B. (1960) Energy Balance Approach to Evapotranspiration from Crops. Soil Science Society of America Journal, 24, 1-9.
14. Liu, H. and Foken, T. (2001) A Modified Bowen Ratio Method to Determine Sensible and Latent Heat Fluxes. Meteorologische Zeitschrift, 10, 71-80.

15. Geiger, Z.R. (1959) The Climate near the Ground. 2nd Edition, Harvard University Press, Cambridge, MA, 482 p.
16. Slatyer, R.O. and Mcllroy, I.C. (1961) Practical Microclimatology. UNESCO, Prepared and Reproduced by CSIRO, Australia.
17. Waggoner, P.E., Ed. (1965) Agricultural Meteorology. American Meteorological Society, Boston, 188 p.
18. Evans, L.T. (1962) Environmental Control of Plant Growth. Academic Press, New York, 449 p.
19. Foken, T. (2008) Micrometeorology. Springer-Verlag, Berlin, 306 p.
20. Lehmann, A. and Kalb, M. (1993) 100 Jahre meteorologische Beobachtungen an der S?kularstation Potsdam 1893-1992. Deutscher Wetterdienst, Offenbach, 32 p.
21. Stull, R.B. (1988) An Introduction to Boundary Layer Meteorology. Kluwer Academic Publishers, Dordrecht, Boston and London, 666 p.
22. Liebethal, C. and Foken, T. (2007) Evaluation of Six Parameterization Approaches for the Ground Heat Flux. Theoretical and Applied Climatology, 88, 43-56.
23. Arya, S.P. (2001) Introduction to Micrometeorology. Academic Press, San Diego, 415 p.
24. Jensen, M.E. and Haise, H.R. (1963) Estimating Evapotranspiration from Solar Radiation. Journal of the Irrigation and Drainage Division, 89, 15-41.
25. Van Bavel, C.H.M. (1956) Evapotranspiration, the ET Method, a New Way to Find a Right Time to Irrigate. Better Farming Methods, April 1956, 14.
26. Wright, J.L. and Lemon, E.R. (1966) Photosynthesis under Field Conditions. VIII. Analysis of Windspeed Fluctuation Data to Evaluate Turbulent Exchange within a Corn Crop. Agronomy Journal, 58, 255-261.
27. Wright, J.L. and Lemon, E.R. (1966) Photosynthesis under Field Conditions. IX. Vertical Distribution of Photosynthesis within a Corn Crop. Agronomy Journal, 58, 265-268.

CHAPTER 20

A Study on the Causes of Sediment Accumulation in the Drainage Systems

Cengiz Koç[1], Ebru Yilmaz[2]

[1]*Department of Civil Engineering, Engineering Faculty, Bilecik Şeyh Edebali University, Bilecik, Turkey*

[2]*Bozdogan Vocational School, Adnan Menderes University, Bozdogan, Turkey*

ABSTRACT

Accumulation of sediment and silt in the drainage canals is undesirable, yet inevitable occurrence in the course of the use and operation of any drainage canal network. In this study, D-25 drainage canal group, taking place in the Nazilli irrigation system with an area of 1165 ha is the only system where all planned activities have been completed. It has been determined that the drainage system was constructed according to original drainage project. The depth of accumulated sediment in the drainage canals in the research area was determined from the difference between the measured elevation and the elevation given in the design projects. The reasons for siltation in the D-25 drainage canal group have been studied by looking at the results of the elevation measurements made in 2010-2012. The measurements made in D-25 drainage canal group showed that there were significant differences between the actual structure (bridge, culvert, and conduit) bottom elevations and the elevations given in the design projects. In addition, the length of some canals would not coincide with the design project either. 83.3% (93.3% in length) of the canals had differences in structure bottom elevations. Of the total 55 structures 45 (81.8%) had a 0.10 m or more difference in bottom elevation from the project. Of the erroneous structures 73.3% had an average of 0.40 m, and 26.7% had an average of −0.25 m difference in bottom elevations from the design projects.

Keywords: Drainage Canal; Sediment; Structure; Design Project; Turkey

1. INTRODUCTION

Humans have long used open-air canals for land drainage in Mesopotamia around 9000 BP [1] as well as Egyptians and Greeks around 2400 BP [2] . The long March canal in 1789 was the first recorded canal project in Maryland. By the early 20th century, land drainage was a large-scale endeavor involving state and federal partnerships which focused on the removal of surface water and groundwater [3] . Agricultural drainage canals are essential for the removal of surface and groundwater in order to allow for crop production in poorly drained agricultural landscapes. Canals also mediate the flow of pollutants from agro-ecosystems to downstream water bodies. Stream flow velocities and sediment transport regime in the drainage canals cause accumulation of fine sediments and silts along the canal network. Deposition of sediment in the drainage canal network is an undesirable, yet inevitable occurrence in the course of the use and operation of any canal network system. Apart from the sediment deposited in the bottom and consequently reducing the designed, basic purpose and functional performance of the canal network as well as hydraulic works constructed on them, the physical, chemical and biological properties of these sediments deposits are becoming issues of more immediate concern.

This study was initiated in Nazilli irrigation scheme D-25 drainage canals group of Büyük Menderes basin in Turkey, to determine the effects and causes of silt accumulation and mechanical silt cleaning which are expected to alleviate the problem on the water table, and to evaluate the current situation. This problem needs to be properly addressed considering the total length of the canal network, that is, the total volume of sediment to be removed by dredging if the functional performance of the drainage canal network is to be restored and maintained.

2. MATERIALS AND METHODS

2.1. Material

Nazilli irrigation D-25 drainage canal group, which forms the work area for this study, is located in Turkey's west, in Buyuk Menderes basin. Work area is located at 61°40' - 65°00' east longitude, 41°84' - 42°02' north latitude (Figure 1). Total service area of studied D-25 drainage group is 1165 ha. Slope is in north-south, east-west direction and varies between 0.02 - 0.006. Rainfall shows an irregular distribution by months and seasons. The average annual rainfall is 608 mm. Downstream conditions of drainage system is limited by water surface elevation of Büyük Menderes. When the water surface elevation of Büyük Menderes is high, downstream conditions of drainage canals are not appropriate [4] . Effects which cause the need for drainage in the research area include rainfall, hillside waters, side streams, floods, irrigation and water leaks. Flow direction of water table complies in general with topographic slope. While soil in Nazilli irrigation area usually consists of alluvial soils; light textured soils form the majority. Because of this feature, permeability of the soil is high [5] . The main crops are cotton, corn, vegetables, alfalfa, wheat, citrus and fig.

Total length of the canals located in D-25 drainage group which has been chosen as study area is 29 + 342 km (24 km + 342 m). Properties of built drainage canals are given in Table 1 [6] . Drainage canal length per unit area is 25.1 m/ha. On the drainage channels, 3 bridges, 1 walkway, 1 gallery, 47 culverts have been built.

2.2. Method

In the study area, methods used for calculating sediment amount accumulated in investigated drainage canals and for determining the factors effecting sediment accumulation and drainage problem dimensions are given in Figure 2.

2.2.1. Drainage Systems in the Study Area

In order to determine the drainage systems in the study area, Land Classification Reports, Drainage Planning or Detailed Drainage Reports and Drainage System Projects of the areas chosen are initially provided. By examining drainage reports, canals and qualifications located in recommended drainage system are recorded. As a result of the investigation of drainage projects, which have been prepared based upon the drainage report, canal characteristics, its artistic structure and properties of anticipated drainage system are collected. By comparing drainage systems which have been formed as a result of the application and the ones recommended in the drainage report and drainage project; the differences between them are determined.

2.2.2. Identifying the Dimension of Sediment Problem

In order to determine the dimension of sediment problem in the drainage system, elevation controls were performed before or after mechanical cleaning in the canals. To determine the black elevations of the canal, levelers were built from the canals or different drainage groups with the interval of minimum 30 m up to 60 m. By com paring canal black elevations with the red elevations compiled before from the project and profiles, sediment heights were found. Elevation control works were performed for the years 2010, 2011 and 2012 and one in a year. The extent of work to be undertaken is in most cases determined by visual inspection, followed by detailed measurement of the volume, area, or unit lengths of each task. The most complex estimate is generally that of silt

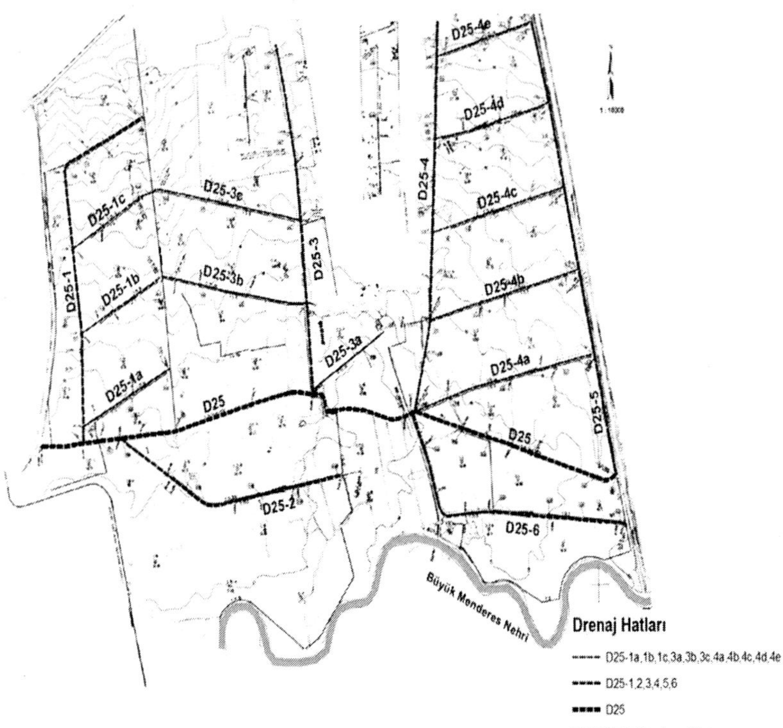

Figure 1. Nazilli irrigation D-25 drainage canal group.

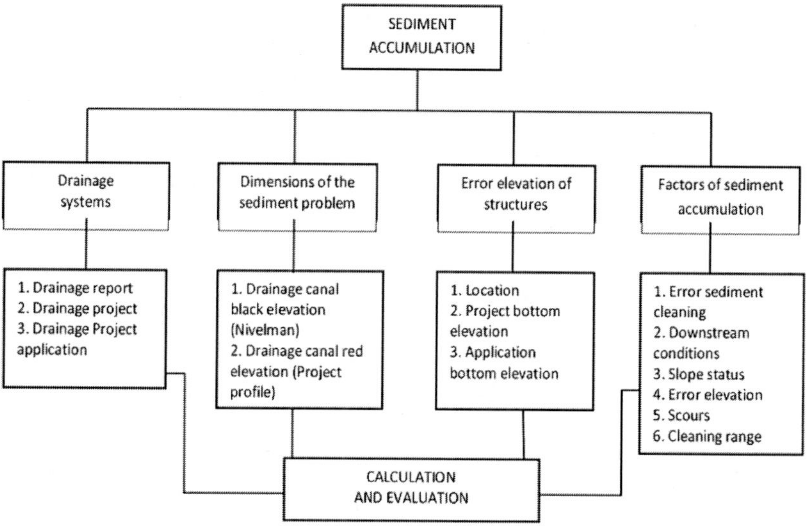

Figure 2. Schematic representation of the method used to determine the accumulation of sediment in drainage channels in the research area.

clearance, since owing to variations in the sections of the canal and consequent differences in water velocity, the silt deposited unevenly. The amount of silt deposition in the different stretches of a given canal may quite well differ 3 - 5 times. In order to determine the amount of silt cumulated, a topographical survey is made by taking the sections every 30 or 60 m according to the required degree of precision. With the data obtained the area of the section is calculated. The volume of sediment (V_P) for stretch of length, summing the volume of trapezoids in the range cross section gives the total end volume of the cross-section.

$$V_P = 1/2\left(A_1 + A_2\right) * L \tag{1}$$

where A_1 and A_2 are transversal section of the sediment and L the length of the section, and therefore for a canal of constant section:

$$V_P = \left\{1/2\left(A_1 + A_n\right) + A_2 + A_3 + \cdots A_{n-1}\right\} * L \tag{2}$$

$$V_P = \left(A_1/2 + A_2 + A_3 + \cdots A_{n-1} + A_n/2\right) * L \tag{3}$$

Table 1. Canal characteristics of D-25 drainage canal group.

Canal name	Length (km)	Drainage area (ha)	Discharge (m^3/s) min.	Discharge (m^3/s) max.	Section no	Bottom width (m)	Depth (m)	Side slopes	Bottom slope	Roughness (n)	Velocity (m/s)	Discharge (m^3/s)	Tractive force (kg/m^2)
D-25	0 + 000 - 0 + 660	1427	1.749	7.000	1	3.00	2.40	1/2	0.0002	0.028	0.546	7.056	0.225
	0 + 660 - 2 + 186	863	1.092	5.200	2	2.00	2.32	1/2	0.0002	0.028	0.508	5.220	0.203
	2 + 186 - 3 + 003	352	0.491	2.400	3	1.20	1.92	1/2	0.0002	0.028	0.421	2.413	0.152
	3 + 003 - 4 + 694	55	0.109	0.290	4	1.20	0.90	1/2	0.00066	0.028	0.386	0.307	0.177
D-25/1	0 + 000 - 0 + 350	181	0.274	1.100	1	1.20	1.35	1/2	0.0004	0.028	0.446	1.099	0.197
	0 + 350 - 1 + 150	130	0.274	1.100	2	1.20	0.96	1/2	0.0052	0.028	1.152	1.124	1.124
	1 + 150 - 1 + 851	88	0.274	1.100	3	1.20	0.89	1/2	0.0096	0.028	1.434	1.107	2.515
	1 + 851 - 2 + 304	48	0.068	0.440	4	1.20	0.95	1/2	0.0010	0.028	0.499	0.471	0.294
D-25/1-a	0 + 000 - 0 + 745	50	0.09	0.490	1	1.20	0.95	1/2	0.0012	0.028	0.547	0.516	0.353
D-25/1-b	0 + 000 - 0 + 281	40	0.055	0.500	1	1.20	0.82	1/2	0.005	0.028	0.931	0.548	1.12
	0 + 281 - 0 + 708	40	0.055	0.500	2	1.20	0.88	1/2	0.0024	0.028	0.707	0.526	0.617
D-25/1-c	0 + 000 - 0 + 281	37	0.058	0.540	1	1.20	0.85	1/2	0.004	0.028	0.874	0.581	0.964
D-25/2	0 + 000 - 1 + 992	323	0.318	0.467	1	1.20	0.97	1/2	0.0009	0.028	0.485	0.488	0.275
D-25/3	0 + 000 - 0 + 544	428	0.509	3.100	1	1.20	1.40	1/2	0.0025	0.028	1.150	3.106	1.293
	0 + 544 - 1 + 543	337	0.509	3.100	2	1.20	1.32	1/2	0.0038	0.028	1.347	3.136	1.820
	1 + 543 - 2 + 295	232	0.509	3.100	3	1.20	1.24	1/2	0.006	0.028	1.600	3.172	2.640
	2 + 295 - 2 + 543	200	0.509	3.100	4	1.20	1.36	1/2	0.003	0.028	1.228	3.084	1.494
D-25/3-a	0 + 000 - 0 + 350	36	0.065	0.500	1	1.20	0.90	1/2	0.0022	0.028	0.682	0.508	0.895
	0 + 350 - 0 + 630	36	0.065	0.500	2	1.20	0.81	1/2	0.0056	0.028	0.968	0.546	1.221
D-25/3-b	0 + 000 - 0 + 600	51	0.082	0.400	1	1.20	1.03	1/2	0.004	0.028	0.345	0.427	1.005
	0 + 600 - 1 + 041	51	0.082	0.400	2	1.20	0.78	1/2	0.005	0.028	0.867	0.959	0.682
D-25/3-c	0 + 000 - 0 + 815	55	0.111	0.920	1	1.20	1.04	1/2	0.002	0.028	0.779	0.959	0.628
	0 + 815 - 1 + 105	55	0.111	0.920	2	1.20	0.87	1/2	0.009	0.028	1.350	0.969	2.268
D-25/4	0 + 000 - 0 + 369	268	0.372	1.400	1-a	1.20	1.01	1/2	0.0054	0.028	1.240	1.404	2.755
	0 + 369 - 0 + 540	149	0.372	1.400	1	1.20	1.13	1/2	0.0024	0.028	0.927	1.437	0.926
	0 + 540 - 1 + 048	99	0.372	1.400	2	1.20	1.08	1/2	0.0032	0.028	1.024	1.402	1.155
	1 + 048 - 1 + 170	57	0.372	1.400	3	1.20	0.93	1/2	0.0015	0.028	1.654	1.465	3.266
	1 + 170 - 1 + 604	20	0.372	1.400	4	1.20	0.93	1/2	0.0074	0.028	1.406	1.459	2.294
	1 + 604 - 1 + 699	20	0.372	1.400	5	1.20	0.43	1/2	0.0111	0.028	1.625	1.439	3.150
	1 + 733 - 2 + 193	20	0.372	1.400	6	1.20	1.00	1/2	0.0062	0.028	1.316	1.448	1.984
	2 + 193 - 2 + 598	20	0.372	1.400	7	1.20	0.98	1/2	0.0074	0.028	1.406	1.459	2.294
	2 + 598 - 3 + 027	20	0.372	1.400	8	1.20	0.99	1/2	0.0066	0.028	1.343	1.434	2.079
D-25/4-a	0 + 000 - 0 + 760	64	0.092	0.590	1	1.20	0.89	1/2	0.0025	0.028	0.732	0.561	0.655
	0 + 760 - 1 + 124	64	0.092	0.590	2	1.20	0.96	1/2	0.0014	0.028	0.598	0.583	0.419
	1 + 124 - 1 + 861	64	0.092	0.590	3	1.20	0.86	1/2	0.0036	0.028	0.841	0.581	0.886
D-25/4-b	0 + 000 - 1 + 103	53	0.104	0.480	1	1.20	0.88	1/2	0.0022	0.028	0.677	0.504	0.565
	1 + 103 - 1 + 661	53	0.104	0.480	2	1.20	0.81	1/2	0.005	0.028	0.915	0.516	1.090
D-25/4-c	0 + 000 - 0 + 370	485	0.065	0.400	1	1.20	0.83	1/2	0.0025	0.028	0.669	0.411	0.573
	0 + 370 - 1 + 057	485	0.065	0.400	2	1.20	0.79	1/2	0.004	0.028	0.790	0.407	0.828
	1 + 057 - 1 + 537	485	0.065	0.400	3	1.20	0.75	1/2	0.007	0.028	0.964	0.410	1.281
D-25/4-d	0 + 000 - 0 + 700	40	0.058	0.400	1	1.20	0.88	1/2	0.0015	0.028	0.559	0.414	0.386
	0 + 700 - 1 +385	40	0.058	0.400	2	1.20	0.75	1/2	0.0066	0.028	0.913	0.389	1.208
D-25/4-e	0 +000 - 0 + 360	35	0.043	0.310	1	1.20	0.83	1/2	0.0016	0.028	0.536	0.329	0.366
	0 + 360 - 0 + 777	35	0.043	0.310	2	1.20	0.78	1/2	0.003	0.028	0.671	0.331	0.603
D-25/5	0 + 000 - 0 + 719	20	0.109	0.290	1	1.20	0.78	1/2	0.0026	0.028	0.625	0.308	0.523
D-25/6	0 + 000 - 1 + 500	53	0.159	0.233	1	1.20	0.85	1/2	0.0007	0.028	0.365	0.243	0.169
	1 + 500 - 2 + 053	53	0.159	0.233	2	1.20	0.77	1/2	0.0017	0.028	0.352	0.250	0.371

2.2.3. Determination of Art Structures and Elevation Errors

In order to determine the difference between the project and the elevations of applications of art structures (structures constructed on the irrigation and drainage canal) which were made during the leveling made in canals, art structure base elevations (application elevations) were measured. Elevations differences were found by comparing these elevations with the project elevations which were determined from the canal project and the profile. According to project elevation, art structures with differences more than +0.10, −0.10 m are considered to be "bad". Art structures were also classified as less than −0.10 - −0.49 m, −0.49 m and bigger than +0.10 - +0.49 m and +0.49 m.

2.2.4. Factors Effecting Sediment Accumulation and Determinations of Effecting Groups

The main factors affecting the accumulation of sediment in canals: 1) doing the sediment cleaning on the red elevation (incorrect sediment cleaning); 2) the inadequacy of the downstream conditions; 3) the low slope; 4) art structure elevation errors (especially, art structure built higher than project elevation); 5) doing the sediment cleaning lower than red elevation (scouring), (scouring caused by the other reasons in canals, higher slope etc., were not added into the group); 6) very long sediment cleaning intervals.

In order to investigate these specified factors, downstream conditions and art structure elevation differences were obtained from the results of leveling studies. Slope values were calculated by weighted average method from the project and profiles. The cases when the sediment cleaning is performed on the red elevation were determined with the examination of the profiles in canals, where elevation controls were performed right after mechanical cleaning. Furthermore, scouring is determined with the investigation of canal profiles as a result of elevation control. Information about mechanical sediment cleaning is obtained from DSI 21 Regional Office records. In order to determine the factors effecting sediment accumulation, canals are classified depending on cleaning intervals first (difference between the last cleaning date and leveling work). Then downstream conditions, art structure project and application elevation differences of the canals located in different cleaning intervals were examined and factor or factors effecting sediment accumulation for each canal were determined.

3. RESULTS AND DISCUSSION

3.1. According to Drainage Report and Drainage Project

Drainage report, drainage project and situation after the application of drainage system in research area are presented in Table 2. According to the drainage report, drainage canal length is 31 + 155 km, and average main canal depth as 3.50 m, average depth of secondary and tertiary drainage canals as 2.50 m, the base width 2.50 m, gradient slope as 1/2 and slope <0.0004 are recommended for the research area. Even though D-25/2 canal is recommended to be 4 + 066 km according to drainage project, this canal was taken as 1 + 922 km in the project.

Table 2. Qualifications of D-25 drainage canal group to the drainage report, drainage project, and project application.

Canal name	According to the drainage report					According to the drainage project					After application	
	Length (km)	Bottom width (m)	Depth (m)	Side slope	Slope	Length (km)	Bottom width (m)	Depth (m)	Side slope	Slope	Length (km)	Bottom width (m)
D-25	0 + 000 - 0 + 640	2.50	3.50	1/2	0.0004	0 + 000 - 0 + 660	3.00	2.40	1/2	0.0002	0 + 000 - 0 + 600	4.50
	0 + 640 - 2 + 175	2.50	3.50	1/2	0.0009	0 + 660 - 2 + 186	2.00	2.32	1/2	0.0002	0 + 600 - 2 + 186	5.00
	2 + 175 - 3 + 020	2.50	3.50	1/2	0.0011	2 + 186 - 3 + 003	1.20	1.92	1/2	0.0002	2 + 186 - 3 + 050	4.00
	3 + 020 - 4 + 670	2.50	3.50	1/2	0.0011	3 + 003 - 4 + 694	1.20	0.90	1/2	0.0007	3 + 050 - 4 + 692	3.50
D-25/1	0 + 000 - 2 + 093	2.50	2.50	1/2	0.0070	0 + 000 – 0 + 350	1.20	1.35	1/2	0.0004	0 + 000 - 0 + 350	5.00
						0 + 350 – 1 + 150	1.20	0.96	1/2	0.0052	0 + 350 - 1 + 150	5.00
						1 + 150 - 1 + 851	1.20	0.89	1/2	0.0096	1 + 150 - 1 + 850	5.00
						1 + 851 - 2 + 304	1.20	0.95	1/2	0.0010	1 + 850 - 2 + 304	3.50
D-25/1-a	0 + 000 - 0 + 780	2.50	2.50	1/2	0.0021	0 + 000 - 0 +7 45	1.20	0.95	1/2	0.0012	0 + 000 - 0 + 745	5.50
D-25/1-b	0 + 000 - 0 + 700	2.50	2.50	1/2	0.0031	0 + 000 - 0 + 281	1.20	0.82	1/2	0.0050	0 + 000 - 0 + 280	3.00
						0 + 281 - 0 + 708	1.20	0.83	1/2	0.0024	0 + 281 - 0 + 708	5.50
D-25/1-c	0 + 000 - 0 + 710	2.50	2.50	1/2	0.0035	0 + 000 - 0 + 585	1.20	0.85	1/2	0.0040	0 + 000 - 0 + 585	4.00
D-25/1-d	0 + 000 - 0 + 650	2.50	2.50	1/2	0.0049							
D-25/2	0 + 000 - 4 + 060	2.50	2.50	1/2	0.0010	0 + 000 - 3 + 992	1.20	0.97	1/2	0.0009	0 + 000 - 1 + 975	3.50
D-25/3	0 + 000 – 4 + 060	2.50	2.50	1/2	0.0110	0 + 000 - 0 + 544	1.20	1.40	1/2	0.0025	0 + 000 - 0 + 540	3.50
						0 + 544 - 1 + 543	1.20	1.32	1/2	0.0038	0 + 540 - 1 + 540	4.20
						1 + 543 - 2 + 295	1.20	1.24	1/2	0.0060	1 + 540 - 2 + 100	2.00
						2 + 295 - 2 + 543	1.20	1.36	1/2	0.0030	2 + 100 - 2 + 543	1.50
D-25/3-a	0 + 000 - 0 +629	2.50	2.50	1/2	0.0043	0 + 000 - 0 + 50	1.20	0.90	1/2	0.0022	0 + 000 - 0 + 350	
						0 + 350 - 0 + 630	1.20	0.81	1/2	0.0056	0 + 350 - 0 + 625	
D-25/3-b	0 + 000 - 1 + 051	2.50	2.50	1/2	0.0018	0 + 000 - 0 + 600	1.20	1.03	1/2	0.0040	0 + 000 - 0 + 500	3.00
						0 + 600 - 1 + 041	1.20	0.78	1/2	0.0050	0 + 500 - 1 + 040	4.00
D-25/3-c	0 + 000 - 1 + 080	2.50	2.50	1/2	0.0032	0 + 000 - 0 + 815	1.20	1.04	1/2	0.0020	0 + 000 - 0 + 600	3.00
						0 + 815 - 1 + 105	1.20	0.87	1/2	0.0090	0 + 600 - 1 + 097	4.50
D-25/3-d	0 + 000 - 1 + 118	2.50	2.50	1/2	0.0047							
D-25/3-e	0 + 000 - 1 + 065	2.50	2.50	1/2	0.0061							
	0 + 000 - 3 + 070	2.50	3.50	1/2	0.0058	0 + 000 - 0 + 369	1.20	1.01	1/2	0.0054	0 + 000 - 0 + 359	2.50

D-25/4						0 + 369 - 0 + 540	1.20	1.13	1/2	0.0024	0 + 359 - 0 + 540	3.00
						0 + 540 - 1 + 048	1.20	1.08	1/2	0.0032	0 + 540 - 0 + 786	3.50
						1 + 048 - 1 + 170	1.20	0.93	1/2	0.0015	0 + 786 - 1 + 048	3.50
						1 + 170 - 1 + 604	1.20	0.98	1/2	0.0074	1 + 048 - 1 + 604	3.50
						1 + 604 - 1 + 699	1.20	0.43	1/2	0.0111	1 + 604 - 2 + 087	2.30
						1 + 733 - 2 + 193	1.20	1.00	1/2	0.0062	2 + 087 - 3 + 027	2.30
						2 + 193 – 2 + 598	1.20	0.98	1/2	0.0074		
						2 + 598 - 3 + 027	1.20	0.99	1/2	0.0066		
D-25/4-a	0 + 000 - 1 + 443	2.50	2.50	1/2	0.0022	0 + 000 - 0 + 760	1.20	0.89	1/2	0.0025	0 + 000 - 0 + 700	2.70
						0 + 760 - 1 + 124	1.20	0.95	1/2	0.0014	0 + 700 - 1 + 100	3.00
						1 + 124 - 1 + 861	1.20	0.86	1/2	0.0036	1 + 100 - 1 + 850	3.00
D-25/4-b	0 + 000 - 1 + 200	2.50	2.50	1/2	0.0028	0 + 000 - 1 + 103	1.20	0.88	1/2	0.0022	0 + 000 - 1 + 100	3.00
						1 + 103 - 1 + 661	1.20	0.81	1/2	0.0050	1 + 100 - 1 + 651	2.30
D-25/4-c	0 + 000 - 1 + 070	2.50	2.50	1/2	0.0043	0 + 000 - 0 + 370	1.20	0.83	1/2	0.0025	0 + 000 - 0 + 350	3.50
						0 + 370 - 1 + 057	1.20	0.79	1/2	0.0040	0 + 350 - 1 + 050	2.50
						1 + 057 -1 + 537	1.20	0.75	1/2	0.0070	1 + 050 - 1 + 534	3.00
D-25/4-d	0 + 000 - 0 + 930	2.50	2.50	1/2	0.0034	0 + 000 - 0 + 700	1.20	0.88	1/2	0.0015	0 + 000 - 0 + 700	3.50
						0 + 700 - 1 + 385	1.20	0.75	1/2	0.0066	0 + 700 - 1 + 375	3.50
D-25/4-e	0 + 000 - 0 + 770	2.50	2.50	1/2	0.0019	0 + 000 - 0 + 360	1.20	0.83	1/2	0.0016	0 + 000 - 0 + 350	3.00
						0 + 360 - 0 + 777	1.20	0.78	1/2	0.0030	0 + 350 - 0 + 766	4.00
D-25/5						0 + 000 - 0 + 719	1.20	0.73	1/2	0.0026	0 + 000 - 0 + 753	2.50
D-25/6						0 + 000 - 1 + 500	1.20	0.85	1/2	0.0007	0 + 000 - 1 + 500	5.00
						1 + 500 - 2 + 053	1.20	0.77	1/2	0.0017	1 + 500 - 2 + 039	3.50
Total	31 + 155					29 + 367					29 + 309	

D-25/3 canal recommended being 4 + 066 km in drainage report is projected as 2 + 543 km, since it is located in the residential area. Even though D-25/5 canal was not recommended in the report, it was projected as 0 + 719 km in project phase. Since D-25/3-d and D-25/3-e canals recommended in the report remain in the residential areas. They were not projected. D-25/1-d canal recommended in the drainage canal is combined with D-25/1 and projected as 2 + 304 km. Canal depths and base widths were significantly decreased in the project phase. Major changes were made in the canal slope in the project according to the report. Total projected canal length is 29 + 367 km. After implementation of the project, canal number and total canal lengths were determined as 18 and 29 + 309 km. respectively. When examined drainage canals of D-25 group are evaluated in terms of their states in the application. It was determined that there are no significant differences in terms of the number of canals or lengths; canals are opened in the application 3 times wider than the values in the project and values in the application are exceeded the values in the drainage report, main canal slope in the project is approximately 1/2 less than the report. As a result of the application, canal length per unit research area is determined as 21.5 m/ha. Art structure identified in the work area after the application is 55 in total containing 5 bridges, 2 passages, 1 galleria and 47 culverts.

3.2. Dimensions of the Sediment Problem

Canal black elevations and sediment heights in D-25 drainage group canals located in the research area were calculated with the leveling works performed during September-October 2010 (before the canals were cleaned). in October 2011 and in 2010-2011 (after all canals were cleaned) and before cleaning of the canals in September 2012. In the study, the results have been shown in Table 3. Classifications of the canals depending on sediment cleaning intervals are presented in Table 4. As a result of the leveling works performed in research area, the study shows that average 0.38 m (at least 0.00m. up to 0.86m) sediment was accumulated in the drainage canals from the years of 2008-2009, when the drainage project was implemented to the fall season of 2010. However, sediment levels in D-25/4, D-25/4-b and D-25/5 canals were calculated as 0.04. 0.10 and 0.00 m. A significant portion of the calculated sediment amount in 2010 was cleaned in the same year. For the research area, total of 35,552 m^3 sediment cleaning was programmed in 2010-2011. 35,002 m^3 sediment cleaning was performed. Average 8817 m^3/ha sediment before cleaning in 2010 (minimum 0.365 m^3/ha, maximum 33080 m^3/ha), 5940 m^3/ha sediment after cleaning (min. 0.24 m^3/ha, max. 20163 m^3/ha) were calculated. In USA, the North-side Canal Company keeps records of the quality of sediment mechanically removed from drainage canals each year. Some canals are cleaned every year, but others are cleaned only after sediment accumulation seriously restricts flow. This may be every second or third year or longer. An estimate of the quality of sediment deposited canals each year was obtained by averaging the quantities removed mechanically over the past 5 years. This quantity was 295,000 metric tons. This is equivalent to 4.5 metric tons/ha [7] . It was observed after the cleaning that re-sedimentation is high. Average sediment height formed in fall season in 2011 after the cleaning in 2010-2011 in all canals is 0.28 m (min. 0.00 - max. 0.74 m), it was formed in D-25/1-c, D-25/4, D-25/4-b and D-25/5 as 0.00, 0.07, 0.03 and 0.12 m, respectively. Then, even though cleaning wasn't performed in the canals, sediment height in fall season in 2012 was calculated to increase a little to reach 0.35 m (min 0.15, max 0.79 m). In this case, D-25 group drainage canals are classified in two main groups, one needs to be cleaned every year and second needs to be cleaned every 2 years (Table 4). As it can be seen by examining the table that after sediment cleaning performed in 2010, since accumulation of sediment is in the level which should be cleaned in 13 canals having total length of 22 + 880 km (77.3%). As a result of leveling made in 2011, cleaning period was determined as 1 year. It was determined that there is no need for sediment cleaning in 4 canals with total length of 6 + 016 km (20.6%). As a result of the leveling operation in 2012, it was determined that sediment in cleaning level in all canals was accumulated (average 0.35 m). In this case, in 4

Table 3. Cleaning range, sediment height, and elevation difference of structure of D-25 drainage canal group.

Cleaning range	No	Canal name	Sediment height (m)	Canal length (km)	Structure: From downstream (km)	Structure: Elevation difference (m)	Bottom slope
		D-25/2	0.25	1 + 975	0 + 007	−0.25	0.0009
					1 + 211	−0.13	
	1	Total	0.25		1 + 213	−0.38	0.0009
		Average	0.25		0 + 309	−0.19	0.0009
	2	D-25/1	0.32	2 + 304	0 + 459	0.23	0.0093
					1 + 051	−0.35	
	3	D-25/1-a	0.61	0 + 745	0 + 110	0.20	0.0012
	4	D-25/1-b	0.35	0 + 708	0 + 005	0.23	0.00343
					0 + 281	0.53	
	5	D-25/3	0.71	2 + 543	1 + 352	0.74	0.00409
					2 + 299	0.65	
					0 + 010	0.75	
	6	D-25/3-b	0.46	1 + 040	0 + 457	0.24	0.00234
					0 + 810	0.35	
	7	D-25/3-c	0.16	1 + 097	0 + 005	0.23	0.00379
					0 + 511	0.11	
	8	D-25/4-a	0.29	1 + 850	1 + 143	0.46	0.00271
	9	D-25/4-c	0.47	1 + 534	0 + 370	1.15	0.00457
First year					1 + 312	0.15	
	10	D-25/4-d	0.27	1 + 375	0 + 070	0.43	0.004
					0 + 550	0.77	
	11	D-25/4-e	0.15	0 + 766	0 + 360	0.23	0.00238
					0 + 580	0.18	
	12	D-25/5	0.21	2 + 039	0 + 010	0.19	0.00101
					0 + 210	0.10	
		Total	3.83	16 + 001	10 + 960	7.97	0.0345
		Average	0.34		0 + 548	0.39	0.0031
					0 + 483	0.91	
					1 + 073	1.03	
					2 + 182	−0.14	
					1 + 684	0.72	
	13	D-25	0.74	4 + 692	3 + 082	0.16	0.00036
					3 + 445	0.20	
					3 + 635	0.22	
					4 + 055	0.17	
					4 + 533	0.24	
		Total	0.74	20693	21 + 990	3.51	
		Average	0.74		2 + 740	0.39	0.00036
	14	D-25/1-a	0.24	0 + 585			0.0040
	15	D-25/5	0.15	0 + 753			0.0026
		Total	0.39	1 + 338			0.0066
		Average	0.19				0.0033
					0 + 359	0.26	
					0 + 787	0.20	
					1 + 048	0.38	
Second year					1 + 170	−0.80	
					1 + 733	0.21	
	16	D-25/4	0.20	3 + 027	1 + 855	−0.20	0.00612
					2 + 020	−0.18	
					2 + 193	−0.13	
					2 + 515	−0.25	
					2 + 550	−0.21	
	17	D-25/4-b	0.37	1 + 652	0 + 603	0.11	0.00312
Second year					1 + 101	−0.30	
		Total	0.57	4 + 673	17 + 985	−0.91	0.00924
		Average	0.20		1 + 498	−0.27	0.0046

Table 4. Classification to the project status and cleaning range of D-25 drainage canal group.

Projected canals	Un-projected canals	Canals cleaned in 2010			Unknown cleaning range
		Cleaned in 2011		Cleaned in 2012	
		D-25	D-25/3-c	D-25/1-c	D-25/3-a
		D-25/1	D-25/4-a	D-25/4	
		D-25/1-a	D-25/4-c	D-25/4-b	
		D-25/1-b	D-25-4/d	D-25/5	
		D-25/2	D-25/4-e		
		D-25/3	D-25-6		
		D-25/3-b			
18		13		4	1
100%		72.20%		22.20%	5.60%
29 + 309 km		22 + 668 km		6 + 016 km	0 + 625 km
100%		77.30%		20.60%	2.10%

canals which don't need to be cleaned at the end of the first year (total 6 + 026 km length, 20.6%), cleaning interval was determined as 2 years.

3.3. Art Structures and Elevation Errors

Differences determined between the application elevation of art structures in the works made in 2010 and project elevations are shown in Table 3. Classification made according to the differences of art structure, types and project elevations existing in the research area are given in Table 5. When Table 5 is investigated, in 45 out of 55 (81.8%) built art structures, 0.10 m and more differences were observed according to the project elevation. When the distribution of the incorrect art structures were observed as 83.3% numerically, and 93.3% in length of canals, 45 out of total 55 art structures are defective according to art structure number, the rate is 81.8%. Average 0.70 m on project elevation (+) (min. 0.14 - max. 1.61 m), average 0.25 m under project elevation (−) (min −0.11 - max −0.80 m) elevation errors have occurred. Elevation difference in 73.3% (33 of them) of defective art structures is less than 0.10 m, average is 0.40 m (min. 0.10 - max. 1.03 m), less than −0.10 min 26.7% (12 of them) and average is −0.25 m (min. −0.11 - max −0.80). These results show that defective art structures are in almost all canals in D-25 drainage group and their elevations on the project elevation structures are common. Investigation made in canal profiles show that some art structures predicted before are built during the applications. Comparing to 72 art structures in the profiles, 55 art structures were determined.

3.4. Factors Effecting Sediment Accumulation and Factor Groups

Classification made according to the factors causing sediment accumulation and cleaning interval related to D-25 drainage group canals in research area is given in table 6. Sediment accumulation reasons in 13 canals (22 + 668 km, 77.3%) with 1 year cleaning time are incorrect sediment cleaning in 1 canal (1 + 975 km, 6.70%), incorrect sediment cleaning in 11 canals (16 + 001 km, 54.6%) with art structures built in average 0.39 m height (min. 0.11 - max 1.15 m) from the project elevations, incorrect sludge cleanings in 1 canal (4 + 692 km 16.0%), lower slope and art structures being higher than project elevations average of 0.45 m (min. 0.16 - max. 1.03 m).

Prolongation of cleaning interval in 4 canals (6 + 016 km, 20.6%) with 2 years sediment cleaning interval; high canal slopes in 2 canals (1 + 338 km, 4.6%) and cavities as a result of excessive sediment cleaning in 2010, art structure application elevations being higher than average 0.23 m (min. 0.11 - max. 0.38 m) from the project elevations in 2 canals (4 + 678 km, 16%), high slopes and beside the art structures on the project elevations are reasoned to the cavities as a result of art structures under the project elevation. In the observations made in D-25 group, the formation of glide and as a result of this deterioration was determined in slopes of D-25/1-b

Table 5. Classification to the types, number, and elevation faulty of structure in the D-25 drainage canal group.

Drainage group	Difference to the project elevation (m)	Types and number of structures														
		Total			Bridge			Culvert			Pipe outlet			Other		
		Total	Error	%	Total	Error	%	Total	Error	%	Total	Error	%	Total	Error	%
	−0.10 - −0.49		11	24.5					11	29.8						
	<−0.50		1	2.2								1	50			
	(−) elevation		12	26.7					11	29.8		1	50			
D-25	+0.10 - +0.49		23	51.1		2	40		20	54					1	100
	>+0.50		10	22.5		3	60		6	16.2		1	50			
	(+) elevation		33	73.3		5	100		26	70.2		1	50		1	100
	Total	55	45	81.8	5	5	100	47	37	73.7	2	2	100	1	1	100

Table 6. Classification to the factors affecting the sediment accumulation and cleaning range of D-25 drainage canal group.

Factors affecting the sediment accumulation and cleaning range							
Canals cleaned every year					Canals cleaned biennially		Other
Error sediment cleaning	Error sediment cleaning + Structure error elevation			Error sediment cleaning + Insufficient slope + Structure error elevation	Error sediment cleaning + Scour	Structure error elevation + Scour	
D-25/2	D-25/1	D-25/1-a	D-25/1-b	D-25	D25/1-c	D-25/4	D-25/3-a
	D-25/3	D-25/3-b	D-25/3-c		D-25/5	D-25/4-b	
	D-25/4-a	D-25/4-c	D-25/4-d				
	D-25/4-e	D-25/6					
1		11		1	2	2	1
5.60%		61.00%		5.60%	11.10%	11.10%	5.60%
		72.20%			22.20%		5.60%
1 + 975		16 + 001		4 + 692	1 + 338	4 + 678	0 + 625
6.70%		54.60%		16.00%	4.60%	16.00%	2.10%
		77.30%			20.60%		2.10%

D-25/1-c, D-25/4-b and D-25/4-d canals. This situation was seen to be as a factor in increasing sediment accumulation.

4. CONCLUSION AND RECOMMENDATIONS

As a general conclusion, there are significant differences between the drainage report with project and the project with application in terms of canal properties. Therefore, facility information form and project area layout plans mostly do not reflect the situation in the application and cause confusion as a result. In terms of cleaning interval processes (although no special study has been done), no difference was observed among main, secondary and tertiary drainage canals. [8] reported that cleaning periods in drainage canals constructed by DSI in Turkey is 4 - 5 year for main canals, 2 - 3 years for secondary canals and 1 - 2 years for tertiary canals. However, this case shows the necessity of performing the evaluation and studies about cleaning period times. While the factors effecting sediment accumulation show variation depending on the special conditions of the drainage systems, giving a canal cleaning time for each drainage network is very difficult. In drainage canals in research area, elevation checking needs to be done every year. However, this is not followed in practice and canal sediment cleanings are not planned according to elevation control results. When leveling and elevation works are performed in canals, canal or canal segments with sediment deposition can be determined clearly. Thus, as a result, unnecessary machine entrance and utilization to the other sections are prevented and improper cavities as a result of incorrect cleanings in canals do not take place.

Important and common elevation errors in art structures are accepted to be one of the important factors preventing the solution of drainage problems. Art structures built above project elevation causes sediment accumulation depending on canal slope, the distance to canal downstream and the degree of elevation differences. Sediment accumulation in the case when the slope is high and art structures are far from the downstream is greater than the canals where the slope is low and the structure is close to the downstream. Art structures built under project elevation causes cavities and changes in canal properties depending on the sediment cleanings and slopes.

Art structures constructed on drainage canals in research area are necessary to be built on the project elevation. Building art structure over the project elevation should be prevented and built ones should be arranged according to project elevation. Studies shows that under some specific conditions, art structures can be built under the project elevation of base elevation (less than 0.50 m), this case might be useful in terms of extending sediment cleaning interval.

Determination of factors effecting sediment accumulation for each canal in research area was tried in the study. These factors are thought to be faulty sediment cleaning, lack of downstream conditions, lower slope (less than 0.0004), elevation errors of art structures and the effect of those together. Lack of downstream, one of these factors, is valid for D-25 canal, which is a main drainage canal of D-25 main drainage group. However, given that each canal has a certain drainage area, almost in every secondary and tertiary canals, downstream conditions caused from the lack of sediment cleaning arise. Therefore, it is necessary to give the priority to main drainage canals in sediment cleanings and performing the cleaning from downstream to upstream.

When the relation between the slope and sediment accumulation is analyzed, it is seen that slopes less than 0.0004 increases sediment accumulation, while slopes bigger than 0.001 in general causes cavities unless structural measures are taken. Since the slope is a factor which is determined by topography, it might not be possible to prevent lower slopes. However, in the possible locations, applying the slope as 0.0004 - 0.001 is useful for decreasing sediment accumulation. In addition, it is necessary to give priority to the canals with lower slopes in sediment cleaning. Due to the topographical structure, [9] reported that various art structures should be projected to not to exceed 6 kg/m^2 for maximum drag force or to decrease slopes in the projects in locations with higher slopes.

In the case where canal cleaning is conducted according to leveling studies, it is possible to prevent remaining of sediments or excessive cavities. However, attention should be given to creating possibilities for leveling studies and then to the training of the machine operator.

Cengiz Koç, Ebru Yilmaz Art structure elevation errors should definitely be avoided during the project implementation. To accomplish this, field inspections should be increased during project implementation and during the final acceptance. Art structure elevation errors are the most important factor which is the most effective in sediment accumulation and prevents the expected benefit from the sediment cleaning. In particular, errors on the project elevation should definitely be avoided. The effect of these structures, especially in the canals with low slopes, increases more.

REFERENCES

1. Van Schilfgaarde, J. (1971) Drainage Yesterday, Today, and Tomorrow. Proceedings of the ASAE National Drainage Symposium, American Society of Agricultural Engineers, St. Joseph.
2. Shinnohammadi, A., Wenberg, R.D., Ritter, W.E. and Wright, E.S. (1995) Effect of Agricultural Drainage on Water Quality in Mid-Atlantic States. Journal of Irrigation and Drainage Engineering, 121, 302-306.
3. Mister, D. (2006) Public Drainage in Maryland. In: Needleman, B.A. and Wills, S.A., Eds., Improved Management of Agricultural Drainage Ditches for Water Quality Protection: Field Tour Guide, University of Maryland, College Park.
4. State Hydraulic Works (DSI) (2007) Irrigation and Drainage Facilities Operated and Constructed by DSI. Ministry of Environment and Forestry, General Directorate of State Hydraulic Works (DSI), Department of Operation and Maintenance, Ankara.
5. State Hydraulic Works (DSI) (2008) Drainage Report of Nazilli Irrigation Area. Ministry of Environment and Forestry, General Directorate of State Hydraulic Works (DSI), Department of Operation and Maintenance, Ankara.
6. State Hydraulic Works (DSI) (2005) Records Related to Irrigation Networks in Department of Regional 21. Ministry of Environment and Forestry, Department of Regional 21, State Hydraulic Works (DSI), Department of Operation and Maintenance, Aydın.
7. Brown, M.J., Carter, D.L. and Bondurant, J.A. (1974) Sediment in Irrigation and Drainage Waters and Sediment Inputs and Outputs for Two Large Tracts in Southern İdaho. Journal of Environmental Quality, 3, 347-351.
8. Bekisoglu, Ş. (1995) Solution Methods and Problems Occurring in Irrigation and Drainage Canals. Ministry of Environment and Forestry, General Directorate of State Hydraulic Works (DSI), Department of Operation and Maintenance, Ankara.
9. Kızılkaya, T. (1988) Irrigation and Drainage. Ministry of Environment and Forestry, General Directorate of State Hydraulic Works (DSI), Department of Operation and Maintenance, Ankara.

CHAPTER 21

Applications of SuDS Techniques in Harvesting Stormwater for Landscape Irrigation Purposes: Issues and Considerations

Andrew B. Shuttleworth[1], Ernest O. Nnadi[2*], Fredrick U. Mbanaso[2], Stephen J. Coupe[3], Joris G.W.F. Voeten[4] and Alan P. Newman[2]

[1] SEL Environmental Ltd, Blackburn, Lancashire, United Kingdom
[2] School of Energy, Construction and Environment, Coventry University, Coventry, United Kingdom
[3] Centre for Agroecology, Water and Resilience, Coventry University, Coventry, United Kingdom
[4] Permavoid Ltd, Amsterdam, Netherlands

ABSTRACT

While urbanization and increasing population has put much pressure on natural drainage channels and resulted in increase in flooding, there is increased pressure on available water resources due to climate change, reduction in frequency of rainfall events and drought. The emergence of a sustainable drainage system (SuDS), also known as best management practice (BMP) and low impact development (LID), has changed the management strategy of drainage from conventional to sustainable. SuDS techniques seek to deliver the three cardinal paradigms of sustainable drainage: quantity, quality and amenity and as such, they can offer an additional benefit for applications such as landscape irrigation. Most SuDS techniques have the potential for water storage with minimal or no modifications required. This chapter, while covering the capabilities of SuDS systems, explores SuDS devices such as pervious pavements equipped with excess storage capacity, cisterns and tanks harvesting roofwater, infiltration systems aimed at supporting the growth of urban plants and green roofs with the potential to store water in order to maintain water demanding planting scheme even during dry periods. It also covers systems where SuDS is the main driver to device installation and address issues and considerations surrounding applications of such systems in water harvesting for irrigation.

Keywords: Irrigation, Drainage, SuDS, Stormwater, rain harvesting

1. INTRODUCTION

Urbanization can initiate undesirable local modifications to the water cycle. In particular, the effects of the spread of impermeable area both within and at the margins of towns and cities have highlighted the need to modify the way that society deals with stormwater. The problems of surface flooding in cities and overloading of foul sewers (when combined systems are in place) are the most headline-grabbing aspects of increasing impermeability but reduction in aquifer recharge can also be important when water demand rises due to increase in local population.

In many western societies, particularly in the United Kingdom, there is a preference for housing schemes with gardens (including, often, very water demanding planting regimes). This increases the demands for water well beyond traditional domestic use. Alongside these domestic developments, hotels, hospitals, office blocks, light industrial units and retail developments often try to show a supposedly green face to the world in the form of extensive landscaped areas around their equally extensive impermeable parking surfaces. We are thus faced with the dual problem of a demand for water for landscaping accompanied by rapid and wasteful run off from impervious parking areas. This can cause flooding problems in response to short heavy rain events. The problem is particularly important in summer months when the demand for watering domestic, commercial and municipal planted areas is at its highest, at the same time as the supply from surface sources is at a minimum.

The term SuDS was originally used in the United Kingdom as an acronym for 'sustainable urban drainage systems' but this term has more recently lost favour and is commonly replaced by SuDS (sustainable drainage systems). In the USA, the terms low impact development and best management practices (LID and BMP) cover the same approach to drainage and in Australia the term 'water-sensitive urban design' is favoured. While all of the above terms are used, sometimes interchangeably, to indicate a holistic approach to stormwater management, they are all dominated in their philosophy by the concept of stormwater source control. This entails controlling both the quality and quantity of stormwater as close to its site of deposition as possible. Commonly, parts of the United Kingdom can be under drought orders (which impose summer hosepipe bans which limit landscape irrigation) and still suffer from localized flooding during short-term storms. These occurrences are becoming an increasing feature in some parts of the United Kingdom and this adds extra incentive to retrofitting of SuDS systems that can also make a contribution to the reduction of water use. The principles of SuDS are often presented as follows:

1. Storing runoff so that it can be released slowly or used beneficially (attenuation).
2. Allowing water to soak into the ground so as to mimic the processes on an undeveloped surface (infiltration).
3. Where necessary to allow the water to move doing so at the surface and at a controlled velocity.
4. Using the processes of collection, storage and transportation to facilitate the removal of pollutants by sedimentation, precipitation, adsorption (in its widest sense) and degradation.

Most SuDS devices have the potential to be used for water storage with minimal or no modifications required. **Table 1** summarises the irrigation water resource potential of a range of hard SuDS devices that have been used for water harvesting, whereas **Table 2** outlines a number of green SuDS devices that themselves place a potential demand on water resources but which can be modified to harvest water for self-irrigation and sometimes for other uses.

TABLE 1. Irrigation water resource potential of a range of hard SuDS devices.

Device	Quantity issues	Quality issues
Underground and above ground barrels, tanks and cisterns harvesting roof water only	The sizing of the system needs to be done carefully and the usage of the water be well established if both stormwater attenuation and water harvesting are to be achieved in the same system. With modern computer control the sizing can be reduced by actively draining tanks in response to predicted rain events.	The quality of water will be affected mainly by atmospheric fallout and the nature of materials used in roof construction. While this is not really considered an issue new UK guidance requires that even roof water should have some treatment before discharge to a watercourse
Underground and above ground storage harvesting both surface runoff and roof water	The additional resource will require a greater storage volume for it to be utilisable. If there is attenuation available upstream of the storage tanks this can be minimized as a problem	Surface water will generally be of much poorer quality than roof water alone (see above) but provided the system is correctly designed to retain day to day pollutant releases it will generally only be an issue if major pollutants releases overcome the pollution attenuation mechanisms. Salt applications in temperate zones can be a problem.
Pervious pavements and similar without off line additional storage	Unless taking runoff from impervious surfaces too the total inputs are limited to the water falling on the surface. Unless designed with extra storage capacity even an attenuation based system will have limited capacity to store water for more than short periods of time	Input of potentially harmful organisms from faecal contamination and chemical pollutants from atmospheric fallout. If exposed to traffic there will also be day to day input of automotive based pollutants and if not provided with upstream protection the water can be subject to contamination from the very rare losses of engine oil and fuel. In temperate areas de-icing salts can be an issue.
Pervious pavements with off- line additional storage	By directing water into off line underground or above ground tanks the storage required for attenuation or to overcome limited rates of infiltration will not be compromised by the need to recover water	

TABLE 2. Green SuDS devices that can be modified to harvest water for self-irrigation and other uses.

Device	Quantity issues	Quality issues
Green roofs with additional off line storage Green roofs with integrated storage.	Both need careful sizing taking into account predicted available rainfall and demand placed by both growing plants and any additional off-roof demands. Storage needs to take into account sufficient temporary storage to provide stormwater source control when substrate is close to saturated. Provision may need to be made for rapid dumping of stored water when large storms are predicted in a 'reservoir full' situation.	If used for self-irrigation of the roof or for additional off-roof irrigation there will be little problem but if, as in some installations, green roof drainage water is pooled with other water and used for such as toilet flushing the colour can become a source of user complaint. If fertiliser is used on the roof, then recycling the water via the roof will retain the nutrients where they are needed.
Infiltrating urban tree planters with subsidiary methods of storage.	Maintaining sufficient moisture for the trees can be a problem. Proprietary systems which incorporate water storing foams (as used by florists) can offer a solution. Species selection for the variability of available water needs to be done carefully.	There will be day-to-day input of automotive based pollutants and if not provided with upstream protection the water can be subject to contamination from the very rare losses of engine oil and fuel. In temperate areas de-icing salts can be an issue.

This chapter, while covering the capabilities of all SuDS systems, gives particular emphasis to those that are particularly suitable for retrofitting. In this chapter, the greatest attention is given to devices such as pervious pavements equipped with excess storage capacity, cisterns and tanks harvesting roofwater, infiltration systems deliberately aimed at improving the growth of urban plants (including trees) and green roofs, which can offer the potential to store water in excess of the capacity of the substrate so as to maintain a water demanding planting scheme during dry periods. Inevitably, there will be some overlap with harvesting for purposes other than irrigation. Competing use for harvested water includes toilet flushing and washing machine use and with appropriate treatment, many of the types of use that normally demand potable water. We must also recognize that rainwater harvesting systems are commonly provided without involvement of the SuDS philosophy, being designed without any attempt to enhance the control of stormwater (but sometimes doing so by accident). This chapter only covers systems where SuDS is the main driver to device installation.

Harvested rainwater from urban environments (even water directly from roofs) will inevitably contain pollutants that would not be expected if the water was collected from rural upland catchments or extracted from protected aquifers. If water is to be used for irrigation after being in contact with pollutant materials inevitably generated by both the daily activities of urban living and by fallout from industry, a first question must be whether the quality of the water is good enough for purpose. Irrigation water limits vary from place to place and not all organisations include all the possible pollutants (see Ref. [1]). A useful comparative table is provided by Nnadi et al. [2] which presents the irrigation water limits provided by a range of authorities.

2. ROOFWATER HARVESTING: RAIN BARRELS AND CISTERNS

While many SUDS systems have been used to harvest rainwater for a combination of both irrigation and indoor uses (such as toilet flushing and washing machine pre-wash) it is the direct harvesting of roof water (and sometimes water from non-trafficked paved areas) into barrels and cisterns that has seen the greatest uptake as a component of the LID programmes in the USA. The relatively high quality of this source of harvest (although well short of potable water standards in most jurisdictions (see Ref. [3]) that makes this source attractive for internal use as well as for irrigation. The primary differences between a cistern or rain barrel which forms part of a SuDS system and one which is simply a water supply device are the way the container is sized and the way the water is managed. Whether managed manually by an individual householder (possibly backed up by an advice and education program) or through an automated system, unless efforts are made to maintain enough available storage to contain at least a significant proportion of a design storm and unless the stored water is managed either by using the water or by releasing a proportion before the next storm then this is not SuDS at all but purely a water harvesting exercise. Jones and Hunt [4] made the point that the greatest problem with roofwater harvesting as a stormwater control mechanism is that many of the cisterns and barrels remain full all of the time because water is not utilized. Managing the cistern by using the water regularly or releasing during a dry period is necessary if roofwater storage is to be seen as SuDS.

North Carolina State University (NCSU) offers online a very comprehensive review of rainwater harvesting [5] and also one of the best sources of information provided for home owners on water cisterns and their use [4]. Their document usefully attempts to guide the owners in relation to the need for additional ground support for cisterns in gardens, an often overlooked factor. NCSU also provides an informative guide to roof water quality [6]. Many other American universities provide a similar extension literature and the various States of the USA represent such a broad diversity of climates that the guidance from either state regulators or that provided through local university extension programs should be selectable if someone is looking for information that suits their climate elsewhere in the world. That is not to say that authors from outside the USA have not been active in reporting their work in a variety of jurisdictions (although in many cases, the stormwater source control element is missing).

Cisterns can be either above ground or below ground, but the land take and unattractive appearance of reasonably sized water cisterns are often a barrier to adoption. Underground tanks can be expensive and offer significant structural problems if they are to be covered with a trafficked surface or even one with a reasonable dead load. Unlike above ground storage, where irrigation use can often be achieved under gravity flow, they will normally need a pump.

Underground tanks are not normally suitable for occupier retrofit even by the most capable householders. An option that addresses this is the Skeletank®, where the thin polyethylene tanks get their structural strength from internal interlocking polypropylene skeleton. This provides structural strength while maintaining a system that can be carried, if necessary, on the roof of a car and can be lifted easily by two people. They are easily linked together to achieve the size required. The system was originally designed to provide stormwater attenuation and soakaway tanks (used upside down) within individual domestic curtilages. The system has since been adopted for rainwater harvesting and lightweight pump and filter chambers are available as accessories. Apart from the wiring for the pumps, the system is totally suitable for installation by an amateur. **Figure 1** shows an installation in Preston (Lancashire, UK) which provides an interesting example of a householder built retrofit water harvesting SuDS project particularly since the primary aims of harvesting the water was to provide irrigation water to landscaped areas and to support a small pond on site while allowing total disconnection of drainage from the combined sewer. The owners of the property in question were undergoing conversion works which included the installation of a new patio area. The tanks were installed underneath the proposed patio area. The concept of the system begins with the collection of rainwater in a downpipe chamber. On this site, three downpipe chambers were installed, two chambers for a 94 m^2 roof area at the main house and the final chamber positioned to collect water from a 55 m^2area (serving an adjacent outbuilding). These chambers are intended to pre-clean the rainwater and allow silt to deposit within the sump of the chamber. Water passes through a fabric filter on the outlet spigot of the chamber before running into the tanks which were connected together using standard underground drainage pipe to provide a combined reservoir. For this site, sixteen tank units, providing a total storage capacity of 4800 litres, were used. The required number of units was determined by a modelling to fit available storage to the various demands including sufficient temporary storage, with controlled discharge to accommodate stormwater during rain events when the temporary storage is full. The rainwater pump chamber is the final component in the chain and is primarily used to pump rainwater on demand for irrigating flower beds and grassed areas.

Figure 1. Left: Skeletanks® in place and connected together. Right: lightweight pump chamber.

When available in excess stored rainwater is also pumped into a header tank for internal household use, such as, flushing toilets and water feed for washing machine pre-wash. The primary re-use of the harvested rainwater is to irrigate the flower beds and grassed areas surrounding the property and also for supporting the flora and fauna of the pond. Since the installation of the system, both use of mains water for irrigation and for topping up the pond has been minimal. Checks have shown the public health microbiological qualities (including legionella) are not a problem.

3. PERVIOUS AND MACRO-PERVIOUS PAVEMENTS AS A WATER HARVESTING SYSTEM

Pervious pavement systems (PPS) have been around as SuDS elements since the 1970s and are extremely well documented (see Refs. [7–12]). In the United Kingdom, their origin is largely associated with Chris Pratt's design which utilized concrete block pavers provided with infiltration channels as the wearing course [13].

The pervious wearing course surface can also be porous asphalt or porous poured concrete. The wearing course is usually laid on a bedding layer of, typically, 10 mm material and there is often a geotextile as a separation and filtration layer between the bedding layers and either the subbase of aggregate (typically a uniform 50 mm, with a void ratio of about 30%) or on a subbase replacement of plastic load-bearing boxes (a typical void ratio of 90%). If the pavement is to be used for water harvesting, it requires a reliable under-sealing layer which can be a folded or welded polyolefin membrane or, particularly within individual curtilages, can be conveniently provided by one or more of the sub-surface tanks discussed in the example above. These can also be used below pervious pavements to capture a proportion of the stormwater with the number of tanks selected in relation to demand with both unharvested and overflow water disposed of by infiltration or at a controlled rate to a surface water body or a combined sewer.

The most common application of PPS has been parking areas, whether they are extensive parking lots or individual house driveways and courtyards. The efficacy of pervious pavements in intercepting and immobilising or degrading traffic-derived pollutants has been the subject of a large body of research in many countries including England [14, 15], Scotland [16], Spain[17], Australia [18, 19] and the USA [20].

One of the most extensive experimental studies of PPS-derived water quality, with respect to irrigation, from pervious parking surfaces was instigated by Nnadi [21]. This was an indoor simulation experiment on a relatively large scale. The experimental outline was to apply oil drops (simulating oil dripping onto pavements from parked vehicles) and street dust to pervious pavement system models built into 1200 mm × 1000 mm × 600 mm heavy-duty polypropylene boxes (used for heavy industrial storage). Models were created with a selection of wearing courses including several cross-sections of new porous asphalt, a 10-year aged asphalt from a quarry parking surface, porous poured concrete and concrete paving blocks with infiltration channels. Thirteen millimetre simulated rainfall events were applied to the models at 15 mm/hour (52 min rain events) twice monthly. These were applied a day before and a day after the monthly application of used motor oil and dust. The effluents were captured and analysed for total suspended solids (TSS), metals and total petroleum hydrocarbons (TPH). The results were considered in relation to water quality as applied to irrigation and other forms of re-use as detailed in Nnadi et al. [2]. The results indicated that the porous asphalt and porous concrete are as good as the block paved permeable pavements in the treatment of stormwater pollution with all surfaces delivering water quality that is suitable for irrigation and clearly demonstrated the capability of the porous asphalt and concrete pavement system to trap hydrocarbons within the system.

While the use of pervious pavements for control of day-to-day automotive-derived pollutants is well established, it must be remembered that the traditional PPS has shown a poor pollution retention performance when challenged by major hydrocarbon releases [22]. Modification of the permeable pavement to incorporate a gravity separator just below the laying course has been reported for both aggregate-based [15] and plastic box-based pavements [23] showing a capability of retaining sufficient oil should every car in a parking lot simultaneously lose the content of their sump.

The first pot trials to investigate the acceptability of water quality from pervious pavements for irrigation originated at Coventry University as a PhD programme by one of the authors [21]. The source of water was a series of laboratory models. The models used were based on the modified designs by Newman [24] which incorporates a gravity separation system to ensure that free oil can never escape from the system and were artificially fertilized by adding Osmocote® slow release fertilizer and brushing it into the infiltration channels. This was to enhance the biodegradation rate of the hydrocarbons. The pot trial investigation showed that where the effluent water could be used for irrigation the excess nutrients provided distinct advantages to the growing plants [25]. In all cases, the plants irrigated with effluent grew much better than those irrigated with tap water. In effect, the system was being used for fertigation.

There has also been more recent work using a live car park, this case based on a highly modified design called a macro-pervious pavement system (MPPS) (**Figure 2**). In 2011, an investigation was initiated [26] on an MPPS installed as a prison car park in Scotland.

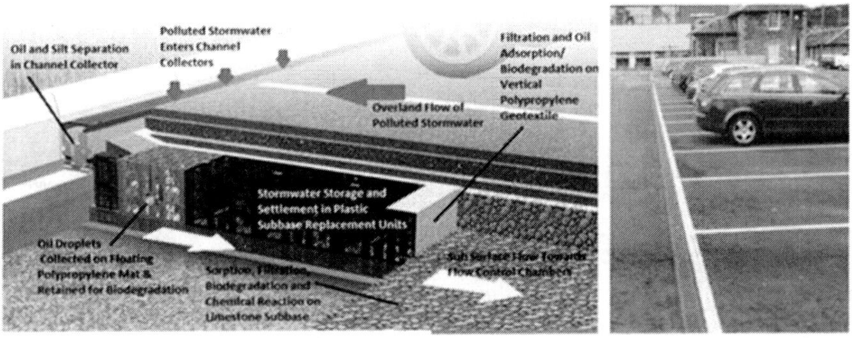

Figure 2. The macro-pervious pavement system as installed at Perth Prison, Scotland. The line of 'Channel Drain' is actually a series of oil and silt separating infiltration points, which direct stormwater into the sub surface storage and attenuation layer.

The parking lot consists of two major sub-catchments of around 1350 m^2 and a minor one of 300 m^2(which was not included in the irrigation water study). The majority of the surface of the parking lot consists of traditional asphalt with stormwater entering the subbase/storage and attenuation layer of crushed limestone through miniaturized, linear, gravity separator units (which look like normal channel drains from the surface) through a chamber containing a floating mat of oil-sorbing textile. It then flows into the subbase, which drains towards separate flow control chambers (with orifice plate flow control) in each of the sub-catchments. These provided convenient sampling points and in-line storage chambers which would be available for landscape irrigation. Samples were analysed for a wide range of determinants.

For the heavy metals, all results were below the most stringent of irrigation water limits. For suspended solids, all measurements were below 20 mg/l, a concentration below which blockage of drip irrigation systems is not a problem. For TPH, the irrigation limits that were adopted are dependent on the solubility of the various carbon chain fractions (except for C_9–C_{14} where a limit of 1.8 mg/l applies). Since no measurements of TPH ever exceeded 1.8 mg/l and in the absence of any observed free product (indicating that solubility was never exceeded for any fraction), it can be concluded that hydrocarbons should have no detrimental effect if this water was harvested and used for irrigation.

BTEX concentrations were all below the 1 μg/l limits of detection for these compounds. The irrigation water limits for benzene and substituted benzenes proposed by the New Zealand Government [27] range from 800 μg/l for benzene to as high as 39 mg/l for toluene. MTBE was also invariably below its detection limit (10 μg/l). Irrigation water standards for MTBE could be found in the literature but since the State of Florida mandates a limit of 50 μg/l in drinking water, this value could be seen as a very conservative irrigation water limit. Boron was always below the 40 μg/l limit of detection and thus well below the 500 μg/l soil pore water limit for the most sensitive plants [28]. Examining the major cation concentrations, it was shown that calcium, magnesium and potassium were low compared to any irrigation water limits and most authors would expect that at least magnesium and potassium would need to be supplemented for optimum plant growth. For most of the year, neither the sodium (**Figure 6**) or sodium absorption ratio (SAR) values were a concern at this site, but in the winter months, following the application of de-icing salt (which will vary from year to year) the values of both these parameters show the water to be unsuitable for irrigation with sodium values in January up to 1200 mg/l. By March, the sodium concentrations were below 100 mg/l but above the 50 mg/l irrigation water limit until May. It was concluded that if water is to be stored for irrigation purposes it would be necessary to divert the meltwater (and the rainwater falling onto a salt contaminated pavement) away from the storage tanks during the winter months after filling them as much as possible during October and November and then topping them up in April and May before the need for irrigation is established in the summer. If the worst of the salt contaminated water can be diverted, the direct effect on plants will be minimized. Another issue from this particular site was the pH that was higher than optimal, but this was due to the limestone used in the subbase and would have been much lower if a granite subbase had been chosen. Adjustment of pH with addition of calcium sulphate was proposed as a potential solution, but species selection for high pH tolerance would be an alternative.

Pot trial experiments were carried out using both ryegrass and tomato plants (using the experimental protocol previously proposed by Nnadi [21] irrigated with effluent collected from this pavement in September with initial results being reported to the SuDSnet conference in 2015 [29] and continued experiments (and tissue analysis) have shown that at this site, the water is perfectly suitable for irrigation, if harvested before the salt application. However, the data represents only relatively short-term use of the water. Hence, further work in this area is required and this provides a good research opportunity for anyone interested in this area of study.

One problem with block paving surfaced PPS is that they can become a habitat for weeds and chemical methods of weed control still dominate as the most preferred in the United Kingdom [30] and most European cities [31, 32]. Although several herbicides are in use, glyphosate-containing herbicides (GCH) are by far the most widely used herbicide for weed control on hard surfaces [32–35]. Recent work on the impact of GCH on pollution attenuation and biodegradation in a PPS indicated that GCH is not retained in the PPS structure and is subject to rapid wash through in response to water movement thus increasing the potential risk of reaching receiving environments [36]. Hence, not only does one need to consider the herbicide itself, but the breakdown of other pollutants such as hydrocarbons within the PPS can be potentially affected [35] with potential impact on the quality of irrigation water. In a recent study to determine the suitability of stormwater harvested from PPS for reuse purposes in conditions where GCH was applied as part of PPS maintenance procedure, Mbanaso et al. [37] observed that effluent from the test models including those dosed with high GCH concentration of 7200 mg/l do not pose infiltration or salinity problems when used for irrigation. They, however, indicated that high dosage of the herbicide could lead to an elevated electrical conductivity of the recycled water. Hence, if PPS-derived water is used for irrigation; chemical methods of weed control should be avoided until the dynamics of breakdown of the particular herbicides are better understood. Further work needs to be done on establishing the lifetime of such compounds in both the PPS and the stored irrigation water.

3.1. Green Roofs

As well as the need to control rapid water runoff in urban areas, there are several other problems to which SuDS can make a contribution. The urban heat island effect [38, 39] and reduction in green space with associated loss of both amenity and biodiversity [40] are important issues. Much of this is being exacerbated by climate change. A contribution to the mitigation of such problems is to apply green SuDS techniques and foremost amongst these is the attempt to utilise the roof spaces of buildings to create green roofs. Green roofs can be either extensive, usually planted with a mat of relatively drought-resistant species such as sedum or intensive [41], or being more 'garden like' where the aim goes beyond the immediate environmental benefits with aesthetic and social aims being contributed to by a more varied and attractive planting scheme with the potential to contribute more to biodiversity.

The potential of green roofs to contribute to the mitigation of rainwater runoff is well established, see Refs. [42–45]. What is less often stressed, however, are the problems in certain climates associated with the fact that after a heavy rainstorm, the water in the saturated substrate can take a considerable time to drain or evaporate and a subsequent storm will be offered significantly reduced attenuation.

The capability of green roofs to contribute to the cooling of buildings and combating the heat island effect has been widely reported [45–50]. Green roofs can also play a role in improving biodiversity within urban areas [40, 51] and, the aesthetic value of a green roof can often be a dominant factor in its adoption [52] even though aesthetic planting schemes can come into conflict with biodiversity aims [53].

Particularly where the green roof is intended to be accessible and provide, in part, the function, for example, of an urban park, it is important that the provision of water, to maintain adequate growing conditions (for a wide range of plants), during dry periods is recognised. This helps to enhance the amenity value. One of the factors is the maintenance of adequate soil moisture content. The limited load-bearing capacity (LBC) of the roofs of (existing) buildings often dictates the amount of water that can actually be safely retained on the roof itself. Designing traditional intensive green roofs to satisfy the load-bearing capacity of a roof thus involves finding a balance between storing as much rainwater as possible, maintaining conditions for plant growth and respecting the LBC of the construction [54, 55]. Adding more substrate typically achieves a water-stored-to-weight ratio (WSWR) of just 0.2 l of water per added kg of soil for loamy sand, or 0.4 l/kg for a typical, specialised, extensive green roof substrate [56].

An alternative to storing the water solely within the substrate is to drain the water into a cistern to allow the water to be used to irrigate the roof during dry periods and, if correctly sized, can offer temporary storage during storms occurring when the substrate is close to saturation. A green roof with a cistern for reuse was the subject of a modelling study by Hardin et al. [57]. Their system included a green roof with its drainage system connected to a cistern which in turn supplied irrigation water to the roof via a pump. A supplemental water source is also connected to the cistern to provide water should there not be sufficient water to perform the irrigation event. It was proposed that the irrigation should be managed via a controller, similar to what is widely used for home lawn irrigation, which only irrigates on the prescribed times unless sufficient rain has fallen within 24 h of the intended irrigation event.

Building rainwater cisterns inside or adjacent to the building and using pumps and irrigation systems is thus an option which will provide both runoff attenuation and on-site rainwater reuse. However, the running costs, capital costs and building space/land take combined with the need for maintenance and the propensity for active systems to break down are factors which would militate against their choice. There is also a need to consider the trade-off between energy used to pump the water from ground or basement level and the loss of space if the cistern is maintained in one of the higher floors of the building.

An alternative to using a separate cistern is to maintain 'ponding elements' under the substrate. Green roofs.com [58] presents a useful summary of some of the systems available, some of which are dependent on a slightly sloping roof, one which incorporates active pumped elements and a system that stores up to 40 mm of water in a plastic drainage layer equipped with an overflow device (although little extra capacity is available for temporary stormwater storage during rain events). However, without a separate cistern, the volume available for storage will be relatively small unless special steps are taken. One such approach has been developed in the Netherlands [56]. This modified green roof stores the water directly under the substrate within modified load bearing plastic void formers originally developed for pervious pavement applications. The water is not required to be pumped up to the substrate, this being achieved by capillary actions through 'capillary cones' inserted into the load-bearing vertical struts [59].

An excellent example of an application of this type of system is adjacent to the elevated railway station at Orlyplein in Amsterdam [56], shown in **Figure 3**. This is a remarkable installation which has transform the former rooftop bus station (vacant for some time after it being moved to ground level) into a popular public park area with a resulting increase in economic development into the form of many additional rooms constructed in adjacent hotels.

Figure 3. The roof park built on the former Orlyplein Bus Station *deck*, before, during and after construction.

The previously lonely and threatening exit from the rail station has been transformed into a place that is popular and well-populated whenever the station is open.

The system of capillary irrigation outlined above has also found application in ground level applications. A very recent example of this is the 'Green Stream' in Zuidas an Amsterdam city district in the Netherlands. Zuidas is a very densely build urban area and therefore prone to urban flooding during intense rain events. With the city district being under re-development, aiming to become the major business hub in the Amsterdam Metropolitan region, the city is developing innovative multifunctional designs to improve urban quality, reduce the urban heat-island effect by improved evaporative cooling by plants and increase water retention (and reuse) to prevent urban flooding. The 'Green Stream' is a project in which rainwater from rooftops and adjacent sidewalks is collected in a 2 m wide planting strip that runs along the houses and sidewalk (**Figure 4**). The strip is deeper than its surroundings, making it a natural water collection point in the street design. To get from the sidewalk to the houses, bridges are used to cross the 'green stream'. This planting strip is designed to be dry throughout the year, but is allowed to flood during rain events. Planting species are selected to be able to withstand occasional flooding. Innovative in the design is the 150 mm high water attenuation system 40 cm below the planting, which is fed with water from the roofs. A continuous chain of plastic void forming units, as used in the Orlyplein roof park, is placed in a waterproof liner to create a subsoil water-tank and features the same capillary irrigation system, capable of returning water to the soil when plants are using water without the use of pumps (and thus energy). The improved water availability for plants maintains their evapotranspiration rates at close to the potential evapotranspiration generated by the local weather, improving their urban cooling capacity. Surplus of water can drain freely to ground water level alongside the water drainage and capillary irrigation system. To prevent the Green Stream from overflowing onto the sidewalk, extra emergency overflows are created at the maximum fill level, connected to the conventional sewer.

Figure 4. Left to right: Schematic of the green stream system, under construction and on the day of completion.

Another application of this technology has been on sports surfaces such as football pitches both on rooftops and at ground level. The system provides water to the growing grass turf, while the void space can be made sufficiently deep to satisfy the most stringent stormwater attenuation requirements for new stadium construction. Even if supplemental water has to be used to maintain the playing surface in very hot dry countries, the application of water from below by capillary action is more efficient than spray irrigation from above. Currently, trials are underway in preparation for the soccer world cup in Qatar.

3.2. Irrigation of Urban Trees

Novel below ground water storage options provide a significant opportunity for the reliable delivery of acceptable water to trees in the urban environment. In a non-urban environment and under natural establishment conditions, a mature tree have grown partly due to a favourable water regime that has provided sufficient water reliably over the life of the tree. Historically, urban trees have faced significant challenges that are not faced by non-urban trees which include:

- Establishment in unfavourable environments
- A deficiency in total water volume provision
- Runoff of water from non-permeable surfaces away from tree roots, even where local rainfall totals are adequate
- Low retention rates of applied water, where water evaporates or percolates away from tree roots before uptake
- Insufficient mature and productive soil/substrate for nutrient regeneration and incomplete establishment of beneficial microbial processes (e.g. limited growth of mycorrhizae)
- Tree removal or inappropriate management resulting in damage to trees if land use changes regard existing trees as an obstacle

Thus, water supply considerations provide some of the major challenges faced by urban trees and depending on the tree species, a mature tree with a 500 mm trunk diameter could require 860l of water to be provided for survival and normal function during drought conditions. It is important to acknowledge the local hydrological benefits of urban trees in times of intense and or prolonged rainfall where the drainage system may be at capacity. The positive impacts of trees include rainfall detention, retention and uptake, which may remove significant volumes of water from runoff totals, or delay local peak discharge in comparison with a location without trees [60].

In theory, it would be acceptable to deliberately divert stormwater from a new development to supplement existing urban trees to meet the total water need. This could include disconnection of downpipes into soil and conveyance to areas where trees are located. However, this retrofit option could result in standing water in the case of intense events if infiltration and percolation are not sufficient and a bypassing of the tree if the water is not retained by the soil. Hence, solutions are required that provide the required volume of water, but allow this to be retained close to the roots and provide useful water. In practice, this is necessarily targeted at new developments and is usually provided by tree pits, where a selected soil medium is placed into a structural chamber. Landscaping solutions can incorporate this type of structure into bioretention schemes as shown by the image below of a parking area for a shopping Centre drained by bioretention cells, which often successfully incorporate trees (**Figure 5**).

Figure 5. A bioretention cell with trees and shrubs in Raleigh, North Carolina, USA.

The medium can be an engineered soil or a replacement medium such as absorbent foam-filled structurally resistant void forming boxes. An ultra-absorbent foam that is available commercially as a component in a tree box system was tested independently in a laboratory for its impact on hydraulic conditions and was shown to be a promising solution for runoff management and water absorption [61]. This material also has the potential for water transfer to woody plants.

The use of engineered materials, including allochthonous soils, in providing water to trees has not been proven to fully replicate the functioning of the biological component of mature natural soils, nor have an equivalent of the nutrient recycling and regeneration of soil. However, it has been established that drainage systems that include trees and soil can improve water quality in sustainable drainage applications, including the absorption and retention of heavy metals, nitrogen and phosphorus [62]. The addition of urban pollutants in the form of sediment moved by stormwater is shown in **Figure 6**. In such circumstances, however, the blinding of the surface can become an issue and a means of pre-filtration or sedimentation to take out the worst of the sediment may be advisable and if not available close control of maintenance may be required.

Figure 6. A tree pit receiving urban sediment from an adjacent parking area, Greensborough, North Carolina, USA.

The understanding of the requirements for successful irrigation of urban trees has improved in recent years, particularly as the recognition of the full range of benefits provided by urban trees has become apparent and the use of trees in the urban landscape has been more readily encouraged. As shown in the plates in this section, care must be taken to prevent waterlogging of supporting soils, partly due to the confined nature of tree pits and also the provision of water in excess from large impermeable areas, which may exclude the air that is found in natural soils and may, if serious enough, lead to anoxic conditions and the production of greenhouse gases. Where possible, water should be directed to a sub-root zone position without passing through the root ball itself. Consideration of aeration and an under drain in design should prevent this, alongside a consideration of local rainfall totals and the nearby landscape conditions.

4. CONCLUSION

SuDS elements, by their nature, store water but it is not always simple to either make a SuDS system provide a suitable supply of irrigation water or to make a water harvesting system contribute to stormwater source control. Both quality issues and quantity issues are important and in areas where road salt is used ground level collection needs to be carefully managed to prevent stored water becoming heavily contaminated with salt that could make the water unsuitable for irrigation and the collection mechanism needs to provide attenuation of automobile-derived pollutants sufficient to allow plants to grow without inhibition. Where the prime aim is to harvest a good quality of water such as on a roof, whether this is used directly to irrigate plants on the roof or is collected for offline irrigation sizing of tanks needs to allow enough excess storage to deal with storm events and the tanks need to be managed to ensure such volume is routinely available.

REFERENCES

1. FAO, (2008), Wastewater Quality Guidelines for Agricultural Use. Food and Agricultural Organisation Corporate Document Repository: Natural Resources Management and Environmental Department [online]. http://www.fao.org/docrep/T0551E/t0551e04.htm <25 September 2016>
2. Nnadi, E.O., Coupe, S.J., Newman, A.P. and Mbanaso, F.U., (2015) 'Stormwater harvesting for irrigation purposes: an investigation of chemical quality of water recycled in pervious pavement system' Journal Environmental Management, 147, 246–256.
3. Meera V. and Mansoor Ahammed M. (2006) Water quality of rooftop rainwater harvesting systems: a review, Journal of Water Supply: Research & Technology, 55 (4) 257–268.
4. Jones, M.P. and Hunt, W. F. (2008) Rainwater Harvesting (2008) 'Urban Waterways: Guidance for Homeowners', North Carolina Cooperative Extension Service

AGW-588-11. https://content.ces.ncsu.edu/rainwater-harvesting-guidance-for-homeowners

5. DeBusk, K.M. and Hunt, W.F. (2014) 'Rainwater Harvesting: a Comprehensive Review of Literature', Report No. 425 NC Water Resources Research Institute. http://repository.lib.ncsu.edu/dr/bitstream/1840.4/8170/3/NC-WRRI-425.pdf
6. DeBusk, K.M, Hunt, W.F, Osmond, D.L and Cope, G.W. (n.d.) 'Water Quality of Rooftop Runoff: Implications for Residential Water Harvesting Systems', North Carolina Cooperative Extension, https://content.ces.ncsu.edu/water-quality-of-rooftop-runoff
7. Booth, D. B. and Leavitt, J. (1999), 'Field evaluation of permeable pavement systems for improved stormwater management', Journal of the American Planning Association, 65(3), 314–325.
8. Gomez-Ullate, E., Bayon, J. R., Coupe, S. and Castro-Fresno, D. (2010). 'Performance of pervious pavement parking bays storing rainwater in the north of Spain', Water Science and Technology. 62(3), 615–621.
9. Klein, R. D. (1979), 'Urbanization and stream quality impairment' Water Resources Bulletin, 15(4), 948–963.
10. Rushton, B. T. (2001), 'Low-impact parking lot design reduces runoff and pollutant loads', Journal of Water Resources Planning and Management-ASCE, 127(3), 172–179.
11. Sañudo-Fontaneda, L.A., Charlesworth, S., Castro-Fresno, D. andrés-Valeri, V.C.A. and Rodriguez-Hernandez J. (2014) 'Water quality and quantity assessment of pervious pavements performance in experimental car park areas', Water Science and Technology, 69(7), 1526–1533.
12. Scholz, M. and Grabowiecki, P. (2007), 'Review of permeable pavement systems', Building and Environment, 42(11), 3830–3836.
13. Pratt, C.J. (1996) 'Permeable Pavements for Stormwater Quality Enhancement' Proceeding Paper Urban Stormwater Quality Enhancement: Source Control, Retrofitting and Combined Sewer Technology American Society of Civil Engineers.
14. Newman, A. P., Pratt, C. J., Coupe, S. J. and Cresswell, N. (2002), 'Oil bio-degradation in permeable pavements by microbial communities', Water Science and Technology, 45(7),51–56.
15. Newman, A., Puehmeier, T., Shuttleworth, A. and Pratt, C.J. (2014) Performance of an enhanced pervious pavement system loaded with large volumes of hydrocarbons. Water Science and Technology, 70(5), 835–842.
16. Macdonald, K. and Jefferies, C. (2001), 'Performance Comparison of Porous Paved and Traditional Car Parks' Proceedings of First National Conference on Sustainable Drainage, Coventry University, ISBN 1 903818 06 0, pp. 170–181.
17. Bayón J.R., Castro D., Moreno-Ventas X., Coupe S.J. and Newman A.P., Pervious pavement research in Spain: Hydrocarbon degrading microorganisms, Proceedings of 10th International Conference on Urban Drainage, Copenhagen/Denmark, 21–26 August 2005. CD-ROM
18. Jayasuriya, N., Kadurupokune, N., Othman, M. and Jesse, K. (2007) 'Contributing to the sustainable use of stormwater: the role of pervious pavements' Water Science & Technology, 56 (12), 69–75.
19. Fletcher, T.D., Deletic, A., Mitchell, V.G. and Hatt, B.E (2008) 'Reuse of urban runoff in Australia: a review of recent advances and remaining challenges', Journal of Environmental Quality, 37(5), 116–127.
20. Brattebo, B. O. and Booth, D. B. (2003), 'Long-term stormwater quantity and quality performance of permeable pavement systems', Water Research, 37(18), 4369–4379.
21. Nnadi, E.O. (2009) 'An evaluation of modified pervious pavements for water harvesting for irrigation purposes' Unpublished PhD Thesis. Coventry: Coventry University.
22. Newman, A. P., Puehmeier, T., Kwok, V., Lam, M., Coupe, S. J., Shuttleworth, A. and Pratt, C. J. (2004), Protecting groundwater with oil-retaining pervious pavements:

Historical perspectives, limitations and recent developments, Quarterly Journal of Engineering Geology and Hydrogeology, 37(4), 283–291.

23. Wilson, S., Newman, A.P., Puehmeier, T. and Shuttleworth, A. (2003). Performance of an oil interceptor incorporated into a pervious pavement, ICE Proc: Engineering Sustainability, 156(1), 51–58.
24. Newman, A. P. (2003), Pollutant containment system, WO03040483 [patent]
25. Nnadi, E.O., Newman, A.P. and Coupe, S.J., (2014), 'Geotextile incorporated permeable pavement system as potential source of irrigation water: effects of re-used water on the soil, plant growth and development', Clean-Soil Air Water, 42(2), 125–132.
26. Newman, A.P., Aitken, D. and Antizar-Ladislao, B. (2013), Stormwater quality performance of a macro-pervious pavement car park installation equipped with channel drain based oil and silt retention devices, Water Research, 47(2), 7327–7336.
27. New Zealand Govt. (2011), Guidelines for Assessing and Managing Petroleum Hydrocarbon Contaminated Sites [online]. http://mfe.govt.nz/publications/hazardous/oil-guide-jun99/module-5.pdf
28. Bauder, T.A., Waskom, P.L., Sutherland, P.L. and Davis, J.G. (2014) 'Irrigation Water Quality Criteria' Factsheet 0.506 Colorado State University: http://www.ext.colostate.edu/pubs/crops/00506.html
29. Newman A.P., Nnadi, E.O., Mbanaso, F.U., Shuttleworth, A., Aitken D., Antizar-Ladislao B. and Theophilus S.C. (2015) 'Potential of Pervious and MacroPervious Pavements as Harvesting Systems for Irrigation of Adjacent Lawns and Flower Beds' SUDSnet International Conference 2015 3rd & 4th September 2015 Techno-Centre, Coventry University, UK. http://sudsnet.abertay.ac.uk/documents/SUDSnet2015_Newmanetal_PotentialofPerviousandMacroPerviousPavementsasHarvestingSystemsfo_000.pdf
30. Tuffnell, N. and Britt, C. (2008) Determining the Usage and Usage Patterns of Amenity Pesticides across the UK (PS2230) [online]. Available from http://www.pesticides.gov.uk/uploadedfiles/Web_Assets/PSD/RPA_Amenity_Report_2008.pdf [08/07/2016]
31. Rask, A.M. (2012) Non-chemical weed control on hard surfaces: An investigation of long-term effects of thermal weed control methods. Forest & Landscape Research No. 52-2012. Forest & Landscape Denmark, Frederiksberg. 156.
32. Kempenaar, C. (2010) Best practice for chemical weed control on hard surfaces. European Glyphosate Environmental Information Source, Wageningen: Wageningen University and Research Centre - Plant Research International.
33. Kolpin, D.W., Thurman, E.M., Lee, E.A., Meyer, M.T., Furlong, E.T. and Glassmeyer, S.T. (2006), 'Urban contributions of glyphosate and its degradate AMPA to streams in the United States', Science of the Total Environment, 354, 191–197.
34. Kristoffersen, P., Rask, A.M., Grundy, A.C., Franzen, I., Kempenaar, C., Raisio, J., Schroeder, H., Spijker, J., Verschwele, A. and Zarina, L. (2008), 'A review of pesticide policies and regulations for urban Amenity areas in seven European countries', Weed Research 48(3), 201–214.
35. Woodburn, A.T., (2000), 'Glyphosate: production, pricing and use worldwide', Pest Management Science, 56, 309–312.
36. Mbanaso, F.U., Coupe, S.J., Charlesworth, S.M. and Nnadi, E.O. (2012), Laboratory-based experiments to investigate the impact of glyphosate-containing herbicide on pollution attenuation and biodegradation in a model pervious paving system, Chemosphere, 90 (2013), 737–746.
37. Mbanaso, F.U., Nnadi, E.O., Coupe, S.J. and Charlesworth, S.M. (2016), Stormwater harvesting from landscaped areas: effect of herbicide application on water quality and usage, Environmental Science and Pollution Research, 23, 15970.
38. Baik, J., Kwak, K., Park, S. and Ryu, Y. (2012), 'Effects of building roof greening on air quality in street canyons', Atmospheric Environment, 61, 48–55.

39. Bowler, D.E., Buyung-Ali, L., Knight, T.M. and Pullin, A.S. (2010). Urban greening to cool towns and cities: a systematic review of the empirical evidence. Landscape and Urban Planning, 97, 147–155.
40. Brenneisen S., (2006), Space for urban wildlife: designing green roofs as habitats in Switzerland, Urban Habitats, 4(1), 27–36.
41. Hunt, W.F. and Szpir, L.L. (n.d.,) Permeable Pavements, Green Roofs and Cisterns Stormwater Treatment Practices for Low-Impact Development, Pub. North Carolina Cooperative Extension Service https://www.bae.ncsu.edu/extension/ext-publications/water/protecting/ag-588-06-permeable-pavements-green-roofs-and-cisterns.pdf
42. Stovin V. (2010), 'The potential of green roofs to manage urban stormwater', Water and Environment Journal, 24(3), 192–199.
43. Carson, T.B., Marasco, D.E., Culligan, P.J. and McGillis, W.R.(2013), 'Hydrological performance of extensive green roofs in New York City: observations and multi-year modeling of three full-scale systems', Environmental Research Letters, 8, online at; http://iopscience.iop.org/article/10.1088/1748-9326/8/2/024036/pdf
44. Fioretti, R., Palla, A., Lanza, L.G. and Principi, P. (2010) Green roof energy and water related performance in the Mediterranean climate, Building and Environment, 45, 1890–1904.
45. Wong N.H. , Chen Y., Ong C.L. and Sia A. (2003), Investigation of thermal benefits of rooftop garden in the tropical environment, Building and Environment, 38(2), 261–270.
46. Palomo Del Barrio, E.(1998) ,Analysis of the green roofs cooling potential in buildings, Energy and Buildings, 27(2), 179–193.
47. Castleton, H.F., Stovin, V., Beck, S.B.M. and Davison, J.B. (2010), 'Green roofs: building energy savings and the potential for retrofit', Energy and Buildings, 42(10), 1582–1591.
48. Alexandri, E. and Jones P.(2008), Temperature decreases in an urban canyon due to green walls and green roofs in diverse climates, Building and Environment, 43(4), 480–493.
49. Takebayashi, H. and Moriyama, M.(2007), Surface heat budget on green roof and high reflection roof for mitigation of urban heat island, Building and Environment, 42(8), 2791–2979.
50. Susca T., Gaffin S.R. and Dell'Osso G.R.(2011), Positive effects of vegetation: urban heat island and green roofs, Environmental Pollution, 159(8–9), 2119–2126.
51. Kadas G. (2006), 'Rare invertebrates colonizing green roofs in London', Urban Habitats, 4, 66–86.
52. Matsuoka, R. H. and Kaplan, R. (2008), "People needs in the urban landscape: analysis of landscape and urban planning contributions", Landscape Urban Planning, 84(1), 7–19.
53. Dunnett, N. (2006) 'Green Roofs for Biodiversity: Reconciling Aesthetics with Ecology', Proc. Fourth Annual Greening Rooftops for Sustainable Communities Conference, Awards and Trade Show Proceedings May 10–12, 2006, Boston, MA.
54. Molineux, C.J., Fentiman, C.H. and Gange, A.C.(2009), Characterising alternative recycled waste materials for use as green roof growing media in the UK, Ecological Engineering, 35, 1507–1513.
55. Snodgrass, E. and McIntyre, L.(2010), The Green Roof Manual, Portland, OR: Timber Press.
56. Voeten J.W.G.F., van de Werken L. and Newman A.P. (2016) Demonstrating the Use of Below-Substrate Water Storage as a Means of Maintaining Green Roofs - Performance Data and a Novel Approach to Achieving Public Understanding, Proc. ;World Environmental and Water Resources Congress 2016, Professional Development, Innovative Technology, International Perspectives and History and Heritage, Pathak CS and Reinhart D (Eds) American Society of Civil Engineers , Reston.

57. Hardin, M., Wanielista, M. and Chopra, M. (2012) 'A mass balance model for designing green roof systems that incorporate a cistern for re-use', Water, 4, 914–931.
58. Greenroofs.com (n.d.) Water Storage & Irrigation. http://www.gree nroofs.co m/Greenroofs101/water-storage.htm
59. Van Raam C.H. Shuttleworth A.B. and Culleton P.D. (2014) Plant surface structure and modules and method for forming the same, Patent. WO2014006180 A1
60. Stovin, V.R., Jorgensen, A. and Clayden, A. (2008) 'Street trees and stormwater management' Arboricultural Journal, 30(4), 297–310.
61. Nnadi, E.O., Coupe, S.J., Sañudo-Fontaneda, L.A. and Rodriguez-Hernandez, J. (2014), 'An evaluation of enhanced geotextile layer in permeable pavement to improve stormwater infiltration and attenuation', International Journal of Pavement Engineering, 15(10), 925–932.
62. Hunt, W.F., Smith, J.T., Jadlocki, S.J., Hathaway, J.M. and Eubanks, P. R. (2008) Pollutant removal and peak flow mitigation by a bioretention cell in urban Charlotte, N.C., Journal of Environmental Engineering, 134(5), 403–408.

Index

A

Adoption, 1, 3, 4, 6, 25, 26

C

Challenges, 113, 117, 402
Constraints, 113, 117
Conveyance Efficiency, 202, 328
Crop coefficient, 54, 55, 60, 67, 69, 74, 260
Crop water requirement, 186

D

Deficit irrigation, 275, 277, 278, 286, 287, 308, 311
Drainage Canal, 360
Drainage Development, 113, 117

E

evapotranspiration, 30, 31, 53, 54, 55, 57, 58, 59, 61, 64, 389
Evapotranspiration, 52, 66, 73, 74, 182, 185, 186, 199, 216, 230, 313, 318, 324, 325, 341, 356, 357

G

Greenhouse, 53, 54, 117
groundwater, 2, 29, 30, 31, 37, 165, 166, 216, 241, 254, 280, 360, 394

I

Irrigation, 1, 6, 25, 29, 30, 34, 234, 236, 237, 238, 239, 255, 397
Irrigation Canal Design, 202, 328
Irrigation Module, 120
irrigation requirement, 55, 70, 166, 172, 174, 179, 181

M

Microclimate, 341, 347

Micrometeorological, 341, 342, 356
Monteith method, 185, 186, 189, 194, 197, 198
Morocco, 166, 167, 168, 170, 171, 174, 175, 177, 178, 179, 180, 181, 182, 183

N

Natural Resources, 1, 277, 393
Normalized Difference, 313, 314
N-P-K balance, 166

P

Pan evaporation, 186

S

salinity, 29, 30, 31, 32, 33, 34, 178, 180, 181, 182, 183, 276, 386
salinity tolerance, 30, 50, 169, 178

T

Technology, 1, 2, 5, 7, 25, 26, 50, 52, 53, 56, 73, 95, 96, 97, 396
Turbulent Transport, 341
Turkey, 119, 120, 121, 122, 123, 131, 132, 288, 359, 360, 361, 372

W

wastewater re-use, 30
Water productivity, 72, 278, 280

Y

Yield Response Factor, 314